# Value Chain Struggles

T0314136

## RGS-IBG Book Series

### Published

*Value Chain Struggles: Institutions and Governance in the Plantation Districts of South India*
Jeff Neilson and Bill Pritchard

*Arsenic Pollution: A Global Synthesis*
Peter Ravenscroft, Hugh Brammer and Keith Richards

*Queer Visibilities: Space, Identity and Interaction in Cape Town*
Andrew Tucker

*Resistance, Space and Political Identities: The Making of Counter-Global Networks*
David Featherstone

*Mental Health and Social Space: Towards Inclusionary Geographies?*
Hester Parr

*Climate and Society in Colonial Mexico: A Study in Vulnerability*
Georgina H. Endfield

*Geochemical Sediments and Landscapes*
Edited by David J. Nash and Sue J. McLaren

*Driving Spaces: A Cultural-Historical Geography of England's M1 Motorway*
Peter Merriman

*Badlands of the Republic: Space, Politics and Urban Policy*
Mustafa Dikeç

*Geomorphology of Upland Peat: Erosion, Form and Landscape Change*
Martin Evans and Jeff Warburton

*Spaces of Colonialism: Delhi's Urban Governmentalities*
Stephen Legg

*People/States/Territories*
Rhys Jones

*Publics and the City*
Kurt Iveson

*After the Three Italies: Wealth, Inequality and Industrial Change*
Mick Dunford and Lidia Greco

*Putting Workfare in Place*
Peter Sunley, Ron Martin and Corinne Nativel

*Domicile and Diaspora*
Alison Blunt

*Geographies and Moralities*
Edited by Roger Lee and David M. Smith

*Military Geographies*
Rachel Woodward

*A New Deal for Transport?*
Edited by Iain Docherty and Jon Shaw

*Geographies of British Modernity*
Edited by David Gilbert, David Matless and Brian Short

*Lost Geographies of Power*
John Allen

*Globalizing South China*
Carolyn L. Cartier

*Geomorphological Processes and Landscape Change: Britain in the Last 1000 Years*
Edited by David L. Higgitt and E. Mark Lee

### Forthcoming

*Aerial Geographies: Mobilities, Subjects, Spaces*
Peter Adey

*Politicizing Consumption: Making the Global Self in an Unequal World*
Clive Barnett, Nick Clarke, Paul Cloke and Alice Malpass

*Living Through Decline: Surviving in the Places of the Post-Industrial Economy*
Huw Beynon and Ray Hudson

*Swept-Up Lives? Re-envisaging 'the Homeless City'*
Paul Cloke, Sarah Johnsen and Jon May

*Millionaire Migrants: Trans-Pacific Life Lines*
David Ley

*Complex Locations: Women's geographical work in the UK 1850–1970*
Avril Maddrell

*In the Nature of Landscape: Cultural Geography on the Norfolk Broads*
David Matless

*Transnational Learning: Knowledge, Development and the North-South Divide*
Colin McFarlane

*Domesticating Neo-Liberalism: Social Exclusion and Spaces of Economic Practice in Post Socialism*
Adrian Smith, Alison Stenning, Alena Rochovská and Dariusz Świątek

*State, Science and the Skies: Governmentalities of the British Atmosphere*
Mark Whitehead

# Value Chain Struggles

## *Institutions and Governance in the Plantation Districts of South India*

Jeff Neilson and Bill Pritchard

WILEY-BLACKWELL
A John Wiley & Sons, Ltd., Publication

This edition first published 2009
© 2009 by Jeffrey Neilson and Bill Pritchard

Blackwell Publishing was acquired by John Wiley & Sons in February 2007. Blackwell's publishing program has been merged with Wiley's global Scientific, Technical, and Medical business to form Wiley-Blackwell.

*Registered Office*
John Wiley & Sons Ltd, The Atrium, Southern Gate, Chichester, West Sussex, PO19 8SQ, United Kingdom

*Editorial Offices*
350 Main Street, Malden, MA 02148-5020, USA
9600 Garsington Road, Oxford, OX4 2DQ, UK
The Atrium, Southern Gate, Chichester, West Sussex, PO19 8SQ, UK

For details of our global editorial offices, for customer services, and for information about how to apply for permission to reuse the copyright material in this book please see our website at www.wiley.com/wiley-blackwell.

The right of Jeffrey Neilson and Bill Pritchard to be identified as the authors of this work has been asserted in accordance with the Copyright, Designs and Patents Act 1988.

All rights reserved. No part of this publication may be reproduced, stored in a retrieval system, or transmitted, in any form or by any means, electronic, mechanical, photocopying, recording or otherwise, except as permitted by the UK Copyright, Designs and Patents Act 1988, without the prior permission of the publisher.

Wiley also publishes its books in a variety of electronic formats. Some content that appears in print may not be available in electronic books.

Designations used by companies to distinguish their products are often claimed as trademarks. All brand names and product names used in this book are trade names, service marks, trademarks or registered trademarks of their respective owners. The publisher is not associated with any product or vendor mentioned in this book. This publication is designed to provide accurate and authoritative information in regard to the subject matter covered. It is sold on the understanding that the publisher is not engaged in rendering professional services. If professional advice or other expert assistance is required, the services of a competent professional should be sought.

*Library of Congress Cataloging-in-Publication Data*

Neilson, Jeff.
Value chain struggles : institutions and governance in the plantation districts of South India / Jeff Neilson and Bill Pritchard.
p. cm. – (GS-IBG book series)
Includes bibliographical references and index.
ISBN 978-1-4051-7393-3 (hardcover : alk. paper) – ISBN 978-1-4051-7392-6 (pbk. : alk. paper) 1. Coffee industry–India, South. 2. Tea trade–India, South. 3. Globalization–Economic aspects–India, South. I. Pritchard, Bill. II. Title. III. Title: Institutions and governance in the plantation districts of South India.
HD9199.I42N45 2009
338.4′76337309548–dc22

2008037693

A catalogue record for this book is available from the British Library.

Set in 10/12pt Plantin by SPi Publisher Services, Pondicherry, India
Printed and bound in Malaysia by Vivar Printing Sdn Bhd

1 2009

# Contents

# List of Figures

# List of Tables

# Series Editors' Preface

The RGS-IBG Book Series only publishes work of the highest international standing. Its emphasis is on distinctive new developments in human and physical geography, although it is also open to contributions from cognate disciplines whose interests overlap with those of geographers. The Series places strong emphasis on theoretically-informed and empirically-strong texts. Reflecting the vibrant and diverse theoretical and empirical agendas that characterize the contemporary discipline, contributions are expected to inform, challenge and stimulate the reader. Overall, the RGS-IBG Book Series seeks to promote scholarly publications that leave an intellectual mark and change the way readers think about particular issues, methods or theories.

For details on how to submit a proposal please visit:
www.rgsbookseries.com

Kevin Ward
*University of Manchester, UK*

Joanna Bullard
*Loughborough University, UK*

**RGS-IBG Book Series Editors**

# Acknowledgements

One benefit of researching South India's tea and coffee industries was that an invigorating brew of these local beverages was never far away whenever we sat down with our interview informants, colleagues and friends. During our eleven research visits to South India spread over the period 2004–8, we consumed literally hundreds of cups of tea or coffee in offices, living rooms, verandahs, restaurants, hotel lobbies, roadsides, and factory floors. They ranged from the archetypal steaming steel cups of milky *chai masala* of everyday India, to fine leaf black orthodox teas served in exquisite porcelain cups, and to foaming lattes in oversized paper cups in western-style 'Coffee Day' and 'Barista' outlets. Sitting over a cup or tea or coffee provided the social micro-climate in which the stories of transformation presented in this book were divulged and debated.

It comes as no surprise, therefore, that in these Acknowledgements we want to recognize the support and generosity of people associated with the South Indian tea and coffee industries. Without their assistance, this book could not have been written. We hope that this book may go some way towards securing a more viable and just future for this regional industry.

Mr Ullas Menon, Secretary-General of the United Planters' Association of Southern India (UPASI), was extremely generous in sharing with us his extensive knowledge of his industry, and warmly introducing us to many of his colleagues. Also at UPASI, R. Sanjith was a fount of information about the industry, and always bore a friendly smile. We cannot name all those in the planters' community who gave us their time in support of this book, but specific thanks are due to the following individuals who went out of their way to help us along our path: T. Alexander, N. Dharmaraj, H. Huq, C.P. Kariappa, D.B. King, B. Mandana, R. McAuliffe, T. Pinto, S. Perreira, A. Ponnapa, R. Bajekal, J.K. Thomas, and V. Ramaswamy.

For assisting our fieldwork with small growers, we wish to acknowledge the former Head Scientist at UPASI-KVK, Dr Ramu, his replacement, Mrs Dhanalakshmi Devaraj and all her staff. We also thank the Directors and staff of the Tea Board of India, especially Mr Nazeem I.A.S., and the Directors and staff of the Coffee Board of India, especially Dr Ramamurthy at the research sub-station at Chettali. The directors and staff of the UPASI-Tea Research Foundation are owed a debt of gratitude, as is Mr Lakshman Gowda, President of the Karnataka Grower's Federation.

In the Indian research community, we acknowledge the assistance of Prof. T. Vasantha Kumaran, Prof. P.G. Chengappa, Dr Ajjan Nanjan, C.P. Kushalappa, Dr C.P. Gracy, Dr A Damodaran, and Dr Claude Garcia (of CIRAD and the French Institute at Pondicherry). Dr Shatadru Chattopadhayay (Partners in Change) and Sanne van der Wal (Dutch Tea Initiative/SOMO) assisted our participation in an NGO Workshop in Darjeeling in 2006, which we gratefully acknowledge.

We enjoyed the good company of Ashish Rozario and family, Steve Rebello and Sanjay Cherian during our fieldwork in India, and also wish to thank the staff at Wallwood Guest House in Coonoor, which became something of an occasional 'home away from home' during parts of our field research.

The project to which this book has contributed was funded by the Australian Research Council. We thank and acknowledge our former colleague within the project team, Prof. John Spriggs, who gave us key insights about institutional economics that now see the light of day with this book. Niels Fold, from the Department of Geography at Copenhagen University, spent a productive sabbatical with us in 2006–7 and helped us to hone our ideas. Lindsay Soutar, Nathan Wales and Jasmine Glover assisted in various technical ways in making this manuscript come into being. Our colleagues in Geography at the University of Sydney and our fellow members of the Agri-food Research Network form the basis of our researcher networks, and it is an honour to be part of these scholarly communities. Kevin Ward at the University of Manchester provided supportive and sound advice in bringing our manuscript to final publication, and we thank two anonymous referees for their vigilance in asking us to fine-tune our arguments.

Finally, this project involved considerable time away from home, to the disruption of our families. We dedicate this book to Relyta, Eden and Jemma, and Kerry and Arizona.

Jeff Neilson
Bill Pritchard

# List of Abbreviations

| | |
|---|---|
| 4C | The Common Code for the Coffee Community |
| ABC | Amalgamated Bean Coffee (company) |
| ABF | Associated British Foods (company) |
| ACPC | Association of Coffee Producing Countries |
| AICEA | All India Coffee Exporters' Association |
| ANT | Actor–Network Theory |
| ASG | Apeenjay Surrendra Group (company) |
| asl | above sea level |
| ATO | Alternative Trade Organization |
| BBTC | Bombay Burmah Trading Company |
| BLF | bought-leaf factory (tea industry) |
| CEC | Center for Education and Communication (Indian NGO) |
| CSA | Commodity Systems Analysis |
| CSCE | New York Coffee, Sugar and Cocoa Exchange |
| CSR | corporate social responsibility |
| CTC | cut, tear, curl (method of tea manufacture) |
| DfID | Department for International Development (UK) |
| EIC | East India Company |
| ETI | Ethical Trade Initiative (UK government programme) |
| ETP | Ethical Tea Partnership |
| Eurep-GAP | Euro-Retailer Produce-Good Agricultural Practices |
| FAO | Food and Agriculture Organization of the United Nations |
| FERA | *Foreign Exchange Regulation Act* (India) |
| FLO | Fairtrade Labelling Organization |
| FOB | free-on-board (shipping term) |
| FUP | Factory Upgradation Programme |
| GBE | green bean equivalents (term for measuring coffee volumes) |

| | |
|---|---|
| GCC | Global Commodity Chain |
| Global-GAP | Global-Good Agricultural Practices |
| GPN | Global Production Network |
| GVC | global value chain |
| ha | hectare |
| HACCP | Hazard Analysis Critical Control Point (system) |
| HML | Harrisons Malayalam Limited (company) |
| ICA | International Coffee Agreement |
| ICAR | Indian Council of Agricultural Research |
| ICO | International Coffee Organization |
| ICTA | Indian Coffee Trading Association |
| IDS | Institute of Development Studies (University of Sussex, UK) |
| IGGoT | Inter-Government Group on Tea (FAO-convened commodity group) |
| ILO | International Labor Organization |
| IMF | International Monetary Fund |
| ISO | International Standards Organization |
| kg | kilogramme |
| KGF | Karnataka Growers' Federation |
| km | kilometre |
| KMFT | Kodagu Model Forest Trust |
| KPA | Karnataka Planters' Association |
| KVK | Krishi Vigyan Kendra (Indian adaptive agricultural research organization) |
| lb | pound (Imperial weight measure) |
| LIFFE | London International Financial Futures Exchange |
| m | metre |
| MRL | Maximum Residue Limit |
| NCA | National Coffee Association (of the US) |
| NGO | non-government organization |
| NIE | New Institutional Economics |
| NILMA | Nilgiris District Tea Producers' Marketing Association |
| NPA | Nilgiris Planters' Association |
| NYBOT | New York Board of Trade |
| OECD | Organization for Economic Cooperation and Development |
| OIE | Old Institutional Economics |
| PDS | Public Distribution System (Government of India program) |
| PES | Payments for Environmental Services |
| PFA Act | *Prevention of Food Adulteration Act* (India) |
| QUP | Quality Upgradation Programme |

| | |
|---|---|
| R&D | research and development |
| RBT | Ram Bahadur Thakur (company) |
| Rs | Rupees (Indian) |
| SAP | Sustainable Agriculture Program |
| SHG | Self-Help Group |
| SMBC | Smithsonian Migratory Bird Center |
| SoP | Systems of Provision |
| SPAA | Specialty Coffee Association of America |
| SPS | Sanitary & Phytosanitary (Agreement of the WTO) |
| TBT | Technical Barriers to Trade (Agreement of the WTO) |
| UK | United Kingdom |
| U-KVK | United Planters' Association of Southern India-Krishi Vigyan Kendra |
| UNCTAD | United Nations Conference on Trade and Development |
| UPA | United Progressive Alliance (coalition government of India, 2004-) |
| UPASI | United Planters' Association of Southern India |
| USA | United States of America |
| U-TRF | UPASI-Tea Research Foundation |
| WTO | World Trade Organization |

## Numbers, Dates and Measurements

In the book we occasionally make reference to the Indian numbering system, which specifies 100,000 as one *lakh* and 10 million as one *crore*. Where data is referenced as occurring over a non-calendar year period (eg., 2005–6), the dates in question follow the Indian financial year, which is 1 April to 31 March. Metric measurements are used throughout, except in some of our analysis of the coffee sector, where we follow the international norm and cite prices in terms of US$/lb.

## Indian Administrative Geography

There are four tiers of political administration in India relevant to the concerns of this book: the national government (usually referred to as 'the centre'); state administrations; districts, and taluks. Thus, we can refer to Kundah Taluk in the Nilgiris District of Tamil Nadu (state).

## Currency Exchange

Wherever relevant, we have expressed Indian rupee amounts in their approximate US dollar equivalents. Prevailing exchange rates for key years referred to in the text (inter-bank rate as of 1 January) are listed below.

2000 US$1 = Rs 43
2005 US$1 = Rs 43
2006 US$1 = Rs 45
2007 US$1 = Rs 44
2008 US$1 = Rs 39

# Chapter One

# Introduction

November 2004, Bangalore. In a downstairs conference room of the four-star Hotel Atria, a special closed session of the 46th Annual Meeting of the Karnataka Planters' Association (KPA) is under way. The KPA is a member organization of the United Planters' Association of Southern India (UPASI), which goes back more than 100 years to the age of British planters' clubs on the subcontinent. A senior economic researcher from one of India's leading universities, just returned from Europe, is setting forth a series of issues to which the Karnataka coffee industry will be forced to respond. In association with the German development agency *Gesellschaft für Technische Zusammenarbeit* (GTZ), the European coffee community is developing what it refers to as a 'common code' for the industry. Under the code, coffee producers wishing to sell to code signatories, which include Kraft and Nestlé, will be required to extensively document the histories of chemical use on their plantations, the environmental conditions under which coffee is grown, and their compliance with labour standards. The code is perceived as essentially a means for defensive brand management by the major coffee companies, and the planters fear that it will soon become a requirement for market access. This being the case, abiding by the code may give the planters an edge in the global marketplace. Yet at the same time, implementing these systems will be costly and time-consuming, especially onerous at a time of low coffee prices when many growers are already struggling to make a living. 'This is just East India Company imperialism in a new guise', says one of the planters. 'The Europeans are setting down new standards, and we have to pay the cost of implementing them.' The planters around the table nod their heads in agreement.

September 2005, a tea factory in the village of Bitherkad, in the Gudalur district of Tamil Nadu. A crowd of 200 smallholder tea growers awaits officials representing the Tea Board of India. Smallholders have been major losers from changed priorities of international tea buyers in local auctions, who have increasingly bypassed the generally lower-quality teas they produce. The associated slump in tea prices received by smallholders is cutting deep into these growers' livelihoods. With average tea plantings of less than one hectare each, the 15,000 local tea growers have seen their farm incomes halved, with most now receiving gross incomes of less than US$600 per year from tea. The officials have come to explain a subsidy payment scheme aimed at alleviating the desperate plight of this segment of the rural population. The scheme has been developed after considerable political agitation by growers but, when it becomes apparent that bureaucratic problems will restrict the eligibility of many growers from receiving these payments, the smallholders' frustrations boil over. Speaker after speaker rails against what they perceive as the evils of globalized markets, industry deregulation, and low tea prices.

The meetings at Bangalore and Bitherkad express situated microcosms in the much wider process of the global restructuring of tropical product value chains. Gone are the days when the tropical products sector was anchored by state marketing boards which arranged sales according to crude quality grades and operated price stabilization schemes. These arrangements have been progressively dismantled, and into this lacuna has emerged a host of emergent forms of market exchange and coordination. As new structures have been implemented, they have reshaped income flows and cost burdens, fuelling intense debate and anxiety within producer communities. Across the world, questions are being asked about how these contemporary global value chain transformations are affecting the shape of these industries, the institutional organization of rural producers, and, ultimately, the fate of the largely impoverished agricultural communities that supply these beverages to be enjoyed by affluent consumers. Is it the case that liberalized engagement with global markets, combined with the forces of consumer activism, can provide a path out of the cul-de-sac of commodity dependence, or is this yet another false dawn in the history of developing country agriculture?

This book brings these questions to the forefront of analysis and argues, from *a geographical perspective*, that these issues reflect a series of *value chain struggles* created as place-based *institutions* negotiate the ability of *governance* structures to determine social, economic and environmental outcomes. Applying these arguments to the issue of one production site (South India), we contend that an appreciation of the significance of these struggles is fundamental to the task of understanding the broader politics of developing country export agriculture. We argue that there is no generic answer to the vital question of whether or not contemporary global market processes are contributing to improved rural livelihoods; rather, this is an outcome of site-specific altercations and intersections between economic actors embedded in varying ways within spaces, networks and social structures. To obtain insights into the pattern of winners and losers from value chain restructuring, therefore, requires an approach to research which digs deep into the questions how and why specific economic actors relate to others in specific ways. In this book, we seek to put into action these perspectives. We deploy a specific brand of Global Value Chain (GVC) analysis – informed by a relational economic perspective and the insights of institutionalism – to emphasize the importance of place and context within the global canvas of developing countries, agriculture and trade.

## Tea, Coffee and the Crisis in Tropical Commodities

The subject matter of this book is set against a backdrop of massive global inequality. Across the world, tea and coffee production have traditionally provided the agricultural mainstay for tens of millions of people living in tropical upland areas. It is commonly the case that producers of these two crops have few viable economic alternatives, and numerous tropical countries have come to rely heavily on these products for export incomes.

For much of the past two decades, low tropical commodity prices have impacted severely on these developing country producers and, in the frank admission of former French President Jacques Chirac, there has been a 'conspiracy of silence' in terms of concrete measures by the world community in dealing with these issues (UNCTAD, 2003: p. 45). This silence has occurred not for want of evidence. The collapse of coffee and tea prices provided impetus for extensive documentation of the distribution of economic returns within tea and coffee value chains. Publications with such provocative titles as *Bitter Coffee* (Oxfam, 2001), *Stolen Fruit* (Robbins, 2003), *Robbing Coffee's Cradle* (Madeley, 2001) *Bitter Beans* (Chattopadhayay and John, 2007) and *There is Blood in the Tea We Drink* (John, 2003a) served to emphasize the plight of farmers. Mostly these studies focused on the fact that coffee and tea growers are at the base of value chains in which the overwhelming proportion of economic returns flow to developed country interests. Accordingly, consideration of the human cost of the crash of tea and coffee prices cannot be divorced from broader analysis of how these sectors are inserted within global value chains. Thus, the transformations in these products tell a story of wider significance for comprehending the global political economy of agriculture and, in particular, whether developing countries face a brighter or harsher future.

Getting to the core of these questions requires some preliminary contextual discussion. Until the 1990s, the international trade in these products was extensively regulated by various bilateral and multilateral agreements that set out terms, conditions and flows of exchange. These structures were advanced to a greater degree in coffee, where the International Coffee

Organization (ICO) negotiated the insertion of 'economic clauses' within a series of multilateral International Coffee Agreements (ICAs). In this regime of managed trade, signatory countries agreed to purchase coffee only from producer countries that complied with export quotas. The effect was to enable producer countries to manage the volume and sources of product reaching the world market at any one time, thus encouraging the maintenance of relatively healthy prices and ensuring (through country-based quota allocations) that all signatory producing countries shared in the export trade. As long as the ICAs were ratified by all major coffee producers and the key consumer countries in the capitalist world, the regime provided a powerful instrument for improving the structural condition of coffee producers in world markets (Talbot, 1997a).

In tea, the first International Tea Agreement was entered into by producers' associations in North and South India, Ceylon and the Dutch East Indies in 1933. (African producers, then only minor producers, implemented only part of the scheme.) Governments were responsible for enforcing export quotas and were subsequently involved in negotiating intergovernmental agreements. However, due primarily to political differences amongst producer countries in the late 1940s, the delicate process of determining appropriate export quotas was never successful and the agreement was abandoned in 1955 (Griffiths, 1967). Nevertheless a *de facto* regime of managed trade emerged in this industry because of the role of Cold War bilateralism, with Indo-Soviet barter trade agreements having particular importance to the subject matter of this book.

Such political arrangements provided the dominant institutional architecture for the tea and coffee trade from the 1950s to the late 1980s, before changing radically in the 1990s. In coffee, the pivotal shift occurred in 1989, when the US administration of President G.H.W. Bush rejected a new ICA. Given the weight of US buying power, this decision effectively brought to a close the era of managed trade in coffee. In tea, the shift was defined by the restructuring of international trading alliances following the collapse of the Eastern bloc. As far as South India was concerned, the end of the Cold War saw the demise of the erstwhile bilateral agreements that benefited Indian tea producers. During the 1990s, the market conditions through which South Indian producers sold tea to the former communist states became progressively less lucrative, with significant impacts on industry viability.

These changing political conditions of trade occurred hand in glove with dramatic shifts in economic power within these industries. Throughout the 1990s there was a spate of mergers and takeovers in the global beverages sector which created new corporate entities with enhanced global reach. This process advanced further in coffee than tea, because global coffee sales are dominated to a greater degree by developed country markets featuring global brands and supermarket sale channels. (As discussed in Chapter 3, global tea consumption continues to be dominated by developing and middle-income countries.) Steady consolidation of the international coffee industry meant that by the mid-1990s, eight traders controlled a majority of the coffee imported into Europe, North America, Japan and Australia (Talbot, 2002a: p. 220). This coffee was then sold to roasters, five of which accounted for 69 per cent of global coffee sales (van Dijk et al., 1998: p. 52). For instant (soluble) coffee the degree of concentration was higher still, with Nestlé alone having 56 per cent of global sales (van Dijk et al., 1998: p. 53). In developed market segments of the tea sector, comparable processes took place. In the UK, three brands accounted for 58 per cent of tea bag sales in 2006 (Mintel, 2007).

The massive buying power of these companies dovetailed with institutional shifts in market exchange. On the one hand, the rise of sophisticated market institutions, based around electronic data exchange and the Internet, effectively globalized the processes of buying and selling tea and coffee. Moreover, the expansion of the futures trade in coffee (for reasons explored elsewhere in this book, futures exchanges have not taken off in the tea sector) has facilitated significant *financialization* of the industry (whereby traders participate in these markets not just to procure product at a given future price, but as part of wider strategies for financial asset management and speculation). This is a far cry from the situation that existed up until the 1980s when the mediations of government-to-government trade (via quota allocations and national marketing board sales) shaped the flow of economic returns to individual countries. On the other hand, the enhanced scope and reach of multinational companies has encouraged new protocols for product grading and certification. Spearheading this latest phase of industry coordination and regulation is a concern by downstream retailers and brand owners to specify key value chain requirements with respect to quality, food safety, and the ethical basis of production. Although mostly developed as 'voluntary' conditions for producers, increasingly these requirements have taken on a life of their own and become *de facto* mandatory global standards for export participation. The Global-GAP scheme (known as Eurep-GAP until September 2007)[1] is a case in point. Established in 1997 as an initiative of European consortia of food retailers seeking to formalize food standards with the primary aim of instilling greater consumer confidence regarding food scares, its scope and breadth of adoption has evolved to the point where it is becoming a regulatory foundation for much international agri-food trade. Entwined within these developments is a new politics of audit, whereby the ability to export is predicated on the ability to document and authenticate. Such private sector initiatives evidence the rise of a system we label *global private regulation*; the enforcement of rules and standards on upstream producers by downstream private sector actors. These rules dictate how farmers gain their livelihoods, how they interact with the environment and how their local production systems and trade networks are structured.

The implied assurances and monitoring capabilities that underlie these varied initiatives bring to the fore the entwinement of global private regulation with the technologies of *traceability* – the imposition of compliance regimes which authenticate production trails from 'seed to supermarket'. Global private regulation and traceability together shape developing countries' capacities both to participate in, and extract benefits from, international agri-food trade. Theoretically, in an economic context of low world market prices for undifferentiated agricultural commodities, the authentication of product standards and credence attributes (the latter relating to the social and economic basis of production; claims such as 'cooperatively-grown', 'organic', and 'no forced labor') could provide defences that act as points of distinction in crowded marketplaces. Whether and how this labelling contributes to improved producer well-being remains, of course, a vexed question. Consumers may pay more for such attributes but it is not always clear whether (or to what extent) upstream producers share in these price premiums. Moreover, from producers' perspectives, developing the capacity to respond to such market signals is often costly and difficult. As we explore in this book, this is precisely where the importance of the institutional environment takes form; the ways that producers are embedded within institutional environments can help or hinder their capacities to participate in these chains.

Tea and coffee are quintessential examples of the type of tropical agriculture that sustains the livelihoods of rural economies across the developing world. Although developing countries have diversified their agri-export baskets over the past decade, tropical commodity exports remain a vital mainstay of countless agricultural communities. This book's attention to tea and coffee, therefore, corresponds to a crucial element of developing countries' participation in world markets. By extension, its conclusions hold meaning for understanding the changing conditions through which developing countries are inserted within the global economy.

## Governance, Institutions and Struggle

Our approach to addressing these questions seeks to bridge key divisions in recent analyses of global value chain restructuring in developing country agriculture. Currently, dominant research approaches into these issues tend to encourage polar opposite interpretations. According to one line of argument, the dismantling of state-centred arrangements and their replacement by global private regulation ultimately benefits producer countries because it removes barriers to the efficient transmission of price signals. The supposed invisible hand of the market weeds out inefficient from efficient operators, and rewards the latter. On the other hand, an alternative line of argument, generally associated with critical traditions of social science, suggests that global private regulation empowers the capabilities of large, globally mobile, corporations to impose their will and thereby exploit spatially grounded producers.

Arguments can be deployed on behalf of either of these positions, but both are prey to the charge of essentialism. Cursory observation of developing country agriculture suggests neither that all producers are being immiserated, nor are all benefiting. The shining successes of global market engagement invoked breathlessly by pro-market advocates are counterposed by dependent enclaves mired in the cul-de-sac of servicing export markets under exploitative conditions. Moreover, enumerating any list of 'winners' and 'losers' from global market engagement is a tenuous exercise, because of the speed at which fortunes can be reversed in response to spatial shifts in chain structures. By any account, the engagement of developing country agriculturists with global value chains reflects a volatile and readily reversible patchwork of apparent successes and failures.

For this book, accounting for such differentiation lies precisely at the heart of the analytical problem. We contend that complexity, differentiation and change should not be air-brushed out of analyses in the quest for narrative elegance. Instead, the challenge for research should be to incorporate these factors integrally to explanatory accounts. The vital question that needs to be asked is how and why economic restructuring reproduces territorial difference; *why* economic activity takes its particular spatial forms, and *how* it accrues advantage and disadvantage in different measure to place-bound interests.

Global Value Chain (GVC) analysis provides an efficacious framework for addressing these concerns. Global value chains represent 'the trajectory of a product from its conception and design, through production, retailing and final consumption' (Leslie and Reimer, 1999: p. 404). The object of inquiry in the GVC approach is the entirety of a product/commodity system. Its core analytical focus is on how product/commodity systems are coordinated, and how economic value is distributed amongst participants.

The GVC approach was formulated and popularized by the research of Gary Gereffi in the mid-1990s (Gereffi 1994, 1995, 1996, 1999; Gereffi et al., 1994). Initially, Gereffi set out a template method for GVC analysis[2] that defined the organization of product/commodity systems in terms of three dimensions: (i) an input–output structure (the configuration of purchases and sales by actors in the chain); (ii) territoriality (the geographical extent of chains); and (iii) a form of governance (the issues of how chains are coordinated and who does the coordinating) (Gereffi, 1994: p. 97). Over time, however, this framework morphed into a fourfold method including the new dimension of 'institutional context'. This inclusion reflected the fact, observed by Sturgeon (2001: p. 11), that value chains 'do not exist in a vacuum but within a complex matrix of institutions and supporting industries'. Correspondingly, Gibbon (2001b) enlarged Gereffi's original 'governance' dimension to the more inclusive category of 'governance and institutional structures', while Humphrey and Schmitz (2002) formalized its relevance to GVC analysis in their research on the roles of local and global linkages. Nowadays, the GVC method is routinely characterized through this fourfold template (for instance, Coe et al., 2007: p. 97).

The consideration of 'institutional context' within GVC analysis adds significantly to its utility as a tool of geographical inquiry. Considered in conjunction with 'governance', the category of 'institutions' provides a useful framing device for the examination of how product/commodity systems intersect with space and place. Issues relating to 'governance' encapsulate the coordinating structures which connect economic actors across space; those relating to 'institutions' represent the multi-scalar contexts that explain how economic actors are embedded within particular geographies.

This mutual interest within the GVC approach for governance and institutions represents an oft-forgotten element of its methodology. During the past decade or so, the GVC approach has been conceptualized all too frequently as being solely about *governance*, leading to the misguided perception that the approach has little to say on the complex questions about why and how particular industries come to be located in particular places. Moreover, this narrow-casting of what the GVC approach actually embodies has inspired many researchers to eschew the GVC approach in favor of alternative frameworks which give the surface appearance of being more sensitive to the nuance of geographical differentiation. As we discuss in Chapter 2, the broad field of product/commodity analysis is now encumbered by a diversity of alternative models each seeking to 'bring in' geography in its own, unique way. We contend that much of this proliferation responds to a fallacious assumption that the GVC approach has a sclerotic insensitivity to geographical considerations, and argue that the main effect of this splintering has been to complicate scholarly endeavour within an over-determined theorization of 'how to do what actually needs to be done'. Of course it is important to note that we are not alone in making this point. Acknowledgement of this problem has been a pivotal message in a succession of influential critiques of the field (see Leslie and Reimer, 1999; Hughes and Reimer, 2004; Friedland, 2005; Jackson et al., 2006). Bernstein and Campling (2006, p. 240) go so far as to claim that the field 'has no common purpose, object of analysis, theoretical framework or methodological approach'.

Reassertion of the importance of institutional analysis within a fourfold GVC approach generates a means to address these rifts. The GVC approach embraces an expansive and internally consistent framework to answer key geographical questions and, very importantly, provides a unified set of terminology accessible to wider audiences. For this to occur, however, further consideration is required with respect to the use of the term 'institutions'. Whereas researchers generally recognize that 'institutional contexts' play a vital role in shaping global value chains, what has gone missing in the literature has been a precise articulation of what 'institutions' actually are, and how they relate to GVC governance. The common shorthand refrain is that they represent 'local, national and international conditions and policies' (Coe et al., 2007: p. 97); that is, the external architecture that chains inhabit. Drawing from the broader field of institutional analysis in the social sciences, however, we can animate a much more encompassing and sophisticated rendition of this concept. Institutions are not just framing devices external to product/commodity systems ('out there'), but exist also as the rules, norms and behavioural vehicles that shape the very essence of how product/commodity systems are organized ('in here'). An institutional perspective contends that economic activity cannot occur in the absence of the social relations in which it is embedded (Granovetter, 1985). As articulated by D.C. North (1990: p. 3), institutions are 'the rules of the game in a society or, more formally, are the humanly devised constraints that shape human interaction'. They can be formal (codified, such as in constitutions, laws and contracts) or informal (conventions, codes of conduct, norms of behaviour, religious taboos, etc.). Nevertheless, whatever form they take, they configure economic and social dynamics. An institutional perspective recognizes that the progress, conduct and outcomes of value chain restructuring are steeped in the weights of history, culture and geography; the 'stickiness' of places. In this way, it provides an explanatory framework for understanding the different pathways of economic change across varying social and geographical arenas.

Invoking these arguments not only generates a more robust incorporation of 'institutional analysis' within the GVC approach but also, crucially, brings to light elevated appreciation of how governance and institutions are necessarily co-produced. Systems of value chain governance intermesh with the institutional life of territorially embedded production arrangements; institutions shape governance forms, and governance is enacted through institutions. The point is: institutional formations and governance arrangements coexist in an iterative nexus within global value chains.

Crucially, this iterative nexus is defined by *struggle*. The interplay of new forms of value chain governance with differentiated institutional environments triggers conflicts and tensions of various kinds. The ways these struggles are played out and resolved configures how producers are inserted within global value chains and, more to the point, the economic returns and level of control producers can exercise within them. The detail of struggle,

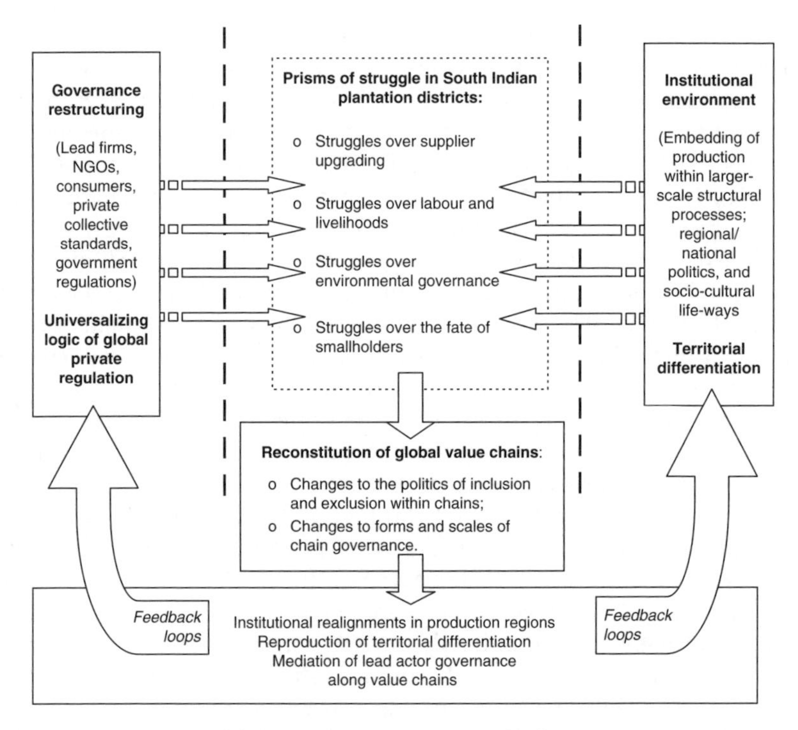


**Figure 1.1** Struggles over global value chains with institutional realignments.
*Source*: Own work.

therefore, becomes a prism through which to observe the broader set of debates about the implications of global value chain restructuring. By focusing on struggles we see in sharpest clarity the significance and implications of value chain restructuring.

The issues of governance, institutions and struggle are brought together in Figure 1.1. The pre-eminent implication of the diagram is to impress the need for caution when accounting for the implications of global value chain restructuring. We contend that there is no 'inevitable way' that developing country producers engage with (generally larger) downstream buyers. The contours of global value chain restructuring are less a finished recipe, and more a continual work-in-progress. This being the case, the vital contribution of this book is to 'get inside' a value chain in the midst of transformation, and to document the varied struggles which are shaping the politics of engagement between producers and downstream actors.

This emphasis on institutions and governance draws on parallel debates in contemporary development theory. During the 1980s and throughout much of the 1990s, the economic policies recommended by development agencies for the Global South were summed up by the 'Washington Consensus'.[3] The mantra of 'getting prices right' – eliminating government interventions in markets to the maximum extent possible – became a catch-cry within development policy for much of this period (Timmer, 1986; Reardon and Timmer, 2007). According to proponents, such policies would 'create space' for private interests to make decisions on production, pricing and marketing, with the effect of generating efficiency gains that would translate to increased economic returns to local producers. Accordingly, a powerful push for agricultural market reforms came from the World Bank, which introduced its first Structural Adjustment Loan in 1980 (Meerman, 1997). Applied to developing country agriculture, this line of argument held forth a vision in which industry structures become aligned solely to the assumed conditions and requirements of global markets, as defined by private sector interests, with domestic resources allocated accordingly and presumed positive flow-on effects for farmers.

As recounted by the former Chief Economist of the World Bank, Joseph Stiglitz (2002), the mid-to-late 1990s was a period of intense debate between the neo-liberal economists in favour of minimizing 'disturbances' to market processes, and 'institutionalists' who argued for a heightened consideration of the fact that markets necessarily take root in specific historical, geographical, political and social contexts. To cite North (1995: p. 23): 'getting the prices right only has the desired consequences when agents already have in place a set of property rights and enforcement that will then produce the competitive market conditions'. Vitriolic disputes over the causes of the financial crisis in Southeast Asia during 1997–98 became a *cause célèbre* for this debate (Burki and Perry, 1998; Pempel, 1999; Nissanke and Aryeetey, 2003). In 2002, the World Bank Development Report (titled *Building Institutions for Markets*) represented a clarion call that these ideas had come of age. It is now increasingly the case that even staunch supporters of neo-liberalism, such as Jagdish Bhagwati (2004), present caveats that emphasize the importance of institutional arrangements, accompanied if necessary by sequencing of reforms, for globalization to generate positive social outcomes. Influential writers and development policy advisors, such as Jeffrey Sachs, are now highly critical of the pre-eminence of the 'getting prices right' mantra within development policy. In his book *The End of Poverty*, Sachs (2005) proposes a new method for the 'differential diagnosis' of economic problems (which he calls 'clinical economics') which appeals to a more geographically nuanced process of problem identification.

These criticisms of the Washington Consensus have enriched the debate on global development and, by way of extension, provide a supportive base for the focus on institutions articulated in this book. It is a danger, we contend, to fall into a market fundamentalist trap whereby the allegedly efficiency-enhancing properties of liberal global markets are envisaged as a *deus ex machina* to raise producers' incomes. Such perspectives, we argue, give insufficient scope to the variability of producers' institutional environments. In the tropical products trade, there is no doubt whatsoever that the current phase of global value chain restructuring is letting loose powerful forces as multinational retailers and branded manufacturers seek to forge a future more amenable to their interests. Yet at the same time, these lead firms deploy their strategies not in a vacuum, but in a real world of spatially-embedded suppliers and consumers with concrete economic and political circumstances. An appreciation of these contexts is vital for a truly comprehensive understanding of the contemporary dynamics of change. As seminally proposed in the work of economic geographers Michael Storper and Richard Walker (1989: pp. 138–53), the mechanics of restructuring inevitably involve 'feedback loops' as the aspirations of companies to improve their profits by developing new strategies meet the 'art of the possible' given the fixities of geography and history, and, in turn encounter and potentially provoke varied responses from affected stakeholders.

Through extensive field-based research in South India, we identify the different ways in which producers are engaging with this set of changes. The rich empirical lode from our field research enables us to tease out broader conclusions on the fate of developing country producers in the contemporary era of global agriculture. As brought together in the book's conclusion, we argue that in the tea and coffee industries of South India, evolved forms of industry governance and associated opportunities for value chain *upgrading* (i.e., improving the returns for producers from participating in chains) are vitally sculpted by the role of the institutional environment. Thus, recent global value chain restructuring is making the differences of place and history more important than ever. This needs to be accorded central consideration in the analysis of who benefits and who loses from the contemporary market-liberal transformations, and in many of the common assumptions underpinning the field of global value chain studies.

## Towards buyer-driven governance in global tea and coffee industries

Before embarking on the detail of recent transformations in South India, it is necessary to provide an overarching assessment of larger-scale shifts in the strategies and structures of major tea and coffee companies. These dynamics provide a vital external frame for recent events in South India, contextualizing how and why large firms are seeking to restructure their engagement with (upstream) South Indian producers, as documented in later chapters of this book.

In both product sectors, the overall narrative of recent change is one of consolidation and forward integration. This is cause and effect of the fact that most of the value-addition in tea and coffee chains occurs in near-consumer segments. In tea, notwithstanding the fact that 'made tea' is exported from developing countries as a finished product, with a price that effectively includes all essential components of manufacture (Talbot, 2002b: p. 707), it has been estimated that average auction prices in producing countries are only 8 per cent of average retail prices for tea sold in Western Europe (van der Wal, 2008, and Table 1.1). The vast majority of the final retail price is accounted for by non-producer interests including shippers, blenders, packagers, owners of brands, and point-of-sale functionaries. In coffee, the price of green beans (the main form in which coffee is sold internationally) comparably represents only a small proportion of the final (retail) cost of coffee in developed countries (Table 1.2). Estimates drawn from ICO data suggest that Arabica growers receive around 20 per cent of the final US retail rice of coffee, whilst Robusta growers receive as little as 6 per cent (Gilbert, 2008: p. 18). As we explore in the following two sections, with such high proportions of value-addition to be captured in near-consumer nodes of chains, it is hardly surprising that these activities have been intensely fought over by large companies during recent years. The consequence has been a shift in industry governance towards buyer-driven arrangements.

Seen in their wider frame, the forward integration of large firms into downstream, brand-centric components of tea and coffee value chains is connected to overarching processes of *financialization* within agri-food production and trade (Pritchard, 1999; Gibbon and Ponte, 2005). Financialization refers to the general set of processes by which financial markets come to exercise a greater pull on economic organization of economies and societies

**Table 1.1** Distribution of value for tea sold into Western European markets

| *Stage of production* | *€/kg* | *Percentage of final price* |
|---|---|---|
| Cultivation, plucking, processing, bulk packaging | 1.25 | 6.91 |
| Auction price | 1.39 | 7.68 |
| Shipment, export taxes | 2.47 | 13.65 |
| Insurance, marketing, packaging/bagging, warehouse | 8.51 | 47.02 |
| Supermarket price (incl. retail taxes) | 18.10 | 100.00 |

*Source*: SOMO et al. (2006: p. 19).

**Table 1.2** Four estimates of the distribution of value for coffee sold into varying markets

| | *Percentage of final retail price* | | | |
|---|---|---|---|---|
| | *Global estimate (undated): van Dijk et al. (1998: p. 10)* | *Sulawesi Arabica, 2003: Neilson (2004: p. 119)* | *Ugandan Robusta exported to Italy, 2001–02: Daviron and Ponte (2005: p. 208)* | *Global estimate, 1994–95: Talbot (1997: p. 67)* |
| Farm gate/producer | 20 | 6 | 6.6 | 19.6 |
| Domestic transport, Processing and export | 13 | 5 | 3.7 | 3.4 |
| International transport | 11 | 9 | 4.0 | 5.0 |
| Roaster | 28 | 80 | 72.2 | 71.9 |
| Retailer | 21 | | 13.5 | |
| VAT | 7 | | | |

*Sources*: As specified.

(Martin, 2002; Epstein, 2005). In the case at hand, financial markets have come to expect that multinational companies appropriately value and protect these intangible assets. Prompting this state of affairs is the fact that brands have become a leading weapon in the struggles between multinational food companies and supermarket chains for profit shares. As the ownership of the supermarket sector has become concentrated in fewer hands, its capacities to dictate terms to suppliers have been strengthened. Supermarkets have had the power and motivation to cull poor-performing proprietary brands, and use own-label product to put additional competitive pressure on branded manufacturers.

## Corporate restructuring among leading global tea firms

The past two decades have witnessed major shifts in the focus and orientation of leading firms in the world tea industry. In the Colonial era and for much of the latter half of the twentieth century, the pivotal players in the global tea trade were merchant capitalists. In India, as we discuss in Chapter 4, most of these companies owed their ancestry to the managing agency system; commission-writing firms which coordinated cultivation, processing and trading activities across a range of commodities. Managing agents were a particular product of the Colonial system, and following the Independence of major tea-producing countries, these firms evolved to become specialist traders. Talbot (1995: p. 146) reports that in 1967, 'four tea exporters controlled about 90 per cent of world exports' (excluding the Soviet trade). A 1977 study by the United Nations Conference on Trade and Development (UNCTAD) found that in 11 out of 20 major tea-importing countries, at least 80 per cent of trade was undertaken by no more than four firms (cited in Ali et al., 1998: p. 16). A recent report by the Dutch NGO SOMO (van der Wal, 2008: p. 25) suggests that seven multinational firms account for 90 per cent of tea traded into western consumer markets.

Over time, these merchant-capital firms metamorphosed into firms with a stronger strategic emphasis on consumer brands (Ali et al., 1998: p. 17). These dynamics are illustrated most visibly in the evolution of the UK market. Before the 1980s, tea blending and branding in the UK was fragmented. A few brands had national distribution and recognition ('Lipton' and 'Tetley', for example), but alongside these was a proliferation of regional-based blender-branders. However, the industry consolidated rapidly as tea bags came to replace loose leaf packaging. Tea bag consistency requires sophisticated blending recipes and the product as a whole is highly amenable to brand visibility; two factors that work to the benefit of large multinational firms over and above smaller national and regional companies.

By the twenty-first century, branded tea manufacturing had become a highly concentrated sector in most national markets, with a common group of multinationals dominating. In the UK in 2006, two brands owned by multinationals ('Tetley' and 'PG Tips', owned by Tata and Unilever respectively) held a combined 48 per cent market share. The next two largest selling brands were also owned by multinationals (Twinings [10 per cent] and Typhoo [7 per cent]). Supermarket own-brands constituted 18 per cent of the market, leaving just 17 per cent to the myriad of other players in the sector (the largest of which was the family-owned Yorkshire Teas [6 per cent]) (Mintel, 2007).

The acquisition of smaller regional companies by brand-focused multinationals executed the demise of many independent 'heritage brands' with ownership associations with individual families and places. Arguably the world's most famous heritage brand is 'Twinings', where historical associations continue to be deployed as a core promotional attribute. The company's packaging styles (particularly its iconic metal boxed teas) invoke tradition and authority. Its flagship London store at 215 Strand acts as an embodied carrier of the company's historical legacy, with its continued operation on this single site since 1706 being a well-publicized draw-card. Less prominently in the public eye is the fact that in 1964 'Twinings' was acquired by Associated British Foods (ABF), a diversified UK agri-food conglomerate with a 2007 turnover of £6.8 billion. This acquisition enabled a fuller weight of national and international marketing resources to be put behind the brand. 'Twinings' now exists as one of a number of assets within the grocery division of ABF, alongside such household names as 'Ovaltine' (a milk drink), 'Ryvita' (dry biscuits), 'Tip Top' (bakery products) and 'Pataks' (curry sauces and chutneys). The combined operations of ABF's grocery division, moreover, contributed just 38 per cent of the company's 2007 revenue, with the remaining majority accounted for through sugar and agribusiness (notably in Southern Africa and China), food ingredients operations, and a retail arm consisting of more than 170 stores across the UK, Ireland and Spain (ABF, 2007).

Another prominent case is 'Tetley', which was established in 1822 by Joseph and Edward Tetley. In the early twentieth century the company passed into the ownership of Joseph Lyons & Co., the famous owner of a chain of British tea shops, hotels, and biscuit factories and then, from the 1960s to the 1990s, it changed hands on numerous occasions until in 1995 it was spun off with a number of other food assets as part of a management buyout. Eventually, in 2000, the business was bought by the diversified Indian conglomerate, Tata, which already possessed significant tea plantation holdings in North and South India, and was a major supplier to 'Tetley'. The acquisition, therefore, saw Tata expand downstream from its Indian production base.

Comparably, in 2005 another diversified Indian conglomerate moved downstream from tea production to acquire a major British tea brand. Exactly one hundred years earlier, in 1905, 'Typhoo Tea' was incorporated in Birmingham by a grocery shop owner and minor tea trader. The company stayed in family hands until 1968 when it was acquired by Cadbury Schweppes, but then was sold off a part of a management buyout in 1986. The new corporate entity holding the divested businesses from the buyout, Premier Foods, was then itself acquired by a group of venture capitalists in 1998 with the intention of being broken up and its individual businesses sold off separately. In this context, 'Typhoo' was bought by the Apeenjay Surrendra Group (ASG), a sprawling Indian conglomerate controlled by the Paul family of Kolkata. At the time of the acquisition, ASG already had existing Indian tea plantation interests and domestic brands through Assam Frontier Tea Ltd.

The most extensive and valuable collection of tea brands, however, has come to be owned by Unilever; a company which alone was responsible for purchasing 12 per cent of the world's black tea in 2006 (Unilever, 2006b: p. 13). In similar fashion to the companies cited above, Unilever has restructured itself downwards along the tea value chain, so that strategic orientation has increasingly been defined in terms of brand stewardship. The history of how Unilever built up its tea brands is traced through its acquisition of two key 'families' of brands; 'Lipton' and 'Brooke Bond'.[4] Backed by Unilever's global reach, scope and commitment to brand development, 'Lipton's' has become the world's most widely recognized tea brand, with sales in 2006 exceeding €1 billion (Unilever, 2006a: p. 2).

Comparable trends of forward integration and a focus on brands are also apparent within higher-value segments of developing country markets, including India. The rapidly growing domestic market for branded and packaged tea within India is similarly structured around high levels of concentration amongst multinational blender-branders. Traditionally, of course, the vast majority of India's tea was sold unbranded and unpackaged, characteristically through the 'tea-wallahs' on the streets and in the office corridors of the country. But with rising living standards among the middle classes, consumption patterns within India have migrated from unbranded to branded market segments (Neilson and Pritchard, 2007a). In tea, sales of branded products grew by 3.3 per cent per annum between 2000 and 2006 (*Indian Business Insight*, 2007), and the same two companies dominating the UK market had a virtually identical grip on branded tea in India. Brands owned by Unilever and Tata accounted for 44 per cent of sales within the formal retail sector in 2006 (*Indian Food and Industry*, 2006: p. 44).

## Corporate restructuring among leading coffee firms

In coffee, the pivotal downstream branding and packaging chain segments have also been intensely fought over by multinational companies. Industry consolidation and restructuring has led to the situation where a small number of diversified agri-food multinationals dominates these value chain segments. These lead firms have engaged in forward integration through their management and ownership of brands and retail outlets, and extended their influence upstream by dictating key terms and conditions of coffee purchases. In their study of coffee value chains emanating out of East Africa, Daviron and Ponte (2005, p. xvi) depict this situation as contributing to an apparent paradox – a 'coffee boom' in consuming markets and a 'coffee crisis' in producing countries – which is perpetuated through the inability of producers to control symbolic and immaterial quality attributes in the specialty coffee sector.

This situation has come into being through industry evolution over the course of several decades. Steady consolidation of the roasting sector occurred during the decades after the Second World War, as larger national and international firms progressively captured market share from smaller rivals. During the 20 years from 1958 and 1978 the four largest coffee roasters in the USA increased their collective market share from 46 per cent to 69 per cent (Talbot, 1995: p. 153). For instant coffee, these same four companies had 91 per cent of the market by 1978. Subsequently, in the 1980s, mergers and takeovers led these firms to become incorporated within global food groups. A 1998 Rabobank report concluded that five roasters (Nestlé, Philip Morris-Kraft,[5] Proctor & Gamble, Sara Lee and Tchibo) accounted for 69 per cent of global roasted bean coffee sales; and one roaster (Nestlé) accounted for 56 per cent of global instant coffee sales (van Dijk et al., 1998: pp. 52–3). The confluence of these arrangements with historic lows in international coffee prices led many researchers, especially those attached to development NGOs, to question the market power of these firms. Whereas in 1980–88 Arabica and Robusta producers received 34 per cent and 24 per cent respectively of the final US retail price of coffee, by 1999–2003 these figures had fallen to 18 per cent and 9 per cent (Gilbert, 2008: p. 18).

In the lead-up to the establishment of the Doha Round of multinational trade negotiations, Oxfam was especially critical of these outcomes. Its flagship report *Rigged Rules and Double Standards* called for 'A new institution to oversee global commodity markets [in tropical products], and a new system of commodity agreements' (Oxfam, 2002a: p. 14), because: 'Low world prices give the handful of transnational corporations (TNCs) that dominate world markets for products such as … [coffee] … access to cheap resources which produce enormous profit margins' (Oxfam, 2002a: p. 150).

Elaborating upon these arguments in a specific report on the coffee crisis, the aid agency commented:

> Two years ago, an analyst report on Nestlé's soluble coffee business concluded: '*Martin Luther used to wonder what people actually do in heaven. For most participants in the intensely competitive food manufacturing industry, contemplation of Nestlé's soluble coffee business must seem like the commercial equivalent of Luther's spiritual meditation.*' Referring to Nestlé's market share, size of sales and operating profit margins, the same author said: '*Nothing else in food and beverages is remotely as good.*' The report estimates that, on average, Nestlé makes 26p of profit for every £1 of instant coffee sold. Another analyst believes that margins for Nestlé's soluble business worldwide are higher, closer to 30% ... How do these roaster companies manage to be so profitable while farmers are in such deep crisis? They gain from the volumes they buy, from the strength of their brands and products, from cost control, from their ability to mix and match blends and from the use of financial tools that give them even more buying flexibility. (Oxfam, 2002b: p. 26)

As summarized by Daviron and Ponte (2005: pp. 141–2):

> Roasters have complete information on quality when they buy coffee and release next to no information to their clients. This factor, together with increasing market concentration, has allowed them to gain a driving seat in the global value chain for coffee. While supermarket chains have a predominant power position in other agro-food chains such as fresh fruit and vegetables – and dictate quality and logistics standards to other actors upstream – coffee roasters have been able to use the asymmetry of quality information on coffee to their advantage. They have downgraded the quality of their product to increase their margins.

In light of the extensive documentation of roaster dominance (see Talbot, 1996, 1997a, 1997b, 2002a, 2002b, 2004; Ponte, 2002a, 2002b; Daviron and Ponte, 2005; Gibbon and Ponte, 2005), there is little reason here to embellish these points further. Suffice to say that, as lead firms within the global value chain, multinational roasters wield considerable power in this industry.

## Local flavours: An introduction to tea and coffee production in South India

The vital empirical concern at the core of this book is how these emergent buyer-driven forms of industry governance have intersected with institutional environments associated with tea and coffee production in South

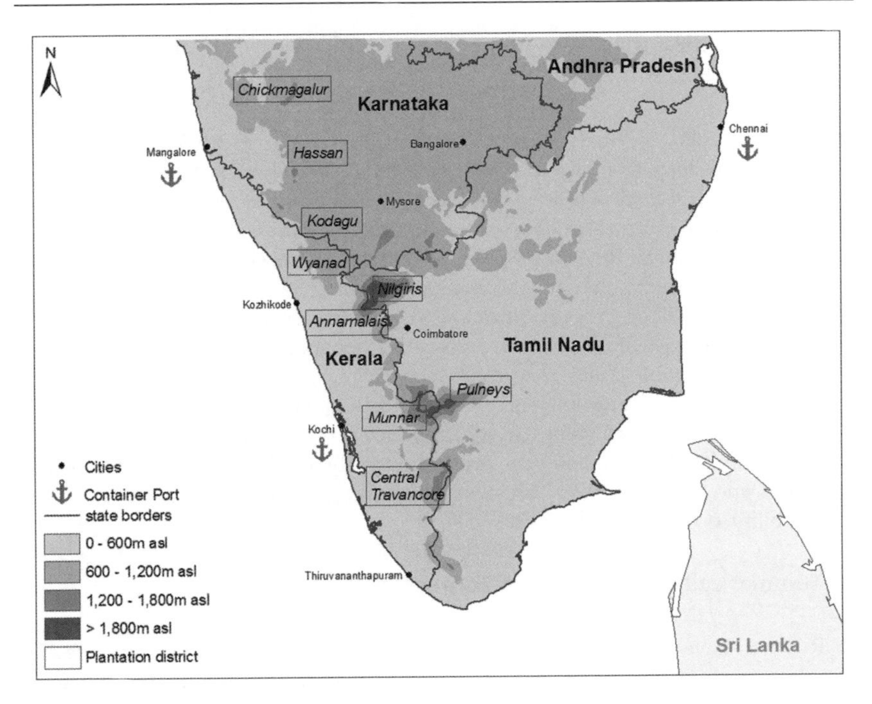


**Figure 1.2** Plantation districts in South India.
*Source*: Own work.

India.[6] This focus reflects our geographical concern that the South Indian tea and coffee value chains cannot be adequately understood without a deep appreciation of the complex and interacting social, economic and agro-economic mosaics in which their production takes place. India is a diverse, multi-ethnic and multi-linguistic amalgam in which discourses of regional identity remain ever-present. These geographical contexts translate to a set of embedded realities whereby tea and coffee production take on particular and different forms in specific locales. These differences include issues of economic structure, the cultural and social ways in which tea and coffee industries are incorporated within local economies and local environments, and the connections between tea and coffee industries and particular political jurisdictions. Tea and coffee in South India is grown across the three states of Karnataka, Tamil Nadu and Kerala, and exists within a number of different regions or, as they are known in India, 'plantation districts' (Figure 1.2). Comparative assessment of these industries across these geographical contexts brings into sharp focus the intersections of history, geography, politics and culture at local, regional and national levels. This book pays keen regard to these issues, concentrating on the different evolutionary trajectories and disparately shaped dilemmas that currently face these spatially-grounded production systems. Such an analysis sharpens an appreciation of how a sector's inherited position in the world economy and its institutional capacity to adapt and respond to global challenges vitally affects the concrete issues of 'who benefits, how much and why' within agrarian communities.

To set the scene for later chapters, it needs to be recognized that both coffee and tea have lengthy, though quite different, histories within South India. Coffee's introduction to South India is widely attributed to the Muslim pilgrim and saint, Baba Budan (Ukers, 1935; Pendergrast, 2001; Kariappa, 2002). It is believed that Baba Budan brought seven coffee seeds from Yemen to the princely state of Mysore in the seventeenth century, from whence cultivation spread amongst indigenous growers throughout the Western Ghats. However, it was British colonization of the Ghats that resulted in the widespread expansion and modernization of the industry in the early nineteenth century. The allure of profits from the spice trade may have inspired the Age of Exploration and the accompanying politics of European Colonialism, but it was the cultivation of tea and coffee within strategic upland sites that extended and consolidated the European agri-export presence within the Indian subcontinent during the nineteenth century. Starting around 1820, the British began to show an interest in the 'wild' coffees of Mysore, and coffee soon became the first major plantation crop cultivated in the South (Kariappa et al., 2004). By the 1840s, commercial coffee plantations were being established in Kodagu and further north in the Baba Budan hills of Chikmagalur.

With respect to tea, the British had aspired to grow the crop on the subcontinent since the late eighteenth century, but it was to take several decades for those dreams to be realized. According to Sir Percival Griffiths in his iconic *The History of the Indian Tea Industry* (1967), one of the leading proponents of establishing tea cultivation in India was the eminent botanist Sir Joseph Banks who recommended in 1788 that tea cultivation be developed in Bengal. Five years later he supported efforts as part of the Macartney Embassy to procure information about tea cultivation.[7] Meanwhile, in 1823 a Major Robert Bruce, whilst exploring the Brahmaputra River basin of Assam, reported back to Calcutta that the locals drank a form of 'native tea' which he thought might be the same as Chinese tea (Moxham, 2003: p. 93). Confusion then ensued as to whether the indigenous Assamese tea bushes were the same species as the established Chinese variety, which was resolved only in 1834 with the report of the 'Committee on Tea Culture'.[8] In May 1838, the first tea was exported from India to London. During the 1830s, tea varieties imported from China were distributed contemporaneously to regions in both North and South India and, over time, the endemic Assam tea was accepted as being of comparable quality as the imported Chinese varieties. Tea was successfully first grown in South India in 1832, at Ootacamund in the Nilgiris, and the first commercial tea estate was established in 1854. However, tea cultivation did not significantly expand in South India until the 1870s when many of the coffee plantations were severely struck with leaf rust and planters converted to tea.

British Colonialism provided the political and institutional vehicle for the expansion of plantation-based cultivation of tea and coffee and, as far as workers were concerned, conditions on estates could be highly exploitative with debt bondage, low wages, inadequate food, substandard housing and nonexistent medical care (Moxham, 2003; Macfarlane and Macfarlane, 2003). As pressures for Independence fermented within colonial India, disquiet about the working conditions on British plantations formed a cause of complaint integral to nationalist struggles. In the years around Independence, the return to capital on tea estates in British India (mostly British-owned, of course) was second to none (Tharian, 1984: p. 41).[9] Potent images of the English planter and the impoverished plantation worker became symbolic of the Imperialist regime and the anti-Colonial struggle. With independence in 1947, tea and coffee estates became bastions for unionized workforces, which asserted a newfound set of rights promulgated in the *Plantation Labor Act* of 1952. Still in existence, this legislation has had important ramifications for the status of Indian plantation labour in the post-Independence period (Bhowmik et al., 1996: p. 13).

As regards the ownership of India's tea and coffee plantations, decolonization was mostly quite orderly and took the form of a progressive replacement of British for Indian interests over the course of three decades. Through these arrangements, Indian entrants to the tea and coffee industries inherited an extant legacy of cultural, agronomic and professional institutions attached to 'plantation life'. Importantly, this provides a vital point of difference separating the Indian experience from that of a number of other tropical countries, such as Indonesia and Viet Nam, where the continuity of productive systems did not survive the transition from Colonialism to Independence. As described by the editors of a collection of articles from the *Planters Chronicle* (a prestigious industry publication now in its 110th year): 'Starting around 1820, pioneer British planters progressively 'opened up' the South Indian planting districts for tea and coffee, applying a blueprint from which a thriving industry emerged, and survives until today' (Kariappa et al., 2004: p. 2).

In short, British ownership of the tea and coffee industries of South India may be long gone, but its legacies continue into the present. Surviving features of the British 'blueprint' include supportive institutions such as planters' associations, clubs, and a strong sense of professional pride, which continue to facilitate knowledge sharing, innovation and a sense of collective purpose. As discussed in Chapters 3 and 4, this aspect of these industries' institutional environment retains profound implications in terms of capacities to comply and respond to emergent agendas in these sectors' global value chains.

In trading terms, the transformation of these industries in the immediate post-Independence period was undertaken in the context where Britain remained India's largest market for tea and coffee, and that these products were vital export-earners for the national economy. Indeed, given the diversity and sophistication of India's export base in the early twenty-first century it is a surprising fact that, as late as 1967, tea was India's largest export-earning sector, comprising 15.7 per cent of its total exports (Talbot, 1995: p. 148). But through the 1960s and 1970s, Indian dependence on the British market cooled. The economic nationalism of Prime Minister Indira Gandhi created an environment which was increasingly hostile to these export industries and, as tea and coffee plantations expanded in Kenya, Tanzania, Uganda and Malawi, British importers were accorded an alternative source of supply. The *coup de grâce* occurred in 1973 when India enacted the *Foreign Exchange Regulation Act* (FERA) which curbed the 'Sterling trade' (i.e., trade denominated in Pounds sterling), and placed greater restrictions on the appointment of foreign nationals and the ability of foreign companies to maintain holdings in India. Then, the strengthening of diplomatic relations between India and the Soviet Union during the 1970s resulted in the 1978 Rupee–Rouble trade agreement, where bilateral arrangements were made in nonconvertible national currencies (Tewari, 1999). Commonly, Soviet military and industrial hardware was exchanged for various Indian raw materials, notably tea and coffee. From the mid-1970s to the mid-1990s, some 80 per cent of South Indian tea was exported to the Soviet bloc (Ajjan and Raveendran, 2001). The usually favourable conditions of these deals for India, due largely to strategic geopolitical motives during the Cold War, and the relatively relaxed quality requirements of the Soviet market, resulted in complacency amongst Indian tea producers. According to common industry consensus, the quality of South Indian tea deteriorated sharply in the period of Soviet market reliance. When, in 1992, the Rupee–Rouble arrangements were abandoned, the South Indian tea sector had all but totally lost its traditional UK buyers and was no longer guaranteed access to the Russian market, precipitating a major crisis over the following decade.

It is the context of this crisis that this book attends. Largely, the analysis here concerns the period from the mid-1990s when deregulation became a major theme in these industries' governance arrangements. In India, the economic policy pendulum swung sharply towards deregulation in the early 1990s following a crisis in the country's foreign exchange reserves (Dash, 1999). The 1991 national budget of then-Finance Minister Manmohan

Singh (later elected Prime Minister in 2004) is usually credited as instigating the country's dramatic policy shift. During the ensuing decade the general mood of liberal policy reforms began to filtrate through to Indian agriculture (Neilson and Pritchard, 2007a), although the rural economy as a whole still remains subject to considerable regulation when compared to many other developing countries. Without going into detail, the vital point is that the domestic implementation of the liberal reform agenda, coincident with an era of low world market prices and concentration of corporate market power within tea and coffee global value chains, has produced a distinctively new competitive landscape for South Indian tea and coffee. Importantly, this landscape is being actively shaped by the place-specific struggles of institutional actors in their engagement with global value chains.

## Our Blend: An Outline of the Book's Structure

The chapters that follow seek to illustrate how the recent experiences of South Indian plantation districts can be seen to represent a politics of struggle as place-based institutions negotiate global governance structures. This journey begins in Chapter 2, which sets out our conceptual framework. The fundamental premise of this book – that the restructuring of tea and coffee producers in South India can be understood as the struggle of place-based *institutions* to negotiate global *governance* structures – entertains a series of deeper theoretical and methodological questions. Essential to our broader purpose is a need to explore and explain our approach within the larger field of economic geography. In Chapter 2, we introduce the GVC approach; review its merit in light of the array of competing and complementary analytical approaches within the broad field of commodity/product analysis; and then make the case for a refined type of GVC analysis that is more sensitive to the importance of place, space and territory. In our rendition of the GVC approach, emphasis is given to the ways in which institutional environments configure governance structures. This approach is heavily informed from relational perspectives in geography articulated recently by theorists working on Global Production Networks (GPNs). Our adaptation and incorporation of those ideas into the arena of GVC analysis seeks to provide a core geographical contribution to that field.

The remaining chapters of the book then apply this framework to our empirical case. Remaining true to our correspondence with the four-pronged method of the GVC approach, Chapter 3 presents the 'input–output' and 'territoriality' dimensions of tea and coffee GVCs for South India. This chapter focuses on major economic, technological and territorial dimensions of these industries. This material is largely contextual rather than interrogative, reflecting the fact that these two categories provide the foundational basis for generating the geography of global value chains. Without an appreciation of the agronomic, technological and economic issues presented in this chapter, the analytical intent of later discussions would be impossible to sustain.

Chapter 4 then commences in earnest the analytical project of this book, where we fully articulate our own rendering of the 'institutional environment' of GVCs through incorporation of our selected industry case studies. The intent of this chapter is to develop a nuanced understanding of how these industries *actually exist*. Contemporary social, economic and cultural arrangements are presented and analyzed in terms of the multi-scaled geographies in which they are embedded. Specific attention is given to the evolution of institutional environments over time, thus bringing to the fore a path-dependent set of interpretations of these industries' institutional environments.

Insights from Chapter 4 are then further developed in case study form in Chapters 5, 6 and 7, where we cast light on three strategic issues central to recent debates on value chain restructuring. Chapter 5 considers the issue of ethical accountability (essentially, the labour conditions under which products are made), by focusing on various agendas in the South Indian tea sector. Chapter 6 examines environmental accountability, focusing on conservation strategies in the coffee-growing district of Kodagu, in Western Karnataka. Then, in Chapter 7, we address the status of smallholders in global value chains, using the example of tea cultivation in the Nilgiri Hills of Tamil Nadu. The common approach throughout these chapters is to identify the grounded struggles within production districts as externally authored systems of chain governance are sought to be imposed within territorially complex sites of upstream production. The primary focus of Chapters 5 and 6 is how global private regulation is being refracted through the institutional settings of South Indian plantation districts. In Chapter 7, it is how local organizations have sought to ensure smallholder production meets international quality compliance requirements. Together, these chapters affirm our institutionally informed, geographical perspective on value chain analysis; we argue that analyses should be attentive not only to the internal dynamics within chains, but their broader manifestations within production arenas.

These foci give rise to a set of strategically relevant issues about the future of these industries, which we address in Chapter 8. Using the concept of value chain upgrading, we explore the diverse ways in which South Indian tea and coffee producers are seeking to improve their economic status in these global industries. Without pre-empting the detail of those conclusions here, the point we arrive at is one that asserts the importance of place-specific institutional environments in structuring producers' positions within chains and, thereby, their capacities to earn a living in the global economy.

The final chapter then reviews these conclusions in the context of theory and policy. It is our hope that the empirical narratives of South India in this book make a major contribution to geographical consideration of commodity systems. Our deployment of an institutionally enhanced GVC analysis provides a mechanism for more forthright examination of the role of place in the structuring of commodity flows. In terms of policy, what this study shows is that contemporary processes of change have raised the stakes for producer regions. Whether individual producers are jettisoned to the low-value cul-de-sac of commodity production, or whether they can reap some gains out of aligning their operations with the demands of restructured global value chains, will be dependent on the outcomes of an ongoing series of geographically informed value chain struggles.

# Chapter Two

# Re-inserting Place and Institutions within Global Value Chain Analysis

*Making sense of how the global economy is evolving is a difficult task.*
*(Sturgeon et al., 2008: p. 297)*

The emergence of the 'globalization era', following the end of the Cold War, has generated considerable tumult within the social sciences regarding how to theorize, research and frame 'global processes'. Sturgeon et al. (2008: pp. 297–8) contend:

> As national and local economies, and the firms and individuals embedded in them have come into closer contact through trade and foreign investment, the complexity of the analytical problem has increased. The classic tools of economics, especially theories of supply and demand and national comparative advantage, while powerful and astonishingly predictive, have always been too thin and stylized to explain all of the richness of industrial development and economic life.

Economic geographers have been an advance guard within the social science community with respect to the need to grapple with these problems. As Peter Dicken has argued (2003: pp. 10–12), a distinctive element of this contribution has been the way in which economic geography researchers have bristled at hyper-globalist accounts – which have featured prominently in some other disciplines – that imagine globalization as a monolith which flattens geographical difference (see O'Brien, 1992; Ohmae, 1995; Friedman, 1999). For economic geographers, such accounts fail to recognize 'the limits to globalization' and discursively construct the concept in terms of a neo-liberal 'wish list' (Allen and Thompson, 1997; Hirst and Thompson, 1999). Instead, economic geographers have tended to appreciate globalization processes as strategies of time-space compression which 'bump into' spatial variability and *reproduce* territorial difference:

'the specificities of institutions, culture and industry are clearly not eliminated by globalization, but brought into intensified contact, triggering a process of accelerated change that leads to *increasing* diversity over time' (Sturgeon et al., 2008: p. 298, italics in original). (See also Sturgeon, 2007). As such, within the field of economic geography, place, space and territory have been theorized as active ingredients in shaping patterns of global economic change. These research perspectives have illuminated how the emergent arrangements in the global economy are not simply the footprint of comparative advantage – the faceless dynamics of supply and demand – but rather have reflected the *politically cultivated* reconfiguration of economic, cultural and social relationships.

In their initial outings, attempts to assert the spatial complexity of globalization processes were developed through the ungainly terminology of 'glocalization' – literally a hybridization of the global and local (Swyngedouw, 1997) – but over time, researchers seem to have abandoned the quest for such singularly encompassing nomenclature. Increasingly, the theoretical literature has come to understand globalization in *relational terms*; a complex reconfiguration of the ways that places are connected to one another which is inherently uneven in its geographical scope and intensity. The foundational insight of a 'relational perspective' is that geographical scale is constructed from the actions and discourses of economic actors embedded within specific domains. Hence, spaces of action are created out of the ways that actors *relate* to one another; rather than being *prima facie* categories that are 'just there'. Applied specifically to globalization, Amin (1997: p. 133) has suggested: 'globalization in relational terms [involves] ... the interdependence and intermingling of global, distant and local layers, resulting in the greater hybridization and perforation of social, economic and political life.'

Heightened cognisance of the relational complexities of global processes has provided an intellectual impetus for a re-scaling of the research gaze. Rather than dwell on the theorization of global processes, the attention of many researchers has been captured by 'actually occurring globalization': the nitty-gritty of how globalizing processes are reshaping particular industry and commodity systems. From these investigative foci, 'thick' understandings of globalization are constructed. They build an appreciation of globalization through place-based and sector-specific tangents ('the traffic in things') (Jackson, 1999), and then work upwards through comparative analysis to signify and identify wider meanings. In these terms, the interrogation of product/commodity systems has come to 'suit the times' of these wider globalization debates. Sturgeon (2001: p. 9) expresses this argument well:

> Discussion of global-scale economic trends is inherently a large and unwieldy topic. Tools are needed to block out some of the 'noise' and allow us to focus on what is important, but we must choose carefully. Those studies that rely

> solely on macro-level statistics, such as trade and investment, cannot help but render invisible the detailed contours of the world economy. This is especially true when we seek to understand the role of personal and firm-to-firm relationships and the influence of power and politics on the development process, things that I hold to be crucial aspects of political economy and to the crafting of effective economic development policy and business strategy. At the same time, smaller-scale studies of national economies, localized clusters of economic activity and firm strategy forfeit a comprehensive view of the larger, cross-border structures that exist, or are coming to exist, in the world economy. For this reason, my own work has gravitated toward the study of cross-border organizational structures in particular industries. Industry-level analysis of economic activity, especially one that uses a 'value chain' approach, works well in studies of cross-border economic integration, because it takes a significant but still manageable slice of the world economy as its object of study. What is revealed in studies of industry value chains are the concrete actors in the global economy as well as the linkages that bind them into a larger whole.

The research task of following a commodity/product through a chain explains how the connections and disconnections of geography create a world of comparability and heterogeneity; of affluence and poverty, of advantage and disadvantage. It recalibrates the way we think about globalization in order to emphasize the 'here-and-now' specifics of decisions, strategies and actions of social actors (businesses, governments, NGOs, etc.). What we end up with is a frame of understanding not dissimilar to what Burawoy labels *grounded globalization*; that is, a sense that 'globalization is not just a simple ideology but a constellation of ideologies that becomes a terrain of struggle' (Burawoy, 2000b: p. 342). In so doing, this approach serves to provide an antidote to the 'TINA' ('there is no alternative') logic of what Beck (2000) labels the 'linearity myth' which emanates in so-called 'globalization from above' accounts of transformation (as an example, see Friedman, 2005).

The challenge of putting this agenda into practice, however, has stimulated considerable debate among economic geographers and, indeed, within the wider social sciences. Whereas the basic idea of tracing a product/commodity as it progresses through stages of value-addition appears a relatively clear-cut and self-evident *modus operandi*, researchers face the problem of how to organize their empirical investigations in order to tell a story of wider significance. Inevitably, as with any area of scholarly inquiry, the choice facing researchers with respect to how to tell their story, influences the story that is told. Again, to cite Burawoy in the context of describing a research team investigating 'grounded globalization':

> … we set out from real experiences, spatial and temporal, of welfare clients, homeless recyclers, mobilized feminists, migrant nurses, union organizers,

> software engineers, poisoned villagers, redundant boilermakers, and breast cancer activists in order to explain *their* global contexts ... We searched for theories that would help us stretch from local to global. We circled back to where we came from, to our theoretical moorings, only now more conscious of who we were, and what tools we needed to make global sense of our sites. (2000b: p. 341, italics in original)

For the research reported in this book, our chief 'theoretical mooring' has been the Global Value Chain (GVC) analytical framework. In accordance with Burawoy's sentiments, however, the process of applying this approach to field-based settings has led us to revisiting and revising aspects of this tool. Although we organize the arguments in this book in accordance with GVC methodology, we utilize this framework not simplistically or uncritically.

This chapter presents our conceptual thinking on these issues. Befitting a field of inquiry which takes 'global economic change' as its core subject matter, it is hardly surprising that the debates around these issues reflect 'a sprawling, multi-disciplinary field of endeavour' (Coe et al., 2008a: p. 267). Consequently, this chapter takes on board a breadth of analytical inquiry, organized in four major sections, which for reasons of clarity in presentation, we label as 'Part One', 'Part Two', etc. Each individual 'Part' of the chapter corresponds to a major area of investigation in its own right and, to some degree, can be read in isolation. Nevertheless, the wider purpose of this book necessitates that the discussions and analysis in all sections of the chapter 'come together' to create an overarching theoretical and methodological line of attack.

- Part One introduces the diverse, and often competing, schools of thought that take consideration of a particular product or commodity as their primary focus of inquiry, including Commodity Systems Analysis, French-inspired filière approaches, Actor–Network theory and Global Production Networks (GPNs).
- Part Two presents the GVC approach as a specific conceptual framework of analysis within this broader field. Detailed attention is given to the concepts of governance and upgrading, the two flagship notions associated with GVC research. We then make specific reference to the alleged weaknesses of GVC analysis, particularly the persuasive arguments put forward by advocates of a GPN approach.
- Part Three then introduces the distinctive conceptual contribution articulated throughout this book. Crucially, we argue that researchers using the GVC approach need to deploy a more sophisticated and spatially-sensitized rendition of the *institutional dimension* with GVC analysis. We believe that our approach forges a new vision for the potential of GVC

analysis; one that understands systems of value chain governance and the institutional life of territorially embedded production arrangements as enmeshed and co-produced, thus bringing geographical concerns of place, space and territory to the forefront of analysis.

- Taking these conceptual arguments forward requires, finally and not insubstantially, consideration of methodology. Part Four addresses these concerns. Somewhat surprisingly, there is a relative dearth of analytical work on questions relating to 'how to do GVC research'. This last section of the chapter brings these issues to the foreground. Firstly, we engage in a generalized account of the practical dilemmas and issues involved in a global chain/network research gaze. Then, we address the specificities of our work in India; the conceptual, ethical and practical questions of how we produced the data and insights revealed in later chapters.

## Part One: Commodity Analysis

The broad field of product/commodity analysis is an exercise in motion, with different researchers seeking to augment, renovate or rebuild particular frameworks in the search for more inclusive and/or purpose-specific frameworks. Assessing the trajectories of such processes of synthesis and integration is a task fraught with difficulty. As Jackson et al. (2006: p. 130) suggest, 'The recent rate of scholarly output [in this field] can make it difficult to give shape to the emerging debates and conceptual fault-lines.' And, as already noted in Chapter 1, during recent years the plethora of different approaches in the field has arguably complicated, rather than simplified, the task of drawing together key conclusions (Leslie and Reimer, 1999; Hughes and Reimer, 2004; Friedland, 2005; Bernstein and Campling, 2006). Trawling through the recent literature reveals at least eight major research strands have been developed in this field since the mid-1990s (Table 2.1). This diversity reflects schisms of discipline (e.g., sociology, geography, economics), research tradition (e.g., Anglophone versus Francophone literatures) and ontology (structuralism versus post-structuralism), amongst others. A prevalent cleavage, however, exists between a range of linear approaches structured around the 'vertical' analysis of product/commodity flows, and non-linear approaches using the alternative metaphors of 'circuits' and 'networks' to emphasize the recursive nature of product/commodity transformations. Brief consideration of these differences is timely, in order to give broader context to the detailed analysis of the GVC approach adopted in this book.

**Table 2.1** Methodological strands in product/commodity analysis

| *Terminology* | *Disciplines* | *Key citations* | *Origins* | *Comments* |
|---|---|---|---|---|
| **Linear metaphors** | | | | |
| 1. Global Value Chains (GVCs) | Economic Geography; Political Economy; Business Studies | Gereffi (1994, 1999); Gereffi, Humphrey and Sturgeon (2005); Ponte (2002a, 2002b); Gibbon (2001a); Gibbon and Ponte (2005) | World Systems Theory (Wallerstein, 1974; Hopkins and Wallerstein, 1986) before being popularized in Gereffi et al. (1994) | Initially associated with the work of Gereffi, and in particular, the question of producer-driven versus buyer-driven governance. |
| 2. Filière | Sociology; Economic Geography | Coste and Egg (1996); Daviron and Fousse (1993); Kydd et al. (1996) | INRA (*Institute National de la Recherche Agronomique*) and CIRAD (*Centre de Coopération Internationale en Recherche Agronomique pour le Développement*) | Original emphasis in the French literature was on processes of price formation within chains. |
| 3. Systems of Provision | Sociology; Economic Geography | Fine (1993); Fine and Leopold (1993); Fine et al. (1996). | The UK Research Council project 'Food consumption: social norms and systems of provision' | Compared to GCC/GVC, the SoP approach prioritizes consumption as a chain entry-point. |
| 4. Commodity Systems Analysis | Rural sociology; Agrarian Political Economy | Friedland et al. (1981); Dixon (1999); Friedland (2005) | William Friedland's highly influential analysis of the Californian iceberg lettuce system | Influential in Rural Sociology of the 1980s/1990s, but fewer applications in 2000s. |
| 5. Supply chains | Agric. Economics; Business Studies | Bourlakis and Weightman (2004) | *Journal of Supply Chain Management* | Associated with business-oriented research. |

| | | | | |
|---|---|---|---|---|
| **Non-linear metaphors** | | | | |
| 6. Commodity Circuits | Cultural geography | Cook and Crang (1996); Crang (1996); Cook et al. (2004). | Post-structuralism and 'cultural turn' human geography. | In these two approaches the emphasis is on the cultural understandings of commodities are heterogeneous over time and space. This seeks to overcome allegedly 'reductionist' foci in linear approaches. |
| 7. Commodity networks | Sociology of Technology; Economic Geography | Latour (1987); Arce and Marsden (1993); Busch and Juska (1997); Murdoch (1997); Whatmore and Thorne (1997) | French Actor–Network Theory (ANT). A primary concern is to collapse the binaries of human: nature and structure: agent | |
| **Hybrid metaphors** | | | | |
| 8. Global Production Networks | Economic Geography; International Political Economy | Henderson et al. (2002); Coe et al. (2004); Hess (2004); *Environment and Planning A Special Issue* 38(7) (2006) | 'Manchester school' of economic geography. The UK ESRC project 'Making the connections: Global Production Networks in Europe and East Asia' | Seeks to 'bring in' geography by augmenting analysis of GVC governance with attention to networks and territorial embedding. |

*Source*: Own work.

The linear frameworks listed in Table 2.1 are all characterized by their focus on charting the stages of transformation and value-addition within products/commodities. The Global Value Chain approach to analyzing commodity systems, which we adopt in this book, is essentially one such linear model. We will describe this particular model in greater detail in Part Two of this chapter.

The filière approach represents a French variant of the field, with origins connected to French agricultural policy. Its emphasis is on the empirical analysis of chain (filière) structures, particularly in Francophone Africa, with minimal attention to broader concerns associated with globalization. Raikes et al. (2000) characterize the approach as 'neutral' in purpose and orientation, signifying its overall lack of interest in drawing implications about global economic processes. This derives from what they refer to as the Francophone tradition in which 'economics was a science of accountancy and not of behaviour' (Raikes et al., 2000: p. 405).

The systems of provision (SoP) framework, which is associated with the work of Ben Fine (Fine, 1994; Fine et al., 1996; Fine, 2002, 2004), draws its inspiration from questions relating to the political economy of consumption. Fine developed the framework coming out of a broader interest in seeking to explain the stratification of diets and consumption by way of income, geography and ethnicity (Fine, 1993; Fine et al., 1992; Fine and Leopold, 1993). His methodology was to work backwards from the proverbial dinner table in order to document the *systems of provision* that enabled populations to be fed.[1]

Thirdly, Commodity Systems Analysis (CSA), associated with the pioneering work of Bill Friedland (Friedland et al., 1981), was influential in the way it reinterpreted the sociology of agriculture in terms of industrial systems. His work on the Californian lettuce and tomato industries uncovered the hierarchical regimes that connected growers with downstream shippers and retailers, and, in ways that pre-empted Gereffi's later work on GVC governance, spoke of the power and coordinating relations enshrined in these regimes. Friedland's contributions have been revived through the work of Jane Dixon (1999), who has 'brought in' consumption to the CSA method.[2]

Finally, the generically titled 'supply chain' approach designated in Table 2.1 refers to a parallel body of chain research emanating from agribusiness studies (Jackson et al., 2006: pp. 131–2). Taken together, the pre-eminent concern of all these linear approaches is the mapping of chain architectures. As a result, they have been all subject to criticism by economic geographers on account of their generally residual attention to spatial processes (Hughes and Reimer, 2004).

By way of contrast, the three non-linear post-structuralist frameworks listed in Table 2.1 are all ontologically constructed in ways that treat geography more centrally within their frames of reference. They possess a more wide-eyed perspective that gives primary consideration not to the product/commodity chain/filière/system itself, but to its interactivity with particular places/spaces/regions. These approaches characteristically emphasize the feedback loops that exist between product/commodity systems and particular places.

'Commodity networks' originated in the post-structural turn which flowed through the social sciences from the late 1980s. In the network perspective, particularly as it was developed in Actor–Network Theory (ANT), human activity is understood to be always localized; it operates in and through real places at specific times. Applied to the analysis of products/commodities, this suggested an approach which emphasized the contingencies that bring 'things' together at particular moments and thus forge an extension of actor networks across space (Law, 1987; Latour, 1987; Arce and Marsden, 1993; Murdoch, 1995; Whatmore and Thorne, 1997; Busch and Juska, 1997).

The second category, 'commodity circuits', is defined largely by its interest in culture. Key works in this field are preoccupied by questions relating to how products and commodities are inserted within cultural spaces and understood by consumers (Crang, 1996; Cook and Crang, 1996; Cook et al., 2004). It is premised on an ontological framework in which events and processes are mutually constituted. At the core of the approach is the principle that commodities do not 'begin' by being produced and 'end' when they are consumed, but operate through recursive circuits in which the dynamics of consumption stand equal to the dynamics of production in determining the shape of product/commodity systems.

Evidently, the post-structuralist commodity circuits/networks approaches foreground the 'lived worlds' of products/commodities as they traverse space. They emphasize the local contexts and *dis*connections that are part and parcel within global exchange. In the research by Cook and Crang (1996) and Cook et al. (2004) on UK consumption of tropical fruits, for example, the vital object of inquiry is *not* the issue of chain governance (and the question of who and what circumstances gains economic value from participating in this industry), but the altogether different question of how local consumers construct particular meanings about these products, and how such meanings differ from how other participants in these systems 'understand' these products. Such interrogations enable detailed consideration of specificity and nuance but, in common with critiques of post-structuralist social theory more generally, the approach is open to the charge that it de-politicizes the analysis of economic activities. Fine (2004: p. 336) asserts that the approach is 'primarily selective description, and lacking in causal theory, or, indeed, a reflective analytical account of what it chooses to describe'. Coe et al. (2008a: p. 287) observe that 'our sense is that these studies perhaps reveal less about the power and value relations inherent to such systems'. Cloke et al. (2004: p. 190) argue that the approach has a rather vague fixation to:

> latch on to specific people, things, metaphors, plots/stories/allegories, lives/biographies, and/or conflicts; and follow them, see what 'sticks' to them, gets wrapped up in them, unravels them and [to see] where that takes a researcher who is (un)able to follow myriad possible leads and to make myriad possible connections.

Such methodologies may valuably bring to light the complexities of product/commodity systems but, as Jackson (2002: p. 15) has commented: 'demonstrating complexity is scarcely a worthwhile end in itself' (also see Hughes and Reimer, 2004: p. 7). Thus, although these circuit/network approaches certainly foreground geography, the danger is that they do so in a way which occludes consideration of economic power.

The concern to 'bring in geography', yet not lose sight of issues of economic power within commodity/product systems gave rise to the Global Production Networks (GPN) concept, a framework of inquiry that associated with a group of researchers known as the 'Manchester school' of economic geography (Bathelt, 2006: p. 226).[3] As articulated by these researchers, the avowed purpose of GPN analysis has been to provide a bridging framework that enables joint consideration of 'vertical' (chain) and 'horizontal' (territorial) concerns. In its formal introduction to the scholarly world in Henderson et al. (2002), it was defined as having a 'focus on the flows *and* the places *and* their dialectical connections' (Henderson et al., 2002: p. 438) (italics in original). Thus, it represents a hybrid approach. According to Hess and Yeung (2006), GPN analysis was constructed from an amalgam of influences including Porter-infused value chain studies (Porter, 1990); network analysis (notably, Yeung's uses of these frameworks to explain the economic geographies of Asian corporations [Yeung 1998a, 1998b, 2005; Yeung and Lin, 2003]); the concept of socio-spatial embeddedness within the 'new economic sociology' of the 1980s and 1990s (Granovetter, 1985; Granovetter and Swedberg, 1992) and GVC analysis.

The hybrid nature of the GPN approach is brought out in Coe et al.'s (2004) development of the concept of 'strategic coupling'. This terminology was developed to describe and account for the *multi-scalar* connections (a term which we borrow and return to in Part Three of this chapter) between the (vertical/chain) governance arrangements of production systems, and the (horizontal/network) ensembles of place-based economic actors. As such, the concept of strategic coupling provided a vehicle which called to attention how regional development trajectories depend *not* on the character of endogenous/local factors *per se*, but the ways these attributes 'couple' with 'a range of institutional activities across different geographical and organizational scales' (Coe et al., 2004: p. 469). Hence:

> the interactive complementarity and coupling effects between localized growth factors and the strategic needs of trans-local actors [propel] regional development. We argue that it is these *interactive effects* that contribute to regional development, not inherent regional advantages or rigid configurations of globalization processes. Despite certain path-dependent trajectories, regional development remains a highly contingent process that cannot be predicted *a priori*. (Coe et al., 2004: p. 469; italics in original)

Advocates of a GPN approach emphasize the benefits of using a 'network' metaphor, as opposed to a 'chain', as it more accurately describes a world of increasingly dense, intricate and flexible interconnections. Networks are understood as 'multidimensional, multilayered lattices of economic activity' (Dicken et al., 2001; Dicken and Malmberg, 2001; Weller, 2006: p. 1250) that breech industry sectors; incorporate ensembles of corporate, state and non-state actors (such as NGOs) and operate flexibly across space. For example, Hess and Coe (2006) use the network concept in their study of the telecommunications sector to trace the shift from 'relatively simple' value chains anchored by national flagship firms, to 'complex networks of firms, states, and institutions'.

Seen in the light of these critiques of linear and non-linear approaches, the evident challenge for researchers in this field is to devise and/or utilize frameworks that retain attention to issues of economic power, yet also give scope to incorporate geographical notions of space, place and territory as active constituents to the construction of social and economic outcomes. To this end, we now ask questions about the capacity for the GVC approach to fulfill these aspirations.

## Part Two: Global Value Chain Analysis

The intellectual origins of what is now known as GVC analysis can be traced to institutionalist and world-historical analyses of global capitalism dating from the 1970s and 1980s. Specifically, the GVC framework has its genesis in the world systems research of Hopkins and Wallerstein (1977, 1986) with its focus on the connections between political regimes and international modes of capital accumulation. In seeking to understand the rise and decline of different economic structures over long-term historical cycles, Hopkins and Wallerstein developed the concept of 'commodity chains' as a means to describe the exchange structures that underpinned the trade of products vital to global development, such as wheat, sugar, coal, cotton, gold, etc.

(Bair, 2005; Gibbon and Ponte, 2005). For Hopkins and Wallerstein, commodity chains were the 'warp and woof of the commodity system'. During expansionary phases of economic development, 'chains are expanded and become more vertically integrated [...] in phases of contraction the reverse is the case' (Raikes et al., 2000: p. 392).

This historically focused framework was first applied to the investigation of contemporary patterns of global trade in a chapter by Gereffi and Korzeniewicz (1990) on the footwear sector, and then, in 1994, became the organizing principle for Gereffi and Korzeniewicz's landmark edited collection of 'Global Commodity Chain' (GCC) studies, which established the field as a discrete arena of inquiry. Contributors sought to apply the concept to the contemporary, globalizing, phase of the world economy. Their intent – which seems hardly revolutionary today but was certainly path-breaking in 1994 – was to analyze production systems explicitly outside of the realm of national economic boundaries. Global Commodity Chain analysis (as it was then called) thus provided a framework that was not rooted to national economies as arenas for analysis. As Gereffi and Korzeniewicz specified, the approach could be used for probing: 'above and below the level of the nation-state to better analyze structure and change in the world-economy' (1994: p. 2). This vital innovation paved the way for the widespread application of the concept which followed over subsequent years.

The GCC framework quickly gained wide currency within the social sciences and, along the way, Gereffi's original terminology of 'global commodity chains' began to be replaced by the more-inclusive wording of 'global value chains'. The word 'commodity' was problematic because it implicitly associated the field with raw products ('commodities'), when, in fact, much of the early GCC research focused on manufactures. Researchers from the Institute for Development Studies at the University of Sussex are generally credited with promoting the change in terminology from GCCs to GVCs (Hess and Coe, 2006: p. 1207). The term 'global *value* chains' not only signified a field of inquiry with a greater range of subjects and scope, but also dovetailed with the influential terminology attached to the work of Michael Porter (1990) on competitive advantage, thus opening it to a wider audience in the fields of economics and business studies.[4] Notwithstanding these changes of nomenclature, however, the GVC approach, as understood in this book, retained its compelling characteristic; namely, a research strategy for understanding the operations of industry systems across the world's geography, thereby providing an informed analysis of how capitalist processes generate opportunities and constraints to different people and places in the global economy.

Fundamental to the quick uptake by researchers of this approach was Gereffi's specification of a clear organizing agenda for GVC research. In his contribution to the 1994 edited book, Gereffi charted a straightforward 'how-to' guide which asserted that chains have three analytical dimensions: (i) an input–output structure; (ii) territoriality; and (iii) a governance structure. The following year, he added the category of 'institutional framework' to his analytical approach, defining this as 'how local, national, and international conditions and policies shape the globalization process at each stage of the chain' (Gereffi, 1995: p. 113). We will return to the issue of the 'institutional framework' of chains in later sections of this chapter. Focusing on the initial three categories, however, it is apparent that 'input–output structure' and 'territoriality' reflect relatively straightforward, taken-for-granted criterion within industry analysis, where productive activities along the chain are mapped. Inclusion of the governance dimension, however, provided a key new addition to these debates. It provided a mechanism to assess product systems in terms of the processes of global coordination which mark the modern world economy. The governance dimension of GVC analysis then gained further influential policy impact when attached to the second flagship concept developed by Gereffi; *upgrading*. This notion encapsulated the question of how producers' positions within chains either enhanced or restricted their capacities to improve their livelihoods. As the two flagship ideas that have been central to most GVC analyses, we now discuss the concepts of governance and upgrading, before turning in the next major section to address recent critiques of this overall approach.

## The governance dimension within global value chains

GVC governance can be defined as the parameters that precondition the terms under which actors elsewhere in the chain must operate (Humphrey, 2005). Gereffi argued that in the contemporary global economy, firms are not 'just there' waiting to sell their goods and services, but are constructed as interdependent entities connected to other upstream and downstream chain participants. In specifying the importance of this issue, GVC analysis thus provided a tangible route for connecting debates on economic power with those on the international coordination of economic activity.

Gereffi's delineation of the concept of 'governance' led to his path-breaking articulation of the idea that globalization was executing a shift in power within chains from producers to buyers. He contended that 'two distinct types of governance structures for GCCs have emerged in the past two decades' (Gereffi, 1994: p. 97), and that 'an important trend in global manufacturing appears to be a movement from producer-driven to buyer-driven' chains (Gereffi et al., 1994: p. 5). The essential distinction between these two types of chains was said to lie in the type, and function, of the firm driving the chain. Producer-driven chains are driven by 'large integrated industrial enterprises' and are most readily seen in 'capital- and technology-intensive sectors such as automobiles, computers, aircraft and electrical machinery' (ibid.). Traditionally, a central element of these industries has been the need for large-scale production facilities making use of scale economies. The necessary sunk capital cost associated with such activities has tended to provide a prohibitive barrier to entry and, therefore, has tended to give these firms considerable powers within chains. In contrast, buyer-driven chains refer to industries 'in which large retailers, brand-name merchandisers, and trading companies play the pivotal role' (ibid.). The industries where these arrangements proliferate, such as clothing and footwear, characteristically have greater ease of entry for manufacturing firms, encouraging intense (and often) unstable conditions of competition. Following from this, the implication is that producer firms can be subordinated more readily by large buyers. Led by Gereffi's own research, the apparel, footwear and consumer goods industries became the exemplary cases from which to observe these processes at work. Gereffi's initial exploration of these issues (Gereffi, 1994) became a fount for succession of later studies (Gereffi, 1999; Bair and Gereffi, 2001, 2002, 2003; Gereffi et al., 2002; Abernathy et al., 1999; Jones, 1998; Mortimore, 2002).

Evidently, GVC analysis captured a simple, but compelling argument about the organizing principles of contemporary capitalism; namely, that as borders opened and procurement strategies began to be played out on a global canvas (aided and abetted by emergent information technologies), downstream buyers gained strategic advantages over their upstream suppliers. At the same time, however, the dualistic characterization of chains as either buyer- or producer-driven evidently represented an ideal-typical portrait to which many sectors did not entirely comply. As suggested by Raikes et al. (2000: p. 397), the dichotomous framework of buyer-driven versus producer-driven arrangements represents a 'rather fixed and rigid' portrayal of global economic processes. Correspondingly, recent years have seen researchers develop a series of variants to this dichotomy. Especially influential here has been the work of Timothy Sturgeon, who articulated the intermediary notion of *modularity*, referring to a category of arrangements in which power was not held unambiguously at either the upstream or downstream ends of a chain, but was shared with 'middle-agents'; highly competent suppliers coordinating large assemblages of semi-processed inputs for further processing (Sturgeon, 2001, 2002, 2003). As documented in studies of production systems in the Southeast Asian electronics and automotive sectors (Sturgeon and Lester, 2004), the emergence of modular systems involved the rise of global-scale supplier firms positioned centrally within internationally fragmented production systems.

Also working on the problem of embellishing the initial buyer/producer dichotomy during this same period, Humphrey and Schmitz (2000, 2002) proposed a fourfold classification schema involving: (i) market relations; (ii) networks; (iii) quasi-hierarchy; and (iv) hierarchy. This approach then formed the backbone for the 'marquee paper' by Gereffi, Humphrey and Sturgeon (2005) which brought together these research strands and spelt out a fivefold categorization of global value chain governance:

(i) *Markets*. Arrangements characterized by spot price or repeat transactions in which the costs of switching partners are low for both parties;
(ii) *Modular value chains*. Suppliers in modular arrangements make products according to customers' specifications;
(iii) *Relational value chains*. Arrangements dominated by complex interactions and mutual dependencies between suppliers and customers;
(iv) *Captive value chains*. Small suppliers held commercially captive to larger buyers because of transactional dependence. Because of market and/or technological constraints, suppliers face considerable costs in switching to new customers, and thereby are placed in a subordinated position, and
(v) *Hierarchy*. Systems of vertical integration within chains.

Through this framework, categories of governance are brought more into line with micro-economic realities. Most market arrangements do not mirror ideal-typical extremities of perfect competition or monopoly, but fall somewhere in the middle (what micro-economists label imperfect competition). Comparably, the framework above is dominated by three governance formations (modular, relational and captive) which fall in-between the ideal-typical models of market and hierarchy forms of governance. Therefore, the development of this fivefold categorization can be seen as marking a major evolution in this field. The dualistic heuristic of producer- versus buyer-driven chains has been recalibrated to match more realistically the findings of studies that apply the concept of governance to real-world cases.

Theorizing the conditions under which each of these governance arrangements may arise, Gereffi, Humphrey and Sturgeon propose three factors as critical: the complexity of information and knowledge transfer; the extent to which such information can be codified, and the capabilities of existing and potential suppliers. As they suggest:

> The complexity of information transmitted between firms can be reduced through the adoption of technical standards that codify information and allow clean hand-offs between trading partners. Where in the flow of activities these standards apply goes a long way toward determining the organizational break points in the value chain. When standards for the hand-off of codified specifications are widely known, the value chain gains many of the advantages that have been identified in the realm of modular product design, especially the

> conservation of human effort through the re-use of system elements – or modules – as new products are brought on-stream. (Gereffi, Humphrey and Sturgeon, 2005: p. 85)

Representing a major renovation in the conceptualization of GVC governance, this fivefold categorization gives flexibility and nuance to the approach, and connects it more directly to economic theories of the firm (Table 2.2). As Gereffi, Humphrey and Sturgeon acknowledge (2005: p. 80), attention to these factors owes an intellectual debt to transaction-cost economics. This approach is generally traced back to Ronald Coase's (1937) *The Nature of the Firm* (which explicitly recognized the firm's existence as a response to transaction cost minimization) and Oliver Williamson's (1975) work on the economics of organization. As Williamson asserted: 'economics should expressly address and help assess the transactional properties of alternative modes of organization' (Williamson, 1975: p. 2). Applying these notions to the question of GVC governance, Gereffi, Humphrey and Sturgeon make the point that forms of governance evolve in relation to shifts in the complexity of information and the ability to codify these into known benchmarks (standards, grades and codes of conduct, etc.). This perspective brings into focus the issue of *conventions*, which have been introduced into GVC research through the efforts of the Danish-based development economists Stefano Ponte, Peter Gibbon, and their colleagues (Giovannucci and Ponte, 2005; Ponte and Gibbon, 2005; Gibbon and Ponte, 2005; Daviron and Ponte, 2005). Originating in the work of French heterodox economists studying risk (Boltanski and Chiapello, 1999; Favereau, 2002), convention theory calls to account how economic exchange depends on the concept of 'equivalence', defined as 'communal agreement on the quality or character' (Biggart and Beamish, 2003: p. 456) of a product or activity. Applied to GVC analysis, the concept of conventions provides a mechanism to explain how the entry barriers that define participation in chains, and the processes by which forms of codified knowledge comes into being, arbitrates 'quality'. As Gibbon and Ponte (2005: p. x) suggest: 'Convention theory enriches GVC analysis' preoccupation with chain governance through a better understanding of the normative dimensions of governance and its consumption-related aspects.'

## The upgrading dimension within global value chains

The concept of upgrading adds an important contribution to the overall utility of the GVC approach. It provides a framework to chart the ways that participants can alter their positions within chains. From their earliest days, GCC/GVC studies had an explicit focus on questions relating to upgrading,

**Table 2.2** Key determinants of global value chain governance

| *Governance type* | *Complexity of transactions* | *Ability to codify transactions* | *Capabilities in the supply-base* | *Degree of explicit coordination and power asymmetry* |
|---|---|---|---|---|
| Market | Low | High | High | Low |
| Modular | High | High | High | \| |
| Relational | High | Low | High | \| |
| Captive | High | High | Low | \| |
| Hierarchy | High | Low | Low | High |

*Source*: Based on Gereffi et al. (2005: p. 87).

with Gereffi (1995) using the GVC framework to develop a model of the sequencing of export composition in East Asia.[5] Defining 'upgrading' in relation to the question of how firms gain entry into 'international design, production and marketing networks', Gereffi and his collaborators asked how economic actors generated the requisite 'skills, competences and supporting services' necessary for this participation (Gereffi, Humphrey, Kaplinsky and Sturgeon, 2001: p. 2). Arguing that: 'participation in global commodity chains … puts firms and economies on potentially dynamic learning curves' (Gereffi, 1999: p. 39), this focus coalesced with debates in the 1990s on export-oriented development and the task of explaining the 'East Asian miracle'.

For economic geographers in the 1990s, Gereffi's focus on the role of vertical inter-firm linkages within value chains was provocative, because it appeared to challenge the influential view that horizontal inter-firm dynamics, within industrial clusters, provided the key to industrial learning and economic development. Reconciling the apparently contradictory claims of 'vertical' upgrading through GVCs and 'horizontal' upgrading through cluster development, Humphrey and Schmitz (2002) made a key contribution by arguing that what was important was not participation in a GVC *per se*, but the relations through which producers were connected to downstream chain actors. This perspective rejected both 'the naïve optimism that prevails in much of the local development [clustering] literature', and also 'the pessimism of the globalization critics who fear that local strategies are rendered irrelevant by global forces' (Schmitz, 2004: p. 2). Evidently, the perspective sought to address the reality that: 'Many producers, especially those of small and medium size, find that participating in and gaining from the global economy do not always go together' (Schmitz, 2004: p. 1).

Coming out of the debate over these issues was a fourfold classification of upgrading (Humphrey and Schmitz, 2002; Schmitz, 2004, 2006):

(i) *process upgrading* (transforming inputs into outputs more efficiently by reorganizing the production system or introducing superior technology);
(ii) *product upgrading* (moving into more sophisticated product lines);
(iii) *functional upgrading* (acquiring new functions in the chain (or abandoning existing functions) to increase the overall skill content of activities); and
(iv) *inter-sectoral upgrading* (using the knowledge acquired in particular chain functions to move into different sectors).[6]

The question of how and why particular subordinate (upstream) producers follow any one of these upgrading paths has long held particular interest for GVC scholars. As evidenced in the edited collection of Schmitz (2004), a key theme of research has been the question of whether certain forms of upgrading are connected to particular forms of chain governance and, more specifically, whether the rise of captive forms of chain governance constrains or encourages upgrading. Humphrey and Schmitz (2004: p. 354) summarize the state of knowledge on this topic by advancing the proposition that in instances where downstream lead firms exercise some power on upstream participants: 'developing country producers experience fast product and process upgrading but make little progress in functional upgrading'. The reason for this is that whereas local producers tend to learn 'practical' things from participating in chains, such as how to improve productivity or make a superior product, downstream lead firms tend to studiously protect key components of value-addition within chains, such as the ownership of patents, brands and strategic know-how. As described by Schmitz (2006: p. 566):

> ... integration into global captive chains is often a double-edged sword. On the one hand, it facilitates inclusion and rapid enhancement of product and process capabilities and enables developing country firms to export into markets which would otherwise be difficult for them to penetrate. On the other hand, it can lead to producers being tied into relationships that prevent functional upgrading and leave them dependent on a small number of powerful customers.

In conclusion, this attention to issues of governance and upgrading has been centrally important to the predominant application of GVC analysis, since its inception in the mid-1990s. As discussed, key concepts and categories have been substantially refined and developed during recent years, marking a new stage in the evolution of the GVC approach. Yet notwithstanding the important recent contributions of scholars such as Sturgeon (with respect to governance) and Schmitz (with respect to upgrading), this recent period has also been a time during which the GVC approach has come under sharp criticism from economic geographers. We now turn our attention to the basis for these critiques.

## Critiques of Global Value Chain analysis

According to proponents of competing schools of thought, such as GPN, the GVC approach provides a good basis for the analysis of governance structures, but its linear framework diminishes its capabilities to 'bring in' geographical complexity (the issues of space, place and scale) within the examination of product/commodity systems. Hence, GPN analysis has been promoted as an approach which goes 'beyond the very valuable but, in practice, more restricted' GVC approach (Coe et al., 2008b: p. 272). Seminally, the 'Manchester school' articulated its critique as follows:

> A major weakness of the 'chain' approach is its conceptualization of production and distribution processes as being essentially vertical and linear. In fact, such processes are better conceptualized as being highly complex *network* structures in which there are intricate links – horizontal, diagonal, as well as vertical – forming multi-dimensional, multi-layered lattices of economic activity. For that reason, an explicitly relational, network-focused approach promises to offer a better understanding of production systems. (Henderson et al., 2002: p. 442)

Furthermore, Coe et al. (2004: p. 469) typecast GVC analysis as saying little 'about how particular sub-national spaces and their institutions are integrated into, and shaped by, transnational production systems'. According to Hess and Yeung (2006: p. 1196):

> The GCC/GVC analysis, however, does suffer from some significant shortcomings that can be remedied ... (see Dicken et al., 2001; Henderson et al., 2002) ... although the chain concept in the GCC analysis brings multiple geographical scales, particularly the global scale, to the forefront of its analysis, the geography of GCCs remains weakly developed and under-theorized – no doubt a reflection of the origin of the framework in sociology. The issue of territoriality is highly aggregated in the GCC framework, identifying the spatial units of analysis as either core or periphery.

These claims that GVC analysis goes *only goes so far*, as a conceptual tool, are rooted deeply in geographical ontology. As noted earlier in this chapter, *space matters* to economic geographers because it is understood as an active agent in the production and reproduction of economic, social and political relations. As Yeung (1998a: p. 110) has argued:

> Space is thus simultaneously a product of its constitutive things and relations *and* a means of production, producing spatially embedded things and relations; it is more than a medium or 'container' of social events and activities as conceptualized in Cartesian geometry. (italics in original.)

Consequently, it is no surprise that economic geographers have observed shortcomings in the propensity of many GVC studies to discuss governance in non-spatialized, top-down ways. From a geographical perspective, the evolution of buyer-driven forms of governance is not just something 'which impacts upon' particular locations; but is a relational construct of how places and spaces are transformed in a globalizing world. Capturing the nuance of this transformation (how space actively shapes the evolution of governance formations) is core business to an economic geographer.

The GPN concept shifts the frame of attention in commodity/product analysis from organizational forms and structures, to the specific relations (in time and space) that connect particular economic activities to particular places (Dicken et al., 2001: p. 91). Influenced by the theoretical work of Massey (1993a, 1993b, 2005), these perspectives eschew the conception of places and spaces as passive receptacles of broader-scale economic logics (as the abstract spatial determinism of location analysis would suggest), whilst also avoiding the pitfalls of spatial particularism (i.e., emphasizing the uniqueness of place at the expense of wider considerations of power and agency) that potentially befall post-structuralist perspectives (Yeung, 2005).

Flowing from these insights, GPN theorists argue that each (linear) stage in the addition of value within an economic system is understood to be embedded within an interconnected web of (non-linear) network upon network spanning social, economic, political and environmental (human and non-human) considerations. What this generates, in practice, is a potentially wide and rich inquisitorial menu. Fostering a worldview in which individual activities/agents are positioned as being embedded in multitudinous networks, rationalizes many and varied entry points for researchers, and justifies research that branches off in manifold directions. A logical consequence of the shift from a 'chain' to a 'network' frame of reference, therefore, is the diffusion of research foci. For its advocates, the clear merit of this transformation is that it facilitates research which 'sees the unseen'; investigations which follow network connections and make visible the complexity and contradictions of global economic life. As Coe et al. (2008: p. 272) state: 'GCCs/GVCs focus narrowly on the governance of inter-firm transactions while GPNs attempt to encompass all relevant sets of actors and relationships'.

There is no doubt whatsoever that the overall focus of the GPN approach speaks to a crucial shortcoming in the traditional articulation of GVC analysis. Quite correctly, these scholars point to how an emphasis on governance and transaction costs within GVC analysis occludes consideration of how such processes are outcomes of economic actors embedded in all kinds of spatial networks. In the next section, we address the problem of the under-theorization of space within GVC analysis from a different angle and, through this tack, work towards incorporating the insights of GPN theorists into the GVC framework. In line with arguments by Sturgeon et al. (2008) we see merit in the nomenclature and taxonomy of the GVC approach,[7] however recognize that if the GVC approach is to substantially address its critics, it needs refinement. In this quest, the 'institutional dimension' of GVC analysis comes to centre stage.

## Part Three: The Institutional Dimensions of Global Value Chains

There is an important, but rarely acknowledged disjuncture between the framework and prevailing mode of application in GVC analysis. As discussed above, the GVC approach is defined through a fourfold method: attention to the input–output, territoriality, institutional and governance dimensions of chains. Yet as also already mentioned, most studies adopting a GVC frame of reference have tended (either exclusively or predominantly) to focus on the single dimension of governance. As this concept is a flagship idea within the approach, it is hardly surprising that research has clustered around it. However, our contention is that such focus has delimited perceptions of its potential capacity to address complex questions relating to the spatial organization of economic activities. The key to unlocking this potential, we argue, is contained in closer attention to the *institutional dimension* of GVC analysis.

The institutional dimension of GVC analysis has not received detailed analytical consideration by GVC theorists. As acknowledged by Hess and Yeung (2006, p. 1196), 'the institutional dimensions of the GCC/GVC analysis seem to be hijacked by its privileging of governance structures'. As a result, to the extent that it occurs, discussion of 'institutional context' from within the GVC literature tends to appear wooden and simplistic. Gereffi gave passing and tangential treatment to this analytical dimension in his early publications, and not much of significance has happened since. For example, Gibbon (2001a: p. 347) has commented: 'According to ... [Gereffi (1999)], the fourth dimension (the institutional framework surrounding a chain) mainly comprises conditions under which control over market access and information are exercised on a global plane'. Thomsen (2007: p. 757) suggests that Gereffi has 'provided little indication of the exact meaning of this term'. From the perspective of contemporary economic geography, such conceptions of 'the institutional framework' are extremely problematic.

Hess and Coe (2006: p. 1208) observe that they perpetuate a framework within GVC analysis in which: 'Culture and non-firm institutions are still treated as externalities'. As we explore in the sections below, this belies recent theorization in the social sciences that conceives a richer and more complex articulation of the role of institutions in economic life. Our goal, therefore, is to re-energize the 'institutional imagination' of GVC studies, and thereby address the criticisms that it provides an inadequate framework for *geographical* inquiry.[8] We do this in the next section by examining the twinned concepts of 'institutions' and 'embeddedness' and then draw together the key points of this discussion. An institutionally enriched form of GVC analysis, we argue, transcends the criticisms levelled on the approach on account of its allegedly sclerotic preoccupation with governance categories forged out of transaction cost dynamics.

## The meaning of institutions

The essential starting point for these arguments is to review what we mean by the various terms associated with the concept of 'institutions' (e.g., 'institutional framework', 'institutional environment', 'institutions', 'institutional capacity', 'institutional arrangements' etc.). In a general sense, as alluded to in Chapter 1, the concept of 'institutions' has emerged to become a key buzzword in the debate on globalization. Its essential contribution has been to refocus attention towards the social, economic, political and environmental settings in which reforms and interventions are enacted. At the same time, however, the burgeoning usage of 'institutional analysis' has led to diverse understandings of concepts that are not always consistent with one another. Because the analysis of 'institutions' has an extensive heritage across a range of social sciences, contemporary deployments of its associated terminology can mean many and varied things to different authors and readers. As Hirsch and Lounsbury (1996: p. 882) suggest, it is perhaps the case that the study of institutions 'necessitates such a wide variety of approaches and methods because of the complexity of the phenomena.' Such diversity behooves us to define precisely how we understand and deploy this framework.

Institutional analysis has figured prominently in economic geography since the mid-1990s, leading many scholars to embrace the notion that the discipline has been engaged in an 'institutional turn' (see, *inter alia*, Amin, 2004; Amin and Thrift, 1995; Barnes, 1999; Macleod, 2001; Martin, 2000; Philo and Parr, 2000). The central question of this 'turn' has been: 'To what extent and in what ways are the processes of geographically uneven capitalist development shaped and mediated by the institutional structures in and through which those processes take place?' (Martin, 2000: p. 79).

In asking such a question, economic geographers have found common ground with like-spirited theorists across the disciplines of sociology, anthropology, political science and development studies, and among some (though not all) institutional economists. Despite differences in terminology, emphasis and orientation, the central concern within these literatures has been to argue that geographies, historical contexts, institutions and organizations *matter*; that economic outcomes can never be disentangled from the concrete social worlds in which they lie.

The slant of economic geographers' embrace of such 'institutional analysis' owes a direct debt to the seminal insights of Karl Polanyi (1944) and their invigoration by Granovetter (1985; Granovetter and Swedberg, 1992). Polanyi's *The Great Transformation* (1944) presented a social history of the economy, thus critiquing the myth of a self-regulating market. Yet, as Trevor Barnes (1999) has pointed out, whilst the publications of Polanyi and Granovetter provided the major intellectual pathway for these ideas to inspire the 1990s 'institutional turn' in economic geography, their core ideas tap into an even earlier tradition of institutional analysis within economics. This so-called '*old* institutional economics' (OIE) (to distinguish it from '*new* institutional economics, or NIE)[9] was associated with early twentieth-century economic theorists such as Harold Innes, Thorstein Veblen, Wesley Mitchell, and John R. Commons. Veblen, in particular, sought to interrogate the meaning and significance of institutions through his work on 'habits of thought'. In defining institutions as 'settled habits of thought common to the generality of men [sic]' (Veblen, 1919: p. 239; see Hodgson, 1993: p. 125), Veblen constructed a framework which posited these as being rooted in cultural and social norms:

> institutions have a stable and inert quality, and tend to sustain and thus 'pass on' their important characteristics through time. Institutions are seen as both outgrowths and reinforcers of the routinized thought processes that are shared by a number of persons in a given society. (Hodgson, 1993: p. 253)

Comparably, Barnes (1999: p. 1) characterizes Innes as a 'dirt economist', practicing 'an economics that was grounded in the nitty-gritty of the institutions, technology and ecology of particular places'. Within economics, the 'methodological holism' in OIE (Toboso, 2001) was rekindled in the later decades of the twentieth century through Douglass North's Nobel Prize-winning work (North, 1990, 1995; North and Thomas, 1973).

North's work has hardly figured in recent economic geography scholarship on institutions but, we contend, offers a potentially important contribution. From our perspective, one of the chief shortcomings in recent economic geography scholarship associated with the 'institutional turn' has been a lack of specificity when it comes to nailing down precise conceptual meanings. As argued by Wood and Valler (2001: p. 1140), the tendency in economic geography is to equate institutions with a 'rather narrow reference to sets of formal concrete organizations'. It is as if, as hinted by Trevor Barnes (1999), geographers have rushed to embrace the notion of 'institutions' ultimately for reasons of pragmatism and convenience; by offering a meso-scaled theoretical middle ground (Amin, 2001: p. 1238) that can give emphasis to the role of place, it provides a justificatory 'warm inner glow' to arguments about the importance of a *geographical* perspective. In the work of North, however, we see a fuller and more finely-honed articulation of key concepts.

In his ground-breaking book *Institutions, Institutional Change and Economic Performance* (1990: p. 3), North identifies institutions as 'the rules of the game in a society or, more formally, the humanly devised constraints that shape human interaction'. North's work on institutional change and economic performance presents institutions not just in the neoclassical sense of reducing transaction costs (c.f. NIE), but as being central to the foundations of economic theory itself:

> Institutions are the humanly devised constraints that structure human interaction. They are made up of formal constraints (for example, rules, laws, constitutions), informal constraints (for example, norms of behaviour, conventions, self-imposed codes of conduct), and their enforcement characteristics. Together they define the incentive structure of societies and, specifically, economies. (North, 1998: p. 248)

For North, there is a clear distinction in his analysis between institutions and organizations, which he analogues as the difference between the rules and the players. However, there is evident lock-in between these two categories: 'both what organizations come into existence and how they evolve are fundamentally influenced by the institutional framework. In turn, they influence how the institutional framework evolves' (North, 1990: p. 5). Unlike the implicit economic Darwinism of transaction-cost economics (that operates on the assumption that the evolution of institutional arrangements reflect 'the survival of the fittest institution'), lock-in is described by North in terms of path-dependent processes as particular interest groups struggle over the perpetuation or restructuring of the status quo. In this regard, the North-inspired field of institutional economics also makes a distinction between the *institutional environment* and *institutional arrangements*. The institutional environment refers to the 'set of fundamental political, social and legal ground rules that establishes the basis for production, exchange and distribution', whereas institutional arrangements are 'an arrangement between economic units that governs the way in which these units can cooperate and/or compete' (Davis and North, 1971: pp. 6–7).

Perceived in these terms, (i) institutions have critical importance as an underlying determinant of the long-run performance of economies; and (ii) institutional path dependence is the key to understanding economic change. For North, historical change in a society is thus gradual, path dependent and endogenous where informal institutions play a fundamental role, both by shaping the institutional structure and by conditioning the mental constructs of individuals and organizations who in their turn generate incremental changes.

North's defining contribution to the field of institutional economics helps to explain why societies exhibit divergent paths of historical change. His analysis of the rise of the western world, for example, understands economic change in terms of the evolution of different institutional environments locking-in different path-dependent trajectories of change (North and Thomas, 1973). Yet notwithstanding these contributions, North's framework takes us only so far in addressing the issues central to this book. He contends that 'institutions matter', but does not specifically come to terms with the more detailed *geographical* questions of their existence; how and why particular institutional arrangements coalesce in individual places. To make progress on these questions, we need to consider Granovetter's reworking of Polanyi's notion of embeddedness.

According to Polanyi, economic arrangements (such as market exchanges) are *embedded* in social and cultural structures:

> The instituting of the economic process vets that process with unity and stability; it produces a structure with a definite function in society; it shifts the place of the process in society, thus adding significance to its history; it centers interest on values, motives and policy. Unity and stability, structure and function, history and policy spell out operationally the content of our assertion that the human economy is an instituted process. The human economy, then, is embedded and enmeshed in institutions, economic and non-economic. (Polanyi, 1957: pp. 249–50)

In the 1980s, Granovetter reintroduced this concept in the guise of the *new economic sociology*. He argued that 'the behavior and institutions to be analyzed are so constrained by ongoing social relations that to construe them as independent is a grievous misunderstanding' (1985: p. 482). According to these lines of argument, economic relationships are never fully commoditized, but depend always on interplay between the economic and the social (Murdoch et al., 2000). As such, the concept of embeddedness brings the sociological setting of the economy back into the fulcrum of economic thought. The approach asserts the mutuality of social and economic processes within specific, temporal and place-based settings. Inspired by these conceptual advances, researchers applied the precepts of new economic sociology to a rich variety of thematic concerns (*inter alia*, Christopherson and Storper, 1986; Cooke and Morgan, 1993; Chari, 1997; Lee and Wills, 1997; Wilkinson, 1997; Callon, 1998; Murdoch, 2000; Murdoch et al., 2000; Wrigley et al., 2005; Kalantaridis and Bika, 2006).

In the 1980s and 1990s, geographers latched onto the obvious territorial components of this concept. Reviewing geographers' contributions to the field, Barnes (1999: pp. 1–2) identifies the attributes of the institutional approach in terms of heightening an appreciation of 'the importance of local knowledge, institutional embeddedness, historical specificity, technological dynamism, and critical reflexivity'. The role of locally embedded institutions and inter-firm relationships, of course, has a long history in industrial geography, going back to nineteenth-century Marshallian industrial districts (Markusen, 1996) and the more recent resurgence of cluster analysis (starting with Porter, 1990). Influential within this broad set of ideas is Amin and Thrift's (1994, 1995) articulation of the concept of 'institutional thickness'. This is a 'multifaceted concept' (Amin and Thrift, 1994: p. 14) encapsulating four core principles. Firstly, it is signified by a dense presence of organizations (such as Chambers of Commerce, trading agencies and associations, local authorities, marketing boards, etc.). Secondly, there must be: 'high levels of interaction amongst the institutions in a local area. The institutions involved must be actively engaged with and conscious of each other, displaying high levels of contact, cooperation, and information exchange' (Amin and Thrift, 1994: p. 14).[10] Thirdly, this interaction must lead to well-defined structures of domination and/or coalition; essentially, actors know the common rules and via this, a constraint is placed on potential rogue behaviour. And fourthly, there is a shared commitment to a common enterprise. As Amin and Thrift (1994: p. 15) indicate, this need not be formal; 'usually [it is] no more than a loosely defined script'. There are certainly some parallels here with ideas of social capital, as popularized by Robert Putnam (1995), and subsequently adopted by the World Bank as the 'missing link' in development (Grootaert, 1997). According to the World Bank's Social Capital Website: 'Social capital refers to the norms and networks that enable collective action. Increasing evidence shows that social cohesion – social capital – is critical for poverty alleviation and sustainable human and economic development' (World Bank, 2007b: not paginated).

Research on these processes in the 'associational economies' of the Third Italy (Emilia-Romagna) and other regions provided the bedrock for assertions that 'local*ness* matters' when it comes to issues of institutions, scale and global competitiveness (*inter alia*, Scott, 1988; Turok, 1993; Cooke and Morgan, 1998). A common thread throughout these studies was discussion of regional economic governance by way of an embedded network of trade associations, chambers of commerce, research centres and the like, whose collective intelligence institutionalizes trusting associations between economic actors.

In a similar analysis, Storper (1997: p. 21) speaks of the *untraded interdependencies* which underlie development in a particular region, 'attaching to the process of economic and organizational learning and coordination'. The transaction costs of untraded interdependencies cannot easily be accommodated and necessarily possess a regional character (Storper, 1997, 1999). Storper (1997: p. 52) goes on to present an account of:

> the role of territorial proximity in the formation of conventions; the role of conventions in defining the 'action capacities' of economic agents and, hence, the economic identities of territories and regions; the economic status of regional conventions of production as a type of regionally specific collective asset of the economy; the status of conventions as untraded interdependencies in economic systems; and why it is so difficult for some places to imitate or borrow conventions and institutions from other places.

In such accounts, there is an explicit recognition of the geographically bounded world of conventions, which Storper (1997: p. 38) defines to include 'taken-for-granted mutually coherent expectations, routines, and practices, which are sometimes manifested as formal institutions and rules, but often not'. These insights into the relationship between regional institutions and economic development lead to policy implications that the creation of public institutions must be based on facilitating a common context of coordination.

Assertions within the geographical literature on the importance of territorially defined notions of embeddedness, however, have come under increased scrutiny during the 2000s. Hess (2004: p. 166) contends that 'geographers have theorized and used the concept [of embeddedness] from a distinct spatial point of view, namely to explain – in addition to economic theories of transaction costs and agglomeration economies – the evolution and economic success of regions built by locally clustered networks of firms' and that, in the process: 'most work in economic geography has been prone to use what I will call an "over-territorialized" concept of embeddedness by proposing "local" networks and localized social relationships as the spatial logic of embeddedness, which might result from "spatial fetishization"' (Hess, 2004: p. 174). Attacking what he sees as 'lack of clarity or "sloppy theorizing"' (Hess, 2004: p. 166) in advancing these arguments, Hess (2004: p. 175) argues that local relations 'by no means [represent] the only spatial logic of embeddedness'. This reiterates Granovetter's influential delineation of the concept, which was defined in the ultimately broad terms of the 'structure of the overall network of relations' (Granovetter, 1990: p. 98, cited in Grote and Täube, 2006: p. 1289). Appreciation of the multi-scalar and socially diverse nature of embeddedness represents an important aspect of the concept's application within economic geography, however, the exact framework for undertaking these tasks remains in flux. As Hess (2004: p. 174) suggests: 'The question of what the firms are embedded in has generated multiple meanings.'

At present, applications of the concept of embeddedness within economic geography seem to be being composed around two different taxonomies. On the one hand, some researchers (e.g., Weller, 2006) reference their use of the concept through Zukin and DiMaggio's (1990) categorization of four types of embeddedness: (i) Cognitive embeddedness (based on bounded rationality, place-based knowledge and associated with proximity); (ii) Cultural embeddedness (which has a place-specific or group-specific flavor, derived from collective understandings of the way things are done); (iii) Structural embeddedness (generated through the incorporation of economic, social, and cultural relations in regulatory and cultural relations), and (iv) Political embeddedness (relationships of actors to rule-making powers). Alternatively, Hess (2004) proposes a threefold categorization of territorial, network and societal embeddedness. Whatever the taxonomy, perhaps the most important considerations are that: (i) the concept of embeddedness implies *more* than just the ways economic actors are connected to (Cartesian) geographical space; and (ii) it is the inter-linking of these various aspects of embeddedness that gives rise to a holistic perspective of embeddedness, in its most complete, Polanyian, sense (Hughes et al., 2008: p. 348).

In a broader sense too, moreover, the concepts of embeddedness, the institutional environment and governance can be understood as constitutive elements of what Williamson (2000: p. 597) labels the 'economics of institutions' (see Figure 2.1). Williamson perceives the embeddedness of society (though customs, traditions, 'settled habits of thought' etc.) as so-called 'Level 1' (foundational) social attributes which have evolved over hundreds, if not thousands of years. These 'Level 1' factors shape the so-called 'institutional environment', which, consistent with North's usage, is defined as the 'rules of the game', especially with respect to laws about property rights and commercial conduct. The institutional environment then prefigures a system of governance, which refers to the 'plays' in the game. Finally, systems of governance then contextualize resource allocation processes; the 'getting the prices' right dynamics of supply and demand. As rightly pointed out by Williamson (2000), the narrower band of New Institutional Economists (and neo-classical economists, as well!) focus on this last level of analysis whilst taking for granted (or ignoring completely) the other three levels on which it sits. As geographers, however, we take the view that an understanding of social outcomes can be derived only from a more fulsome appreciation of how these 'levels' interact. Consequently, in much of the text that follows in subsequent chapters, we give deliberate attention to the place-specific historical circumstances that position and

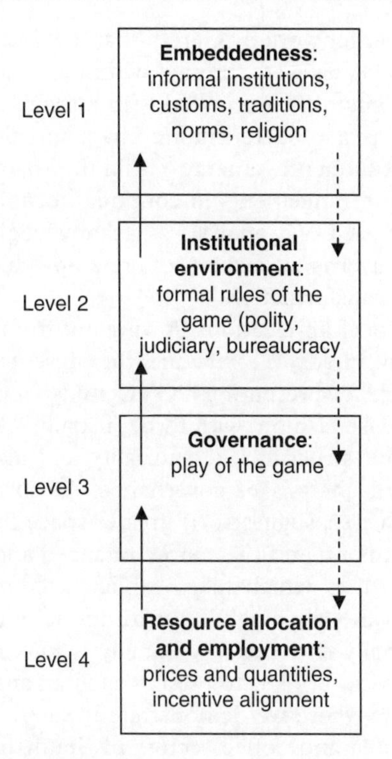


**Figure 2.1** The economics of institutions.
*Source*: Williamson (2000: p. 597).

explain social outcomes; and then we return to this framework in the final chapter of this book.

## An institutionally enriched GVC approach

So far, this chapter has: (i) asserted the need for analyses of product/commodity systems to be sensitive to geographical concerns of space and place; (ii) reviewed arguments about the shortcomings of a governance-preoccupied GVC analysis under these terms; and (iii) proposed that an institutionally-enriched GVC method will adequately address these concerns. In this section, we pull together these various strands of argument.

The pre-eminent question that needs to be confronted is how a more sophisticated kind of institutional analysis might be incorporated within the GVC method? Our review of the field of institutional analysis, above, provides key structuring elements of an answer to this question. In contrast to implied notions in earlier work by Gereffi that the 'institutional framework' was something that GVCs were 'framed within', we argue that the institutional dimension of chains is insinuated within their very core. The institutional environment is a predetermining characteristic of the governance structures which subsequently emerge within the chain and which, in turn, then act upon those arrangements in continual feedback. Thus, we see an institutionally enriched GVC analysis as acknowledging that institutional arrangements and governance structures are co-produced and in a state of perpetual dynamic transformation.

There are two critical implications flowing out from such an elaboration to the GVC method. Firstly, it *complicates* the presentation of analysis. The archetypical template for presenting a GVC study is to introduce data and information in stepwise fashion, with three (usually relatively brief) discussions of input–output structures, territoriality and institutions paving the way for an extended analysis of governance. An institutionally enriched GVC approach, however, suggests: (i) greater space for the analysis of the institutional environment; and (ii) a more nuanced and contextual critique of governance structures which takes explicit account of how these are shaped (and go towards shaping) the institutional environment. So what might an institutionally-enriched GVC analysis look like in practice? The primary objective of this book is to present such an analysis, exemplified in Chapters 5, 6 and 7, where we demonstrate the multi-scalarity of institutional forms in the tea and coffee sectors of South India, and how these forms are being co-produced with constantly evolving chain governance structures. We then return to our model for an institutionally-enriched GVC framework in the concluding chapter.

Secondly, through elevated attention to these facets of economic relationships, some of the vital elements of difference between the GVC and GPN approaches appear to collapse. Notably, introducing an enriched notion of institutions into GVC analysis countermands the charge that it lacks the requisite capacity to address a nuanced appreciation of the active role of spatial relations in shaping social and economic outcomes; and negates assertions of a preoccupation with governance alone. It may be that an institutionally enriched GVC approach is better equipped to address some research questions and industries, whilst GPN frameworks appear better equipped in other cases. As a case in point, research into industries such as tea and coffee (where economic activities are strongly structured around the trade and transformation of a distinct single commodity) would appear to be better suited to a chain metaphor. Advanced manufactures (such as telecommunications and automobiles) which comprise large and intricate clusters of parts suppliers and assemblers, on the other hand, seem better suited to being described through the more open-ended metaphor of a network.

In any event, the subtleties of the conceptual distinctiveness between GVC and GPN approaches are sometimes lost, in any case, in the more harried and messy world of empirical research. Revealingly, Coe et al. (2008b: p. 274) cite Levy (2008) to the effect that despite GPN analysts' 'lofty ambitions, most of the studies spawned by the GPN framework to date are, in practice, very similar to those generated using GCC analysis'. This co-mingling of the two approaches when it comes to the reporting of empirical research, moreover, is manifested concretely in the fact that a 2008 special issue of the *Journal of Economic Geography* focusing on the GPN approach actually includes two articles (Sturgeon et al., 2008; Nadvi, 2008) which explicitly deploy the language and epistemology of GVCs.

## Part Four: From Words to Deeds – Methodological Considerations in GVC Research

The problem of 'how to do' GVC research remains a vexing problem for researchers. There are questions to be asked about the scope of inquiry within GVC studies; the way information is to be gathered, organized and presented; and the positionality of the researcher. It bears important significance to note that the task of undertaking GVC research is methodologically taxing. Information has to be collected from individual places and from individual actors for the purpose of understanding the nature of connections and disconnections with distant others. Value chains typically involve large numbers of actors located in different places with varying (and often competing) perspectives and knowledge about 'how things operate'. Research strategies need to weave in and through these actor-relationships, if a holistic set of perspectives is to be developed.

To the present time, a number of 'how to' manuals have been developed to guide GVC research (such as Kaplinsky and Morris, 2001; McCormick and Schmitz, 2001; Humphrey, 2005).[11] With an intended audience of both academics and practitioners (albeit the often-blurred lines between these communities), these manuals seek to provide broad directional advice for GVC researchers. First and foremost in both manuals is recognition of the imperative of multi-method tactics. In most value chain research, the obvious starting point is desk research using 'grey literatures' such as industry analyses,[12] trade journals, statistical data and the reports of inquiries, etc. Such data enable researchers to undertake preliminary *value chain mapping*; indicative schematic representation of key processes and players. Inevitably, however, as Kaplinsky and Morris (2001: p. 52) admit: 'the theory of value chains suggests simplicity and an easy clarity of focus. However, the real world can be much messier.' Consequently, the collecting of primary data through industry surveys or interviews tends to form an essential second-stage component of research in this field. In terms of timing, however, it is frequently the case that desk- and field-research stages can usefully overlap. McCormick and Schmitz (2001: p. 45) recommend early immersion in the field via preliminary ('scoping') interviews as a means to establish pathways for later (more comprehensive) field research.[13]

Embarking on the quest to collect primary source materials clearly raises a host of methodological concerns. Vital questions relate to the entry point for research; the tools (survey and interview designs) to be used; the processes of respondent recruitment ('cold-calling', or use of gatekeepers for snowball recruitment), and the positioning of the researcher within the fieldwork process. Within the field of value chain scholarship, researchers have confronted these questions in various ways. Common to all, however, is an overarching appreciation of the importance of qualitative research to elicit the nuance of circumstance for individual firms, communities and other stakeholders (NGOs, unions, etc.). As argued by Sturgeon (2006: p. 8): 'An understanding of ... industry-specific factors, and their interaction, requires deep knowledge of specific industries and occupations that *can only be gained* through qualitative research methods' (italics added).

Yet the use of qualitative research brings with it an array of difficult-to-resolve methodological concerns. Qualitative data are shaped explicitly by the procedures under which they are amassed. A subjectivity–objectivity dilemma inevitably haunts qualitative-based field research. As Burawoy (2000a: p. 4) acknowledges, researchers 'cannot be outside the global processes they study. They do not descend *tabula rasa* into villages, workplaces, churches, streets, agencies or movements.' What is necessary, therefore, is to build an appreciation of what Haraway (1991) labels as 'embodied objectivity'; the process of acknowledging and reflecting on how a researcher's interpenetration into other lived worlds inevitably shapes the data which are collected. This approach recognizes knowledge as *situated* and *partial*, and built up through reflexive exchange between (active) researchers and respondent-participants (not subjects) (Burawoy, 1998). Along with this worldview comes heightened analytical concern for the micro-geographies of where, when and under what circumstances particular modes of information gathering (interviews, for example) take place (Elwood and Martin, 2000).

Cross-cutting this entire area of concern is the highly political issue of the insider/outsider status of researchers. This status has widespread implications for the conduct of research, including notably the extent to which respondent-participants choose to share insights, information and attitudes. Evidently, the question of whether or not researchers are perceived as insiders or outsiders is intimately connected to gatekeeper issues; the mechanisms through which researchers gain access to respondent-participants. In value chain scholarship, it is commonly the case that researchers' entry into economic 'lived worlds' is contingent upon some kind of third-party endorsement from entities such as industry organizations, NGOs, or unions etc. This is because data on industry participants are not always public and, in any case, would-be respondent-participants are often inclined to be wary of, and uncooperative towards, unknown researchers. Coming into research on the basis of such relationships, however, can bring with it unwanted baggage. From the perspectives of respondent-participants, alignment issues cast prejudicial pre-judgments onto researchers, creating data openings and closings. Within the scope of a large study, moreover, such issues would be expected to work in a host of different ways, to different effects. Mohammad (2001) speaks of how there can be considerable fluidity in the perceived status of researchers (insider/outsider, etc) as they move between respondents.

Furthermore, once qualitative research techniques have been animated, there comes the question of how to interpret outputs. In line with comments above, dominant social scientific research practice since the 1990s has understood interview testimonies as partial narratives emanating from specific spatial-temporal settings (Miles and Crush, 1993). Extrapolating from individual interviews to generate incisive analyses remains more art than science, though researchers can deploy various strategies to build coherent understandings of common themes across multiple respondent-participants. These include triangulation (asking the same [or similar] questions to different respondent-participants) and the use of visual tools such as Likert scales which seek to translate respondent-participants' views/attitudes to numerical scores (Kaplinsky and Morris, 2001; McCormick and Schmitz, 2001).

The task of extrapolating results even further – using insights from one value chain case to inform theory – poses even greater challenges. The problem for value chain researchers is comparable to that which has faced global ethnographers such as Michael Burawoy, namely, how does a global scale of reference render 'working in the local?' (Burawoy et al., 1991, 2000). Coming out of this debate is heightened appreciation of multi-scalar interactivity. Referring to his team of Berkeley researchers who contributed to the edited book *Ethnography Unbound* (Burawoy et al., 1991), Burawoy notes: 'To use Habermas's language, the Berkeleyites emphasized the way the external 'system' colonized the subject lifeworld and how that lifeworld, in turn, negotiated the terms of domination, created alternatives, or took to collective protest' (Burawoy, 2000a: p. 25).

This recognition of how the space-time rhythms of place reverberate back and forth through multi-scalar systems subverts 'global/local' dualisms in favour of a relational perspective that Massey (*inter alia*, 1993a, 1993b, 2005) calls a 'global sense of place' (see the discussion on this point in Part Three of this chapter). Such multi-scalar interactivity leads to what Tsing (2005) labels as 'frictions'. In this book we use the terminology of 'struggles' to refer to this same process. Whatever the nomenclature, however, the key point is that a vital methodological agenda within value chain research is to forge parsimonious interpretations of how different scales of activity connect with one another, that don't *a priori* privilege on over another.

The operational tactics to achieve these objectives, of course, are rarely clear-cut; although recognition of the ideal frameworks is well known. Coe et al. (2008: p. 290) observe that: 'We need carefully designed and constructed but essentially *grounded* research into the entire structure of GPNs. This implies that a 'lone researcher' approach will not get us very far. The very nature of GPNs – their organizational complexity, their spatial extent and geographical diversity – necessitates multi-national research projects' (italics in original). Burawoy (2000a: 4) argues: 'we need to rethink the meaning of fieldwork, releasing it from solitary confinement, from being bound to a single place and time'. Sturgeon (2006: pp. 8–9) takes these arguments further, suggesting that multi-national/multi-research collaborations are essential to generate *systematic* insights from diverse 'pools' of qualitative research:

> Julia Lane of the National Science Foundation has likened the current state of qualitative industry research to the study of the natural world in the 16th and 17th centuries. Curious researchers made detailed notes and drawings of what they could see of the vastness and variety around them, but there were few mechanisms for compiling the findings of individual researchers into larger pools of knowledge that could reveal broad patterns.

These points underline the broad-ranging nature of methodological concerns attached to value chain research. In the next section, we discuss their application to our own study of the tea and coffee chains emanating from South India.

## Our approach

Prior to presenting the research method employed in this study, it is necessary to answer a few basic questions about the scale of analysis performed. Why tea *and* coffee? Why South India? The world tea and coffee industries share many structural similarities, and the introduction of these commodities into South India occurred within remarkably similar political, economic and social institutional structures. However, as explained in greater detail in Chapter 3 of this book, increasingly divergent international markets for these products provide an interesting study of the influence of global institutions on social and economic outcomes in otherwise very similar producer communities. For similar reasons, our focus on South India is explained by the need to answer questions on the emergence and maintenance of institutions in a real geographic place, and by our desire to more adequately incorporate these dimensions within a reinvigorated GVC framework. However, as demonstrated through the analyses in the case-study chapters of the book, our approach explicitly incorporates the role and influence of economic actors and stakeholders residing in other segments of the chain, such as coffee drinkers in affluent countries, multinational tea trading countries, and multilateral industry organizations. However, as many other GVC studies of these commodities have already comprehensively addressed the downstream segments of these chains, the greater empirical focus of this book resides with the lesser-known upstream sites of production in the chain.

The practicalities of researching tea and coffee in South India have trained our analysis to the critical substance of the contemporary struggles in these industries. Over the course of four years (2004–08) we documented and observed a complex story of value chain restructuring that did not unambiguously advantage or disadvantage the region as a whole. Instead, what we saw was that as buyers sought to alter the terms by which they engaged with tea and coffee producers, they provoked broad-ranging arenas of contestation over the economic, social and environmental constitution of these industries.

Uncovering these dimensions of transformation required a research approach that was sensitive to the detail of change. Corresponding to inductive and grounded methodologies within human geography, where research is organized for the primary purpose of uncovering the relationships between actors in order to account to inform and refine theoretical conceptions of 'how the world operates' (Cloke et al., 2004), our three-pronged methodology entailed: (i) the analysis of published industry material such as technical reports, government reports, news media, and industry journals; (ii) participation in various industry conferences, events and association meetings in South India and internationally; and (iii) most importantly, a comprehensive series of approximately 150 key informant interviews with an extensive range of participants in South Indian tea and coffee industries. Informants included plantation owners, representatives of import/export firms, auction brokers, managers of producer cooperatives, transport, warehouse and shipping agents, representatives of NGOs, government bureaucrats, extension officers, representatives of certification agencies, environmental and agronomic advisers, the leaders of industry associations, journalists, local academics and researchers, trade union representatives, and economic consultants. Interviews with these key informants were generally constructed around a series of open-ended focus questions. Detailed note-taking during interviews was combined with personal observations and were later transferred to a set of field diaries, the combined mass of which became vital primary data used for writing this book. When specific mention in this book is made to information gained from these interviews, a footnote reference is used which records the date and place of the interview, and the type of actor (e.g., 'plantation owner', 'smallholder', 'factory manager', etc.). Mostly these are recorded in ways to ensure the anonymity of the informant, although on occasions where appropriate (and with the consent of the informant), specific names are noted. Key informant interviews were undertaken during 11 separate research visits to South India. Among other things, the fieldwork itinerary was prompted by the fact that South Indian tea and coffee production takes place within a series of separate planting districts and, in the case of coffee, has a highly seasonal pattern. Arranging the fieldwork via a series of separate visits provided a strategy by which each trip would focus on one or two particular districts and ensure an appreciation of seasonal cycles.

Interviews were designed to promote and elicit wide-ranging discussion of contemporary industry dynamics. The methodological praxis we adopted sought to understand the industry through the open-ended testimony of interview participants. Informants were encouraged to tell their own stories of how they perceived the restructuring of their industries, with the view that this strategy would tease out a series of narratives that described the detail of global value chain restructuring. The tactic sought to emulate the technique described in Cloke et al. (2004: p. 123) as 'interviewing by informal conversation'. Not surprisingly, this approach necessarily confronted sensitivities relating to the legal and economic practicalities of business operations, which on occasions seemed to encourage some informants to be evasive. We faced the same difficulties as did Friedberg (2001: p. 354) in her work on green bean value chains, where she wrote: 'I do not make light of the obstacles to research inside industries where investigators of any kind are considered suspect, and where deception and secrecy are standard tools of the trade.' Nevertheless, through amassing a large volume of like-formulated narratives from a wide diversity of informants, individual interpretations of various processes were open to cross-referencing and informed review. Such strategies of triangulation had the merit of bringing into focus the extent of differences in interpretation among industry participants, emphasizing the complex nature of the processes under investigation.

Complementing this approach was a flexible strategy for recruiting informants. We share the observations of King and McGrath (2004: p. 10) who found that: 'comparative research across agencies makes fieldwork very intensive, and it is imperative to be able to abandon certain lines of inquiry and to embrace others based on an awareness of what has been said in the interviews during a visit'. To do the field research for this book required the ability (encompassed in time and financial budgets) to visit locations which were often geographically isolated; to expeditiously capture knowledge from diverse social settings; and to cogently, professionally and diplomatically bring together insights from differently-situated research informants. The course of these endeavours, not surprisingly, presented us with difficult ethical and practical choices about how to immerse ourselves within the various situated contexts we were seeking to document. As discussed above, the very nature of GVC chain research – which is all about understanding the linkages and struggles of economic and social interests across space – tends to demand that researchers align themselves with particular social actors if they are to gain access to arenas of economic activity. Institutional gatekeepers play a pivotal role in connecting researchers with lead informants. Research tends to hinge on gatekeeper dependencies, and research orientation (and, indeed, success) can hang on the uncertain terrain of participant recruitment through snowball techniques. In these circumstances, questions relating to how and in what context research is structured exercises considerable weight. In our case, these issues were managed through the establishment of open channels of communication early on with a number of influential organizations representing these industries. These included: UPASI, which represents the interests of plantation owners; UPASI-Krishi Vigyan Kendra (U-KVK) a Nilgiris-based agricultural outreach and adaptive research agency which works with the smallholder sector; the Tea Board of India, and the Coffee Board of India. In addition to this highly focused regional analysis of production dynamics within South India, our research has also necessarily involved a series of interviews with key industry actors in the major consuming countries to where these global value chains extend. Industry actors, such as coffee roasters, trading houses and tea packers from Australia, Japan, the United Kingdom, the United States and the Netherlands, were interviewed at various times.

## Conclusion

This chapter began with the observation that the analysis in this book of tea and coffee value chains in South India entertains a series of deeper theoretical and methodological questions. In making the argument that the fate of these industries can be understood in terms of the struggle of place-based *institutions* to negotiate global *governance* structures, a wide constellation of ideas and questions are opened up. In terminology and approach this book's analysis is attached to GVC analysis, but this is done neither simplistically or uncritically. The vital point this chapter makes is that the GVC approach needs refinement. Specifically, the institutional dimension of the GVC approach needs a heightened presence in GVC studies. Deployment of such amendments to the GVC method bring it in line with the useful contributions to theoretical debate put forward by other researchers, such as the so-called 'Manchester school'.

Yet (and this is the rub), GVC analyses that incorporate such considerations increasingly take on the substance (if not the precise form) of alternative approaches (such as GPN frameworks) which were designed to 'go beyond' the alleged limitations of the GVC approach. Developing and articulating these arguments has required a breadth of material to be addressed. As we noted at the outset, product/commodity analysis has an ambitious ambit. Individual chains/networks are examined not in isolated terms, but as broader research strategies to document (what Burawoy labels as) *grounded globalization*. In Part One of this chapter, we explored the contextual framework justifying this frame of reference. This paved the way, in Part Two, for an assessment of the GVC approach. As discussed, the key innovation of GVC analysis was its transcendence of the nation-state as the supposedly 'natural' unit of analysis. By focusing on international product/commodity chains, the GVC approach brought into question the role of governance, and the capacity for upstream producers (especially in developing countries) to upgrade their capacities.

For geographers, however, questions remained about the capacities of governance-dominated GVC research to capture the roles of space, place and territory in shaping chain structures. Consequently, alternative approaches have been proposed as 'newer, improved' successor models to the GVC framework. Central to the claims of these approaches is the relevance of a *relational* perspective, whereby economic, social and political structures are seen as co-produced through multi-scalar interactivity. Through these lenses, 'chain governance' is seen not simply as an external structure imposing its will upon passive, spatially bound, economic actors (workers, communities, upstream producers, etc.), but as an artifact of what Coe et al. (2004) call 'strategic coupling'. Recognition of these factors brings to the forefront of analysis Polanyian perspectives of how the economy is embedded in socio-spatial formations.

Part Three of this chapter presented our particular contribution to these debates by providing the intellectual foundations for a broader incorporation of institutional theory within the GVC framework. From Gereffi onwards, GVC researchers have tended to give little attention to the 'institutional dimension' of the approach. It has been crudely typecast as representing external 'framing constructs' in which governance regimes take form. However, if a more sophisticated rendition of the concepts of institutions and embeddedness are incorporated into GVC analysis, the 'institutional dimension' takes on greater importance. Crucially, such an approach responds sharply to accusations that the GVC framework is insensitive to the role of place and the mutually reinforcing aspects of spatial relations.

These arguments, finally, take us into the question of methodology. Undertaking value chain studies provides a unique set of methodological difficulties, with which the existing literature has not grappled fully. However, important insights into these problems are provided through consideration of the field of global ethnography, which has a comparatively strong analytical focus on multi-scalar processes.

These conceptual debates and issues are now brought to life through the empirical presentations that follow. The following six chapters apply the framework developed here to various aspects of the tea and coffee chains emanating from South India. A recurring theme in these chapters being the interaction of scalar institutions through a series of struggles marked by the constantly evolving dialectic between institutional arrangements and economic governance structures along throughout the chain. Then, in Chapter 9, we return to key issues raised in this chapter to reinvigorate our conceptual understanding of value chain struggles.

# Chapter Three

# How to Make a (South Indian) Cup of Tea or Coffee

As presented to consumers, tea and coffee value chains have a whiff of romance. In tea, images of verdant hillsides smattered by smiling tea pluckers provide fodder for worldwide tea advertising campaigns. In coffee, consumers are told narratives about the rich ecological settings of cultivation. In both industries, discourses abound in relation to the intricate, artisanal backroom skills of blenders, roasters (in the case of coffee) and tasters who supposedly dedicate their lives to the noble and exotic art of 'making a good cuppa'.

Tea is usually grown on verdant hills, and coffee (often, at least) in lush ecological settings. Both industries abound, moreover, with professional tasters and blenders who are passionate in their pursuit of a quality product. But in economic terms, such attributes are hardly the defining features of these value chains. Betraying the romance cast upon unsuspecting consumers by advertisers and gourmet travel writers, these industries at their core represent international commodity systems. Tea and coffee are cultivated, processed and traded in line with product-specific requirements, norms and standards that are defined and understood through an array of technical and commercial concepts. Gaining an understanding of these systems represents an essential precursor to the complex analytical tasks (undertaken in future chapters) of documenting the interplay of institutions and governance in these industries. It is therefore incumbent upon us, at this stage, to ground our investigations in preliminary tasks of explanation. Specifically, we need to ask 'what are the agronomic, technical, economic and territorial contours that define these value chains?'

This question is encapsulated by the GVC concepts of 'input–output' and 'territoriality' – the preliminary process of mapping the value chain. Input–output refers to the configuration of purchases and sales by actors in a chain. Territoriality corresponds to the geographical extent of a chain.

As articulated by Gereffi and other GVC researchers (and as noted in earlier chapters), these two prongs of the GVC method are essentially descriptive. Their assessment is meant to provide the foundational basis for generating analytical insights into issues of coordination and control within chains. In this chapter, we address these issues of the input–output and territoriality components of tea and coffee value chains, with a view to establishing the contexts of these industries' expressions in South India.

## The Input–Output Dimensions of South Indian Tea

For the tea grown in South India to ultimately find its way to consumers' cups, it must firstly go through a series of processes: manufacture, warehousing, auction and/or private sale, blending, packaging, branding and retailing (Figure 3.1). At each of these stages, the product's economic form is transformed in some way and value is added to it. The following sections critically review these processes.

### Cultivation

There are two varieties of tea planted commercially in the world: *Camellia sinensis var. sinensis* (generally referred to as the China type) and *Camellia sinensis var. assamica* (the Assam type). As a perennial bush, *Camellia sinensis* grows best in tropical upland locations and requires regular plucking, and so was well suited in the Colonial era to plantation-style cultivation. Shade trees are sometimes planted amidst the tea bushes, as higher shade levels increases chlorophyll levels in leaf and thereby assists the production of more intense tea flavours (Hudson, 2000) (Figure 3.2).

Typically, the upland areas that became sites of tea cultivation in India were forested areas inhabited by tribal communities, which Colonial interests were readily able to expropriate. The planting of tea bushes provided a steady income stream and work schedule that gave a permanent social and economic structure to Colonial life. Decolonization in key producing regions across the world was associated with various transformations to these arrangements. In a number of cases, newly independent governments wrested control over these industries through direct means. In Indonesia and Sri Lanka, plantations were nationalized, and in Kenya, plantation lands were carved up and redistributed to smallholders. In India, by way of contrast, there was a progressive replacement of Colonial ownership with national ownership. Understood in economic terms, the prevailing Colonial form of the tea industry – that is, an industry dominated by large plantations – reflected

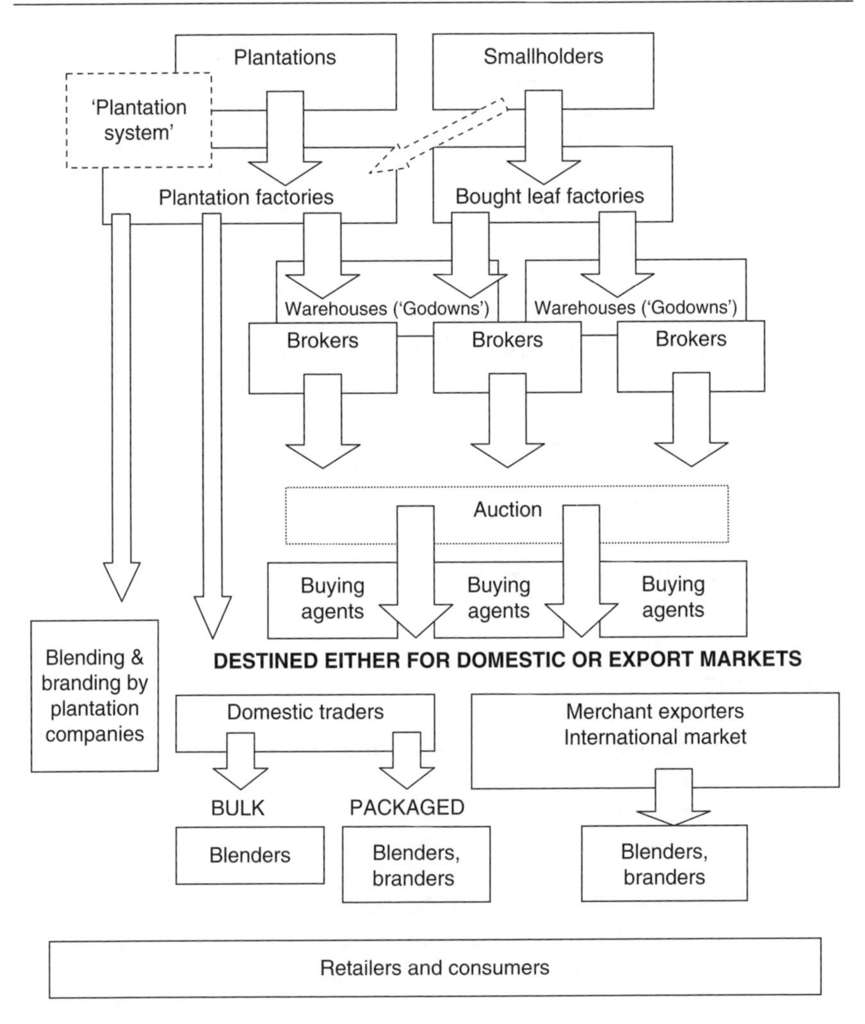


**Figure 3.1** Basic composition of South Indian tea global value chains.
*Source*: Own work.

the fact that its establishment in remote locations favoured enterprises with significant economies of scale (Hayami and Damodaran, 2004). Over time, however, these natural advantages of scale economies came to exercise a waning influence over the industry. As previously remote areas were incorporated into regional and national economies by way of infrastructure development, and as their population densities grew, non-estate forms of cultivation became more viable. Accordingly, in the 1980s and 1990s another wave of transformations occurred as variant models of tea cultivation took root within

**Figure 3.2** Shade cover at Silver Cloud tea estate, Gudalur, Tamil Nadu.
*Source*: Photo by the authors.

producer countries (Table 3.1). For reasons of convenience we use the label 'smallholder' to describe these variant models, however in practice this is a fluid term denoting a wide range of social and economic formations.[1] For some, tea cultivation represents the mainstay of family income, whereas for others it is one element in diversified family incomes that contain both farm and non-farm activities. Whatever the differences, however, it is now the case that the tea industry worldwide is constituted both by plantation and smallholder cultivation.

There is usually a three-year delay between planting and first harvest (Wilson, 1999), and full maturity is obtained after five years. The tea bush itself survives for many decades, though its productivity wanes in old age. Pest and disease can be problematic, and producers need to adopt management techniques that take into account the increasingly stringent

**Table 3.1** Indicative estimates of tea production by smallholders, major countries, 2005

| *Country* | *Production (and exports) ('000 tonnes)* | *Estimated percentage share of production by smallholders* |
|---|---|---|
| India | 831 (175) | 20 |
| China | 941 (283) | 80 |
| Kenya | 295 (284) | 60 |
| Sri Lanka | 308 (299) | 62 |
| Indonesia | 171 (99) | 29 |
| Vietnam | 110 (99) | 70 |
| Tanzania | 30 (24) | 10 |

*Sources*: Production and export data from FAO (2007). Smallholder data from the following sources: *India*: Tea Board of India (2006); *China*: Etherington and Forster (1993); *Kenya*: Tea Board of Kenya (2006); *Sri Lanka*: Central Bank of Sri Lanka (2003: Chapter Three); *Indonesia*: Dirjen Bina Produksi Perkebunan (2004); *Viet Nam*: Nguythitiencu and Nguyentanphong (2005); Tanzania: Baffes (2005).

specification of Maximum Residue Limits (MRLs) imposed by importer countries, especially in the West. In the tea industry, recent consideration of these issues was prompted by a 1995 consignment of Darjeeling tea to Germany which was found to contain traces of the pesticides tetradifon and ethion in quantities much higher than allowed for under German legislation.[2] This MRL issue was taken up by the Inter-Governmental Group on Tea (IGGoT) during its 13th Session in 2001 and, after a few years of back-and-forth disputation between members, a decision was made during the 16th Session (2004–05) to establish a Working Party specifically charged with promoting MRL harmonization. To the time of writing, the most recent public document of this Working Party is its report to the 17th Session of the IGGoT (at Nairobi, Kenya in 2006), which noted no substantive progress on this issue. Thus, producer countries do not yet have the luxury of being able to sell tea into an international marketplace with consistent and mutually-agreed MRL rules.

For both plantation owners and smallholders, a key attraction of tea cultivation is that, with the exception of the pruning period,[3] tea has a capacity to generate a regular, week-in week-out income stream. Plucking can usually take place for 12 months of the year, although, in some regions with cold winters, like Darjeeling in North India, there is a three-month non-plucking season.[4] A general axiom in the industry is that tea quality 'is produced in the field', affected by climate, soils, bush age, variety and growing altitude. Agronomic management, especially plucking frequency and technique, is also critical. The frequency of plucking (the plucking cycle) varies according to the leaf growth rate (which is highly seasonal), with optimal quality gained when tea is plucked in line with a ten-day cycle (during South India's wet season). Frequent plucking ensures capture of the newly emerging green buds before they become fibrous; the golden rule of tea plucking being 'two leaves and a bud'. Young buds have more concentrated levels of polyphenols (chemicals acting as anti-oxidants) which are crucial in giving tea its unique flavours.

Hand plucking best ensures the quality imperative of capturing young buds is met, though it comes at the price of low plucking volumes per worker. With labour being the most important single cost on a tea estate, plantation companies in South India have invested heavily in attempts to increase labour productivity through technical and organizational means. This has included mechanical harvesting, which Wilson (1999: p. 209) describes as techniques such as the use of mechanized shears by one person, reciprocating cutters long enough to cover a row of tea and held by two workers, and self-propelled harvesters which are mounted on wheels or even tracks. In South India, a number of estates have introduced degrees of mechanization, from alternating manual plucking with mechanized shears at various stages of the plucking cycle (a practice which is now quite widespread) through to experiments with reciprocal cutters. Malankara Estates in Kerala, for example, reported it has accommodated a significant reduction in its permanent workforce over the past decade through mechanization.[5] Mechanical plucking, however, faces significant barriers for adoption. Most importantly, the quality of leaf extracted by mechanical plucking is generally lower than that obtained through hand plucking, meaning that there is an important productivity–price trade-off in these decisions.

In addition to decisions regarding plucking technique, tea estates face complex issues in relation to worker deployment across estates. For plantation owners, having the workers 'in the right part of the estate' and providing them with incentives to optimize their plucking volumes, is a crucial element of effective labour management. At the same time, managerial agendas to optimize labour productivity are contextualized by the ongoing plight of the plantation tea worker. There is little doubting that the mainly female workforces of tea plantations have traditionally had major reason to voice complaint about their working conditions. Work as a tea plucker is hard and repetitive, and during the Colonial era in particular, workers had minimal recourse to civil and labour rights. As we discuss in Chapter 5, the enactment of national legislation (the *Plantation Labor Act*) has meant that much has changed for the better since those days, but it remains the case that labour relations on estates often remain fractious and oppositional. Thus, the implementation of productivity-enhancing labour management systems in South India's tea sector continues to be a highly fraught work-in-progress. One such scheme we observed in Kerala involved a three-tiered incentive payment regime, in which workers were paid significant bonuses for exceeding daily minimum plucking levels. According to management, this has substantially improved rates of labour productivity. From union perspectives, however, such systems would seem to represent the creeping intensification of work. In South India, the arbitration of what is acceptable or not is highly dependent upon political decisions taken by State Ministers of Labour, often subsequently contested in the industrial relations courts of Tamil Nadu and Kerala. Complementing such wage-related initiatives is the development of computer-based plucking schedule models, which involve incorporating local weather conditions into a computerized model that estimates tea leaf growth rates in different parts of the estate. According to one system we observed, the use of this programme means that the day's entire plucking activities can be assessed within an hour of the completion of plucking, thus providing management with real-time, day-to-day monitoring of the dynamics of leaf growth and plucking routines.[6]

## Factory processing

The imperative to deliver tea to factories the day it is plucked has pivotal implications for the shaping of tea value chains. Tea plantations frequently operate as sites of processing as well as cultivation, which is different to some other tropical crops, such as coffee, where there is functional separation between these two activities. In general terms, there are three different processing methods used to convert green leaf to made tea, resulting in black, green, or oolong tea. Both black and green tea is produced from the same tea plant, *Camellia sinensis* (from either the China or Assam type), the primary difference being that black tea is fermented, whilst green tea is not (major green tea producers are China, Japan, Viet Nam and Indonesia). Oolong tea is a Chinese tea (properly only prepared from the China variety) which undergoes an abbreviated fermentation process (semi-fermentation). Furthermore, there are two manufacturing processes for black tea. The 'orthodox' method rolls the leaf prior to processing, whereas CTC (Cut, Tear, Curl) tea is cut prior to rolling and is now the most common type of tea sold on international markets (Figure 3.3). The CTC method produces small granule-like particles and is associated with a stronger flavour that, in general, produces a higher cuppage (number of cups per kilogramme). In practice, the rise of CTC tea has been synonymous with the growth of tea bags as the major mode of consumption, for which the smaller granules are well-suited. Whilst it is generally true that African countries produce CTC and Asian countries produce orthodox, much of South India's teas are actually CTC.

Another axiom in the tea industry is that if quality is produced in the field, it is sustained in the factory; that is, that the role of processing is

**Figure 3.3** The difference between orthodox and CTC teas.
*Source*: Photo by the authors. Two orthodox teas are in the top-right of the picture; three CTC teas are in the bottom-left.

subservient to cultivation in terms of establishing the quality parameters of tea. Most tea industry participants would tend to agree with these sentiments, however, in recent years, the processing stage of tea manufacture has probably gained in importance as a factor shaping the perceived qualities of teas (and, more importantly, for producers, the prices they may expect upon sale). In South India as in other tea-producing regions, there is an increasingly important requirement for factories to gain various kinds of quality-defined certifications in order to optimize their position within domestic and international marketplaces. Most significant amongst these are ISO 9001, the international standard for quality assurance, and HACCP (Hazard Analysis and Critical Control Points). In the tea factory context, obtaining these forms of certification requires managers to develop documented systems that ensure 'best practice' procedures of various kinds are followed. The investments of time and managerial energy to these agendas can be onerous, and in our numerous visits to tea factories in South India, we encountered diverse responses to certification. Though certification potentially facilitates higher prices, for smaller BLF operators it was not at all clear that these benefits outweighed the implementation costs. One factory operator we

**Table 3.2** Tea grades used in Indian auctions

| Grade | Description |
|---|---|
| *Orthodox whole leaf grades* | |
| OP/TGFOP (Orange Pekoe/Tippy Golden Flowery Orange Pekoe) | The top grade. Long twisted leaf without stalks; good brightness and strength; 170–200 cups/kg |
| FOP (Flowery Orange Pekoe) | Smaller size twisted leaf; more color and strength; 170–200 cups/kg |
| FP (Flowery Pekoe) | Round shaped twisted leaf; stronger brightness and flavour; 180–220 cups/kg |
| *Orthodox broken leaf grades* | |
| FBOP (Flowery Broken Orange Pekoe) | Intense colour with good strength and moderate brightness and flavour; 200–250 cups/kg |
| BOP (Broken Orange Pekoe) | Medium-sized round and twisted leaf; 200–250 cups/kg |
| GBOP (Golden Broken Orange Pekoe) | Lighter weight/smaller particle of leaf with some fibre; 200–250 cups/kg |
| BOPF (Broken Orange Pekoe Fannings) | Very small particles of leaf; very strong; 250–300 cups/kg |
| *Orthodox dust grades* | |
| BOPD (Broken Orange Pekoe Dust) | Fine particles well twisted; 300–350 cups/kg |
| FBOPD (Flowery Broken Orange Pekoe Dust) | Mixture of light weight and finer dust with fibres; 300–350 cups/kg |
| BOPFD (Broken Orange Pekoe Fine Dust) | Fine light weight particles, powdery; 350–400 cups/kg |
| *CTC leaf grades* | |
| FP (Flowery Pekoe) | Biggest granular size tea; thin and plain in flavour; 150–200 cups/kg |
| BOP (Broken Orange Pekoe) | Medium-sized granular tea; colour and strength; 250–300 cups/kg |
| BP (Broken Pekoe) | Medium granulation with some flaky particles; 300–350 cups/kg |
| SBP (Small Broken Pekoe) | Small granules with strong colour and brightness; 300–350 cups/kg |
| *CTC dust grades* | |
| PD (Pekoe Dust) | Granular fine particles; 300–350 cups/kg |
| RD (Red Dust) | Finer particles (smaller than PD); 300–350 cups/kg |
| SRD (Super Red Dust) | Powdery appearance, smaller than RD; 450–500 cups/kg |
| FD (Fine Dust) | Finer powdery dust; 450–500 cups/kg |
| GD (Golden Dust) | Powdery form of particles; 450–500 cups/kg |

*Source*: UPASI-KVK (2000).

interviewed suggested that the total costs of implementing ISO 9001 and HACCP amounted to Rs 500,000 (approximately US$11,100).[7] However, as discussed in Chapter 4, tea factory owners may be eligible for subsidies from the Tea Board to defray some of these implementation costs.

In terms of actual processes that take place at the factory stage, both orthodox and CTC manufacturing involve a set of relatively similar procedures for the conversion of green leaf into made black tea. Key stages are:

- *Withering*: This removes between 10 and 15 per cent of moisture content within the leaf in order to facilitate rolling. The withering process involves spreading out leaves over wire mesh in a trough through which air is ventilated. The withering process generally occurs for a period of 12–18 hours.
- *Rolling*: The rolling process has the objective of breaking the cell structures within leaves and thereby mixing together the polyphenols and enzymes allowing processes of fermentation to begin. In the CTC technique, machines crush leaves prior to rolling, creating small granules. In the orthodox system, leaves are rolled to form the distinct 'twisted twig' shape of orthodox tea.
- *Fermentation*: The crushing of leaf in the rolling stage releases enzymes which, mixed with oxygen, enable biochemical changes to take place that create the compounds that give tea its colour, strength, briskness and brightness. Fermentation can take place in the open air ('floor fermentation') or with the assistance of a fermentation drum through which air is pushed. In this stage of the manufacture process, leaves change in colour from green to a copper brown.
- *Drying*: The drying stage removes excess moisture and terminates the biochemical changes associated with fermentation, thereby stabilizing the leaf. Conventional dryers used typically in South Indian tea factories transport the tea through a series of circuits at temperatures which escalate from 50 to 100 degrees Celsius. The drying process usually takes 20–22 minutes.
- *Sorting and grading*: Dried tea is then passed through sorting machines which use wire sieves to separate the tea in terms of particle size, which becomes the basis for grading.

## Grading and auction

Tea is not sold as a standard product. The basic classification schema operating in the tea industry – which defines product in terms of particle size – has given rise to a set of quality descriptions denoted by a quixotic and evocative set of terminology handed down since the early days of the tea industry (Table 3.2). For both orthodox and CTC teas a basic separation is made between 'leaf' and 'dust' grades, the latter passing through a smaller mesh size than the former. In general, leaf grades deliver a more refined flavour than dust grades, which exhibit a coarser and stronger flavour that enables a greater cuppage. However, complicating matters further, the taste characteristics that define particular grades in one producer region or country might differ considerably from those in another region or country. Though tea is a tropical commodity frequently compared with coffee, in terms of the variability of its quality parameters, it is more like wine. Thus:

> Little or nothing about grading tea is standardized. There certainly is no central Body or Authority or any such organization policing the use of terms. Paradoxically, the best teas tend to be sold wholesale by chests, and these too are not standardized by weight. Different estates use chests of different weights. The grading is also performed by self-interested estates and companies. (Altman, 2007: not paginated)

And, according to A.F. Ferguson & Co, consultants to the Tea Board of India:

> Often even the actual dimensions of a particular grade [of tea] vary from estate to estate depending on the type of measure used for grading (mesh or volumetric) or the type of material used in the measure (e.g., steel wire mesh vis-à-vis aluminum wire mesh). Thus even though grading is essentially a measure of the physical size of tea particles, in practice different producers have over the years 'evolved' different nomenclatures. The end result is that sorting of lots takes a longer time and there are frequent buyer complaints either on the grading itself or on the deviation of the delivered lot from the sample. (A.F. Ferguson & Co, 2002a: p. 24)

Corresponding to this intricate variability, the marketing and sales systems that have come to dominate much of the world's tea trade are built around dense knowledge networks that mirror the nuance, quirk and complexity of the product itself. Tea is not subject to the kind of smooth-surface market functions that characterize many other agricultural products. Most major tropical crops – coffee, cocoa, rubber, sugar, cotton, palm oil, and pepper – are featured by commodity-based exchange, e-trading and futures markets. In each of these cases, core economic data (for example, ratios of stocks to production) play central roles in price determination. Though there are specialist 'edges' in each of these product markets, the bulk of production can be traded according to standard categories which provide price-setting benchmarks. The tea trade operates differently. There is an absence of globally meaningful standard grades, and the issue of stocks plays a less central role in price determination because, as a crop plucked and processed on a regular week-in, week-out cycle, there is no commercial logic for large stockholdings. Indeed, in the tea industry, stocks function as a transitory

**Table 3.3** Average yearly prices (US$) at major tea auctions, 1985, 1990, 1995 and 2000–6

| | *Colombo* | *Jakarta* | *Mombasa* | *Sth India (all centres)* | *Nth India (all centres)* | *FAO composite price* |
|---|---|---|---|---|---|---|
| 1985 | 1.44 | 1.62 | 1.59 | 1.73 | 1.93 | n.a. |
| 1990 | 1.79 | 1.78 | 1.48 | 2.21 | 2.56 | n.a. |
| 1995 | 1.41 | 1.06 | 1.29 | 1.27 | 1.57 | n.a. |
| 2000 | 1.75 | 1.20 | 2.02 | 0.99 | 1.57 | 1.80 |
| 2001 | 1.61 | 0.97 | 1.53 | 0.98 | 1.48 | 1.56 |
| 2002 | 1.55 | 1.01 | 1.49 | 0.86 | 1.29 | 1.48 |
| 2003 | 1.54 | 0.95 | 1.62 | 0.86 | 1.32 | 1.52 |
| 2004 | 1.78 | 1.02 | 1.61 | 1.04 | 1.58 | 1.66 |
| 2005 | 1.81 | 1.04 | 1.56 | 0.97 | 1.44 | 1.64 |
| 2006 | 1.91 | 1.34 | 2.02 | 1.12 | 1.59 | n.a. |

*Source*: Tea Board of India (2003: pp. 55 and 109–10) [1985 to 2002]; UPASI (2007: p. 9) [2003–6]. Note: n.a. = data not available.

'pipeline' stage in the supply chain, rather than having any wider economic function (i.e., the hoarding of tea stocks by speculators is unheard of). As indicated in Table 3.3, the overall world average price (the 'FAO Composite Price') masks an astonishing divergence of conditions across different market centres. And, moreover, if one was to drill down further into this data, what is even more apparent is that each of these markets also contains an extraordinary divergence of price-realization between highest-fetching and lowest-fetching teas.

These product attributes have lent sustenance to the dominance of physical product auctions as an exchange mechanism in the industry (Table 3.4). In 1679, tea was traded at public auction in London for the first time and, over the ensuing centuries, these forms of market exchange have evolved to take their own idiosyncratic forms. What is particularly striking about the conduct of tea auctions (vis-à-vis those for other agricultural commodities) is the way they give voice to the heterogeneity of this product. This includes extensive pre-sale cataloguing, sampling and indicative valuation, and the sale of tea in individual 'lots' of different weights. As one 'old hand' tea auctioneer from South India describes it:

> The auction system of selling is ideally suited for tea because of the infinite varieties and grades and the scattered and often remote location of estates. The quality of teas produced by the different estates varies widely not only between districts, but even within the same district or region because of

**Table 3.4** Role of tea auctions in major black tea producing countries

| | *Auction sales ('000 tonnes)* | | *Total production ('000 tonnes)* | | *Percentage of production sold through auction* | |
|---|---|---|---|---|---|---|
| | *1990* | *2000* | *1990* | *2000* | *1990* | *2000* |
| All India | 482 | 507 | 720 | 847 | 67 | 60 |
| Chittagong/ Bangladesh | 43 | 42 | 46 | 53 | 94 | 80 |
| Colombo/ Sri Lanka | 217 | 277 | 234 | 307 | 93 | 90 |
| Jakarta/Indonesia | 33 | 33 | 160 | 157 | 21 | 21 |
| Mombassa/Kenya | 124 | 221 | 197 | 236 | 63 | 94 |
| Limbe/Malawi | 21 | 15 | 39 | 42 | 53 | 36 |
| London/UK | 44 | 0 | – | – | – | – |
| Total of above | 965 | 1,096 | 1,397 | 1,642 | 69 | 67 |

*Source*: Tea Board of India (2003: pp. 103, 110–11 and 116) [1990 and 2000]. Note: 'All India' consists of the six auction centers of Kolkata, Guwhati, Siliguri (in the North) and Kochi, Coimbatore and Coonoor (in the South).

> climatic or soil factors. Teas from the same estate also vary in quality between seasons and just this factor makes it difficult for an estate or group to market its product in branded form. (Ramaswamy, 2003: p. 33)

## Blending and branding

Most consumers are oblivious to the attribute distinctions that drive the auction system because tea sold to consumers usually incorporates blends of different grades and origins, which, when acting together, bring out particular combinations of flavour. A standard tea bag, for instance, can include up to 30 different teas (Traidcraft, 2007a: p. 6). Most tea is sold either through brand names that combine elements of proprietary ownership (e.g., Lipton's, Twinings, etc.) with type of blend. Such blends represent specific recipes owned by/associated with a single company, often built around common-use categories (e.g., 'English Breakfast', 'Earl Grey', etc.).[8]

Critically, however, leading tea brands can produce comparable tastes for these various products through a number of blended combinations of grades and origins. What this means is that tea buyers will go to tea auctions with precise requirements in terms of which types of teas they need for their blending purposes; but if price trends act to make one particular grade expensive, they have the flexibility to recalibrate their purchasing strategies and purchase different combinations of tea within global markets. This generates a peculiar characteristic in the operation of tea auctions. Tea is placed on markets as a highly particularistic product defined by intricate nomenclatures and place-specific grades and subjected to extensive cup-tasting and sampling; yet it is bought on those same markets in ways that are characteristic of commodity products, with buyers having the freedom to chop and change in their purchasing decisions in order to take best advantage of whatever market circumstances are present. As the executive of one leading South Indian plantation company quipped, when we asked him about this process: 'buyers insist on distinct and intricate grading of teas, and then blend these back together and call it "value-addition"'.[9]

This capability to flexibly blend different teas also has important geographical implications. Leading tea blenders possess considerable scope when it comes to sourcing teas on international marketplaces. It is relatively easy for buyers to switch from one source to another, especially for popular tea bag blends, as any change in taste within the source of one tea can be disguised by skillfully blending with another. Correspondingly, major brand-name firms traditionally have had strong incentives to locate their blending and final packaging operations in consumer countries, where they can bring together teas from different origins thus taking advantage of flexible sourcing arrangements. For example, the famous tinned teas of Twinings are blended and packaged in the UK and then re-exported globally to other markets (Altman, 2002). Evidently, these kinds of arrangements set the groundwork for debates on the role of branding in determining value chain governance structures (discussed in Chapter 4) and potential for upgrading (addressed in the concluding chapter of this book).

In summary, this delineation of the input–output structures of tea value chains emphasizes: (i) the place-specific contexts that define upstream components of the chain; and (ii) the global economic geographies associated with flexible sourcing and value-addition that dominate downstream chain components. It also indicates how tea is a truly global product constructed around highly complex relations of production–exchange–consumption. These arguments lead into the next section of the chapter, which utilizes the analytical dimension of 'territoriality' to locate South Indian production within these global relations.

## The territoriality dimensions of South Indian tea

In his early GCC/GVC research, Gereffi defined territoriality as the 'geographic dispersion of production and marketing networks at the national, regional and global levels' (1995: p. 113). Most subsequent analysts have interpreted this to refer to the geographical extent of a chain; that is, does it operate globally, regionally (e.g., the European Union), or nationally? In practice, of course, much of the GVC literature emphasizes the global-*ness* of chains (the 'G' in 'GVC'). Even if the product under consideration is not actually traded across the extent of the world (and very few products actually fall in this category), it is still the case that global economic relations impinge upon chain formations. These points are highly relevant when considering the South Indian tea industry.

Tea cultivation originated in China, which had a monopoly on the trade until plants smuggled out by the British saw a second source of supply be developed in India, followed shortly afterwards by identification of the indigenous Assam variety. Thenceforward, *Camellia Sinesis* was taken across the world. By 2005, commercial tea cultivation occurred in around 40 countries, although just six accounted for more than 80 per cent of world production, and four accounted for more than 70 per cent of world exports (Table 3.5).

The geographical patterning of this production and trade reflects the complex relations between consumption and export. China and India remain the world's largest producers of tea, but much of their output is consumed domestically; both countries have large populations to whom tea is a daily staple. The global geography of the tea industry, therefore, is quite unlike that for the kindred beverage crops of coffee and cocoa, where

**Table 3.5** The world's major tea producing countries, 2005

| *Country* | *Tea type* | *Production ('000 tonnes)* | *Exports ('000 tonnes)* | *Share of world exports (%)* |
|---|---|---|---|---|
| China | G | 934.9 | 286.6 | 18.7 |
| India | O, C | 928.0 | 187.6 | 12.3 |
| Kenya | C | 328.5 | 309.2 | 20.2 |
| Sri Lanka | O | 317.2 | 298.8 | 19.5 |
| Turkey | G | 205.6 | – | – |
| Indonesia | O, C, G | 165.8 | 102.3 | 6.7 |
| Viet Nam | C, G | 104.0 | 89.0 | 5.8 |
| Japan | G | 100.0 | – | – |
| Argentina | O | 73.0 | 66.4 | 4.3 |
| Bangladesh | O | 56.0 | 9.0 | 0.5 |
| Malawi | C | 38.0 | 43.0 | 2.8 |
| Uganda | C | 37.7 | 33.1 | 2.2 |
| Tanzania | C | 30.4 | 23.2 | 1.5 |
| Others | – | 184.6 | 83.0 | 5.5 |
| **World** | | **3503.7** | **1531.2** | **100.0** |

*Source*: FAO (2006a: p. 2). Note: G = Green tea, O = Orthodox, C = CTC.

**Table 3.6** The world's major tea importing countries, 2005

| *Country* | *Volume ('000 tonnes)* | *Share of world imports (%)* |
|---|---|---|
| Middle East/North Africa | 392.3 | 28.2 |
| Egypt | 72.0 | |
| Morocco | 48.0 | |
| UAE | 44.0 | |
| Iraq | 43.0 | |
| Afghanistan | 33.0 | |
| Syria | 27.0 | |
| Sudan | 26.9 | |
| Others | 98.4 | |
| Asia & Oceania | 262.3 | 18.9 |
| Pakistan | 134.1 | |
| Japan | 51.4 | |
| Others | 76.8 | |
| CIS | 255 | 18.4 |
| Russia | 177.4 | |
| Others | 77.6 | |
| Europe | 238.2 | 17.1 |
| UK | 128.2 | |
| Poland | 25.5 | |
| Others | 84.5 | |
| The Americas | 139.5 | 10.0 |
| United States | 100.1 | |
| Others | 39.4 | |
| Sub-Saharan Africa | 101.8 | 7.3 |
| Total | 1,389.1 | 100.0 |

*Source*: FAO (2006a: p. 5).

production is in the tropics but the vast bulk of consumption takes place in the affluent temperate zones. Sri Lanka and Kenya are the world's leading exporters and, despite being major producers, both Turkey and Japan are both significant importers. Further unlike the situation for coffee or cocoa, the affluent populations of North America, Western Europe and Japan are not the most important markets for this product. Well over half of the world's tea exports are now destined for low- and middle-income countries, mainly in the Middle East, North Africa and the countries of the former Soviet Union (Table 3.6). The pull of these markets has profound ramifications for the shaping of governance arrangements in tea value chains, as will be discussed later in this chapter. Rising demand for tea in these low- and middle-income

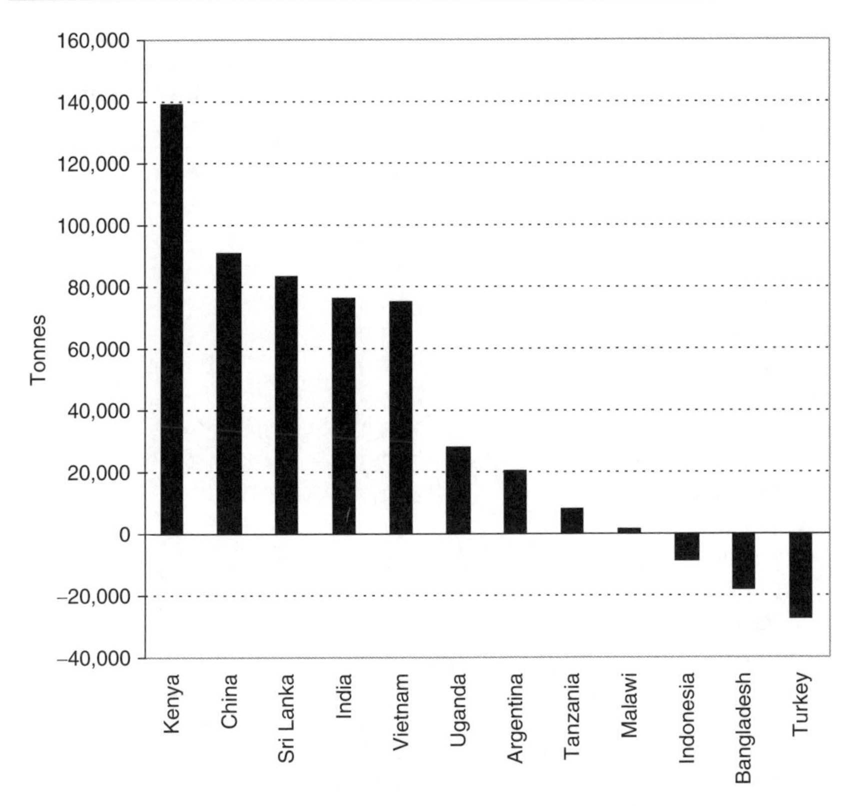


**Figure 3.4** Change in volumes of tea exports between 1990 and 2005, major producing countries. *Source*: Tea Board of India (2004: p. 105); FAO (2006: p. 2).

country markets, fuelled largely by population growth, marks a relatively recent shift in the global geography of the industry. As recently as 1970, the UK accounted for 40 per cent of global tea imports (Baffes, 2004). By 2005, this had fallen to just 9.2 per cent.

Changes to the geography of tea consumption have been matched by significant shifts in the global geography of tea production. Driving the rapid expansion of production has been a desire by producer countries to earn export earnings, with much of the growth in exports between 1985 and 2005 taking place in Kenya, China, Sri Lanka, India and Viet Nam (Figure 3.4). Crucially, these trends are intimately linked to the introduction of neo-liberal and/or structural adjustment policies that gave priority to export-earning sectors. In Sri Lanka, nationally owned tea plantations were privatized in 1992 (Loh et al., 2003) in the context of complementary sector-specific lending programmes arranged by the Asian Development Bank (Ali et al., 1998: p. 52). In Viet Nam, the *doi moi* reforms opened up production by smallholders (in contrast to prior domination by state-owned plantations) and have gradually led to a rise in exports by private tea companies. As in Sri Lanka, moreover, the Asian Development Bank has provided considerable funding for industry expansion, under the terms of its Tree Crop Development Project for Viet Nam (FAO, 2006: p. 1). In Kenya, two decades of policies to encourage smallholder cultivation of this export-earning crop have contributed to significant industry growth. The 1992 decision to end the near-monopoly position of Kenya Tea Packers Ltd. (KETEPA) as the country's largest tea marketing agency was followed by smallholder expansion (Ali et al., 1998: p. 44). In post-Mao China, the loosening of the Ministry of Commerce monopoly, with the issuing of 'Document 75' from the State Council in 1984, ushered in a spate of reforms which saw a more diverse and market-competitive industry emerge (Etherington and Forster, 1993). China's tea exports increased fivefold between 1970 and 1990, with a substantial increase in individual household production and a falling share of production by agricultural cooperatives (from 90 per cent to 55 per cent over the same period, Forster, 1993). The only significant global exporter *not* to have had its tea industry reshaped by the liberal market agenda of recent years is Indonesia, where the national tea industry continues to be dominated by a network of state-owned plantation companies and where there have been no major state initiatives to promote the smallholder sector via marketing coordination or extension services (Neilson and Pritchard, 2006).

Rapid expansion within these producer countries has encouraged the situation where growth in world production has outstripped growth in world consumption. Between 1995 and 2005, increases in global tea production at the rate of 2.3 per cent per annum (FAO, 2006: p. 6) led to conditions of excess supply and softening prices (Table 3.3). Supply responsiveness to these conditions has been mitigated by tea's characteristic as a permanent crop to which smallholders and estate companies owe their survival. It means that 'turning off production' is not necessarily a rational or viable option, notwithstanding the persistence of low prices. Furthermore, the institutional presence of the Asian Development Bank in pumping up Viet Nam's level of tea production by 15 per cent per annum over the period 1995–2005 (FAO, 2006: p. 7) added to global supply notwithstanding what was happening on world markets. World market prices improved somewhat after 2004, but on account largely of the East African drought, which in 2005 pegged back African output by 1.8 per cent (FAO, 2006: p. 2).[10]

## The Input–Output Dimensions of South Indian Coffee

From farm cultivation, coffee cherries undergo a physical transformation which involves pulping and drying; curing (removal of the parchment, cleaning, sorting and grading); warehousing, export, roasting, blending, branding, retail sale and consumption. In this section we review these stages in the addition of value, with a view to develop an understanding of the reasons why particular institutional environments and forms of governance have evolved in the South Indian coffee sector, as described in the following chapters of this book.

## Cultivation

There are two major species of coffee grown commercially in the world today: *Coffea Arabica* and *Coffea Canephora*.[11] For many years, Arabica coffee (originally from the Ethiopian highlands of East Africa) was the only species produced on a commercial scale. This only changed following the emergence of leaf rust (*Hemileia vasatrix*) in Ceylon in 1861, and the subsequent spread of the disease throughout all the major coffee-producing regions of Asia. According to Francis (2001, p. 175), leaf rust had 'devastated whole districts' in South India by 1875. The serious industry decline resulted in the replacement of vast areas planted to coffee with tea, and motivated the search for rust-resistant coffee in its African homeland. This led eventually to the discovery of Robusta coffee in the Congo basin of Central Africa. Robusta coffee was introduced to Java in 1900, South India in 1906 and, by 2006, accounted for 34 per cent of world trade (International Coffee Organization, 2007). As a general comparison with Arabica, Robusta is a hardier plant with higher per unit productivity, which grows well at lower altitudes (generally lower than about 1,000 m altitude in South India), and produces a characteristically harsh, slightly bitter and less acidic tasting coffee. The Arabica and Robusta world markets are only loosely related, with Robusta frequently sold at a significant discount to Arabica.[12] Indian coffee production is dominated by the three southern states of Kerala, Karnataka and Tamil Nadu (Table 3.7), with Robusta constituting approximately 65 per cent (by volume) of national production.

Across the world, coffee is grown in the context of a range of quite different agro-ecological systems. The diversity of coffee agro-ecologies ranges from intensive monocultures to coffee grown within heavily shaded, forest environments. In South India, the expansion of modern coffee cultivation in South India occurred in the nineteenth century through the vehicle of British Colonialism, in the (then) sparsely populated and densely forested hill tracts of the Western Ghats. This ecological system continues to shape coffee cultivation in South India. The crop is overwhelmingly planted and cultivated within shaded, multi-crop ecosystems which support an array of habitat resources (Figure 3.5). The tendency to plant coffee under dense

**Table 3.7** Coffee production by State and District, 2006–7

| *State* | *District* | *Arabica Production (tonnes)* | *Robusta Production (tonnes)* |
|---|---|---|---|
| Karnataka | Kodagu | 21,750 | 83,300 |
| (71.6% of total) | Chikmagalur | 36,800 | 28,600 |
| | Hassan | 17,750 | 8,075 |
| Kerala (20.7%) | Wyanad | 75 | 46,600 |
| | Travancore | 800 | 7,025 |
| | Nelliampathies | 500 | 1,825 |
| Tamil Nadu (6.9%) | Pulneys | 7,900 | 300 |
| | Nilgiris | 1,350 | 3,675 |
| | Shevaroys | 3,600 | 0 |
| | Anamalais | 1,525 | 475 |
| Andra Pradesh & Orissa (<1%) | | 1,775 | 50 |
| North Eastern Region (<1%) | | 175 | 75 |
| Total | | 177,728 | 201,981 |

*Source*: Coffee Board of India (2007).

shade relates to local climatic conditions which include an extended dry season in the Malnad.[13] Without shade, great stress would be placed on coffee plants, leading to overbearing.[14] Traditionally, coffee is planted underneath a multi-tiered canopy which includes a remnant layer of rainforest trees and a medium layer of dadap (*Erythrina lithosperma*). Biodiversity is significant, with the Arehalli smallholder growers' cooperative, in the Hassan District of Karnataka, claiming that their members cultivate coffee in the midst of more than 100 tree varieties.[15] Such levels of biodiversity, however, do not exist universally; in some settings, the varietal diversity of tree shade has been compromised by plantings of *Grevillea robusta* (silver oak), a deep-rooting tree which doesn't directly compete with coffee for nutrients. As discussed in greater detail in Chapter 6, use of this tree as a mono-shade above coffee greatly reduces the diversity of canopy.

As a result of shade-grown cultivation, productivity is low compared to sun-grown monocultures in countries such as Viet Nam and Brazil. However, with the exception of some higher-altitude estates in Chikmagalur District, coffee production in South India is virtually inseparable from pepper cultivation, with pepper vines trained up the trunks of shade trees and producing a vitally important secondary source of income for coffee planters. Nearly 60 per cent of India's pepper production is grown as an intercrop with

**Figure 3.5** Shade-grown coffee, Kodagu District, Karnataka.
*Source*: Photo by the authors. The coffee is in the foreground. The trees are hosting pepper vines, which are a common inter-crop with coffee.

coffee (Raghuramulu, 2007). Citrus was widely inter-planted with coffee in Kodagu in the past, although this has diminished substantially in recent years due to outbreaks of so-called 'greening disease'. Citrus trees are still an important intercrop in the Pulneys and the Shevaroys, with one study estimating that they contributed 19 per cent to the total net income of Arabica plantations across Tamil Nadu (ibid.). The Tamil Nadu Arabica plantations are some of the most diversified in India, with substantial production of cardamom, lemon, banana and avocado. In general terms, however, Robusta plantations tend to be more suitable to inter-cropping, due to the lower altitude at which it is grown and to their lower shade density. Areca nut has been estimated as contributing 24 per cent to the total net income on Robusta plantations in Chikmagalur, and pepper contributes

29 per cent of net income in Kodagu (ibid.). Moreover, India's most widely-known coffee cup taster, Sunalini Menon (2002), has claimed that wide-spread inter-planting with crops such as pepper, cardamom, ginger and vanilla contributes to a characteristic 'hint of spice' in Indian coffee.

Coffee production worldwide is vulnerable to an array of pest and disease threats, leading to a history of extensive agro-chemical interventions. In India, ever since the early planting days and the ravages of leaf rust, the Indian coffee industry has had to deal with serious problems of pest and disease. The most pressing problem facing the Indian industry today (especially for Arabica production) is white stem borer (*Xylotrechus quadripes*). The pest was mostly under control in India until about 1995, after which it resurfaced, apparently due to a prolonged drought and the gradual removal of shade cover. Unlike other pest and disease problems (such as coffee berry borer and leaf rust), white stem borer actually kills the host plant, thereby leading to significant capital losses on plantations. The Coffee Board (2005) reports that average yields for Arabica peaked at 860 kg/ha in 1995–96 prior to infestation, and have since declined to 620 kg/ha. During 2005, some planters we interviewed claim to have lost 50 per cent of their crop to the pest.[16] To prevent the proliferation of stem borer, affected trees need to be quickly identified and removed from the plantation. Coffee berry borer (*Hypothenemus hampei*) and various soil nematodes (sap-feeding micro-organisms) are the other major concerns for Indian coffee production.

Management of these risks needs to be undertaken in the context where, as with tea, importer countries are increasingly restricting chemical use by producers. In the coffee sector, MRLs for 14 pesticides have been specified by Codex Alimentarius in relation to green coffee beans (International Coffee Organization, 2007). This list includes Endosulfan, a controversial chemical currently used in the Indian coffee industry to control coffee berry borer. Endosulfan is classified by the US Environmental Protection Agency as Category I: 'Highly Acutely Toxic'. It is banned in a number of countries such as Germany and the Netherlands, and its use is subject to the Prior Informed Consent (PIC) procedure in other jurisdictions.[17] In 2007, the European Commission proposed to add Endosulfan to the list of chemicals banned under the Stockholm Convention on Persistent Organic Pollutants, which, if approved, could result in efforts to ban all use and manufacture of Endosulfan[18] globally (Umweltbundesamt [UBA], 2007). As a result, there are increasing pressures in both the USA and the EU to ensure the total banning of the use of such chemicals worldwide. Lindane is another controversial chemical widely applied to prevent white stem borer attack. It is a banned chemical (for all uses) in seven countries and is severely restricted in many others (Pesticides Action Network, 2001). Lindane is also subject to the PIC procedure.

A further food safety issue in the coffee industry is that of Ochratoxin A (OTA) contamination. OTA is a mycotoxin produced by fungi (mainly *Aspergillus sp.*) which can be potentially dangerous for human health, such that by the late 1990s, a number of European countries (including Italy, Finland and Greece) had introduced physical OTA limits for green coffee imports. A subsequent multi-country study by the FAO, *Enhancement of Coffee Quality through the Prevention of Mould Formation* (FAO, 2006), has since stalled the introduction of physical limits by the EC in lieu of promoting Good Hygiene Practices along the supply chain. To date, the EC has only set OTA limits for roasted and soluble coffee (EC Regulation No. 466/2001) and not for green coffee. This has effectively placed the responsibility of due diligence on coffee manufacturers, who are applying their own supply chain standards on exporters. OTA levels can be controlled through properly controlling moisture levels along the chain to avoid mould growth and there has been a trend towards HACCP-type certification to minimize the risk of OTA contamination in coffee (Sibanda, 2006). Conformance to these HACCP-based systems is fast becoming a requirement for market access. As discussed throughout this book, a vital contemporary characteristic of agri-food trade is the augmentation of state regulation by a host of private standards that have taken on a quasi-official status, including HACCP and ISO certification, as well as standards such as Global-GAP and 4C. The document *Plant Protection in Coffee: Recommendations for the Common Code for the Coffee Community-Initiative* (see Jansen, 2005) identifies the use of Lindane as unacceptable and Endosulfan as a category red (unwanted practice) on the grounds of worker welfare. Consequently, the Coffee Board has begun to discourage the use of Lindane in the sector.

The flowering of the coffee plant (and by extension fruit set) is determined by the annual monsoon. In the Malnad, the dry months of December through to February are followed by the first rains in late March and the heavy rains of the southwest monsoon from June through to August. The lighter spring rains in March–April are the most important in determining fruit set, with the harvest occurring some 7–8 months later for Arabica, and 9–10 months later for Robusta. As a result, the main coffee harvest generally takes place from November through to February each year. The intensity of coffee flowering is affected by the severity of the dry season, with the clearly defined monsoons of the Western Ghats conducive to distinct seasonality and a relatively fruitful harvest.

Commonly, three rounds of labour-intensive selective picking are performed during harvest: an initial fly-pick to remove early-ripening cherries, followed by the main harvest, and then a final stripping of the tree. Much of the labour for the coffee harvest comes from seasonal migrant workers arriving from the rural communities of the Deccan Plateau in central Karnataka. The coffee harvest in the Malnad coincides with the dry season on the plains and thus offers opportunities for supplementary income for these agricultural workers. On the Kodagu coffee estates, we met with workers from these areas as well as from Tamil Nadu, Kerala, Mysore and the tribal areas of North Karnataka.

On average, the annual labour requirement per hectare is 400 person days for Arabica and 300 person days for Robusta.[19] Whilst the large-scale corporate plantation sector is under a strict obligation to meet the requirements of the *Plantation Labor Act*, mid-sized plantations tend to rely on a minimum of permanent workers, supplemented when required by seasonal workers. Such arrangements mitigate the extent to which the obligations of the *Plantation Labor Act* extend into these production sites. Moreover, it is not uncommon for medium-sized estates to split the legal ownership of their holdings among family members, further circumventing the Act. Some planters we met stated outright that although some form of housing, education and health assistance is usually provided, the small size of their operations make it not at all realistic to expect full compliance with all requirements of the Act. This contrasts with the tea sector. Because of tea's regular week in–week out plucking cycle, estates need to be committed to the employment of permanent workforces. As a crop involving a lesser level of regular maintenance but with sharp peaks in labour demand linked to its annual harvest, coffee is associated more strongly with seasonal workers.

In the Malnad, many coffee pickers (mostly women) are simply paid a flat amount based on the amount of cherry picked, usually measured in 40 litre ('forlit') bins. On one of the corporate plantations in Kodagu, we were told that a picker can harvest an average of about 130 kg of Robusta or 65 kg of Arabica cherries each day, being paid a base wage of Rs 65/day (approximately US$1.48) and a bonus of Rs 0.5 for every kg picked above the daily minimum.[20] The ability to manage labour is probably the greatest challenge to coffee planters, requiring considerable linguistic talent (with up to a dozen languages spoken on some estates) and becoming an increasingly difficult task in modern India. A Kodagu coffee planter explained the hierarchical system of labour management like this:

> The system is pretty well standard across the plantation belt of India, and has been established over a period of more than 100 years. On larger plantations, the owner will employ an estate manager. Although on smaller estates such as mine, I am both the owner and manager. Beneath the manager is the 'writer', who acts as the assistant manager and supervises a number of 'maistre', who are the foreman supervising individual teams of labourers. The bulk of labour on the estate is unskilled, with the exception of borer-tracing and pruning that is.[21]

Across Kodagu and Chikmagalur, as with the case of the tea districts, manual labour has been in particularly short supply in recent seasons. In addition to high labour demands from other sectors of the economy, plentiful rains and improved irrigation on the Deccan can mean that many migrant workers continue farming in their home villages and are reluctant to join in the Malnad coffee harvest. Working on coffee plantations can be a hazardous occupation, especially shade pruning in the upper canopy, and the risk of attacks by wild animals is taken very seriously. This is demonstrated by the tiger attacks on coffee workers in South Kodagu during 2006 (Praveen, 2006), and the fact that more than 51 deaths were reported over a 10-year period due to elephants in Kodagu District alone, most of which occurred on or around coffee plantations (Kulkarni et al., 2007). In February 2005, the labour shortages were such a problem that cherries were going unpicked and drying up on the branches, while informal labour bidding auctions were being held in Chikmagalur to see which plantations were willing to pay the highest price. Labour recruitment and management on the estates has also been further complicated by the delicately poised caste relations and politics of post-*Mandal* India.[22] In Chikmagalur, for example, a planter from the Gouda caste (a progressive farming caste of south-western Karnataka) told us that much of his workforce were Dalits (people without caste), and that one slip of the tongue would lead to him being taken to court for discrimination.[23] Some planters may also live in fear of reprisals from laid off workers, exacerbated by gangs of militant Naxalites roaming the nearby countryside.[24] The upshot of these labor market factors is to seriously compromise the cost-competitiveness of Indian coffee production.

## Pulping and curing

After harvest, coffee cherries are processed in a set of procedures known as pulping and curing. There are two basic methods of coffee preparation used in the world: the dry method (common for Robusta) which involves laying out the coffee cherries in a dry environment until the bean dries 'naturally' within its skin and parchment; and the wet method, which involves initial pulping prior to fermentation and drying (Figure 3.6).

In India, the initial step in processing the cherries is to pour them into a concrete vat of water to separate out over-ripe or otherwise damaged beans, which float to the surface. These 'floaters' are dry processed separately by spreading them directly on the cement drying yards ('barbeques') and then subsequently stored in 'godowns' (warehouses) after about two to three weeks under the sun. In India, such dry processed coffee is generally referred to as 'cherry' coffee. A greater proportion of Robusta production is processed as 'cherry' compared to Arabica. Most ripe, main harvest coffee is wet-processed, and is known as 'parchment' coffee in India. This is conventionally processed by pulping the cherry, fermenting for 12–24 hours and then washing the parchment prior to sun-drying. The washing and fermentation

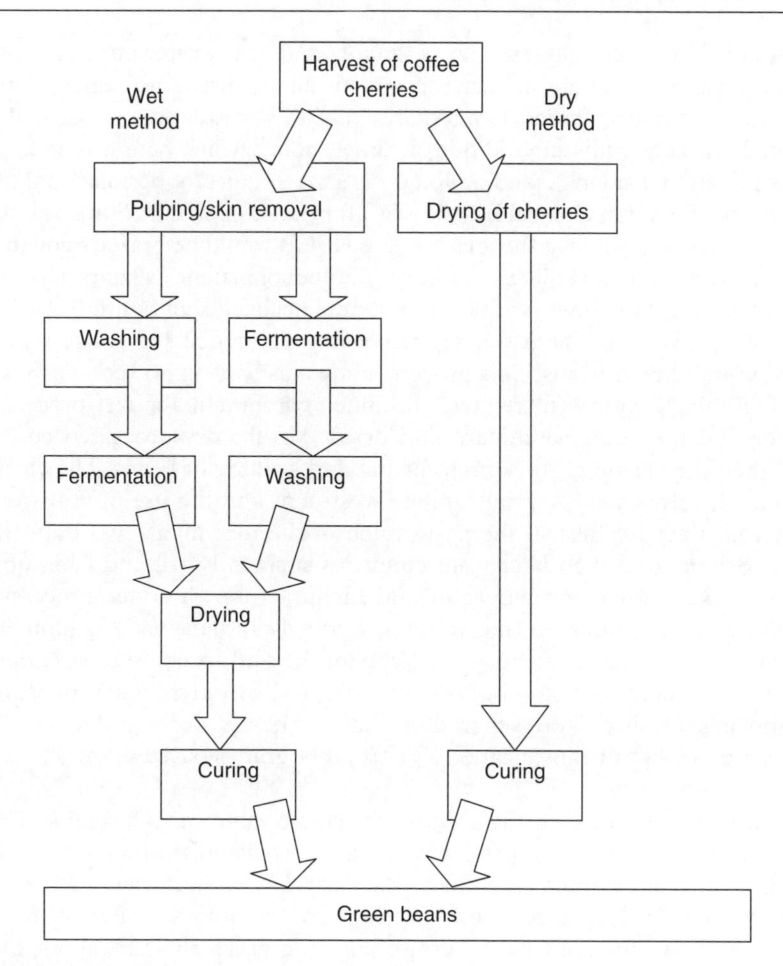


**Figure 3.6** Post-harvest processing stages for coffee.
*Source*: Own work.

process removes the sticky mucilage that adheres to the inner parchment. On some estates that we visited, in an effort to enhance efficiency, the pulped coffee was immediately washed to remove the mucilage and then soaked in water overnight before sun-drying (between 7–10 days, or until the moisture content is about 11 per cent) and then stored in godowns on the estate.[25] In South India, this initial processing is performed on-farm for larger estates, but for small growers, may be sold to independent pulpers (processing at a fee), or pulped at cooperatively-owned operations.

A reliable water supply is required for wet processing, and storage dams are often required on the estates for this purpose. The coffee mucilage itself is considered a potentially serious pollutant, and the wastewater from this process can have severe impacts on aquatic life if discharged directly into streams and rivers. Indian law requires that this wastewater is adequately treated (usually with the addition of lime and enzymes before holding in sedimentation lagoons), and pulping stations require a permit from the Karnataka Pollution Control Board (KPCB) prior to operation. Some planters we interviewed scoffed at the idea that the KPCB would be brazen enough to actually inspect their facilities, however a number of blatant violators have also been successfully charged with environmental negligence in South India.[26]

The next stage for both wet-processed and dry-processed coffee is conversion into green beans. This process, known as 'curing' in India, involves the hulling of coffee (to remove the inner parchment for wet-processed coffee and the inner parchment and dried skin for dry-processed coffee) and then the cleaning, screening, sorting and grading of beans. The curing process involves a substantial capital investment and the use of industrial-type machinery. Whilst in the past, much of the machinery was imported from Britain, and then later from countries such as Brazil and Colombia, today India is a leading innovator and manufacturer of coffee processing equipment and Indian curing works use mostly Indian-made equipment. A common layout for a factory would be for the semi-processed parchment coffee to undergo de-stoning, hulling (removal of parchment), polishing (removing the silverskin), separation of defect beans, deck-grading (which sorts out elephant beans, peaberries and other grades based on mesh size), and then sortation.

In South India, this 'curing' stage of the coffee value chain has undergone significant restructuring during recent years, associated within industry liberalization (see Chapter 4). Prior to the early 1990s there were only a few (highly regulated) curing works within South India's coffee districts. Liberalization, however, led to growers having greater choice about how and through whom they sold their coffee, and in turn this encouraged a rapid increase in the number of curing works across the growing districts. Between 1996 and 2000, the number of curing works in Chikmagalur district alone increased from eight to 20 (Arindampal and Somayaji, 2000). Across the whole of India (remembering that the overwhelming proportion of production takes place in Karnataka and Tamil Nadu), the number of licensed processing units swelled from 39 in 1995 to 85 in 2005 (Coffee Board of India, 2007).

This new competition has given growers the option of selling their parchment coffee to curers at their daily rate, or paying a curing cost (estimated at Rs 2.5/kg) and retain ownership of the product through to its green bean stage. Our interviews across the Malnad suggest that many medium-sized planters simply fix a price with a local town-based agent, who arranges for the collection of the parchment coffee and makes an immediate payment.

Alternatively, the coffee can be held on consignment at the curing works, usually up until the end of October (for Arabica) and November (for Robusta). At this time, the unhulled coffee is either sold to the curing works at prevailing prices or processed for a fee. Town-based agents usually work on behalf of a particular curing house, and there are numerous agents operating in local towns across the coffee districts. With transparent access to global price information, competition is strong and agents appear to operate on tight margins. Agents will receive prices daily from their patron curers and make contracts with planters based on these prices. Most planters we spoke to follow price movements in London and New York closely before selling their coffee to agents. In this sense, Indian coffee is highly commoditized and there does not appear to be much local price variation for different qualities of coffee.[27]

At the larger curing works, truckloads of coffee are initially weighed whilst samples are checked for quality compliance. Each lot is commonly marked by an identification card showing estate name or district of origin, allowing a degree of traceability throughout the supply chain. Some analysts have suggested that the quality of South Indian coffee has deteriorated because of the flush of new curers in the post-liberalization market (Venkatesh and Basavaraj, 1999: p. 65),[28] but our observations and interviews don't lend support to such views. For the premium grades and for coffee to be sold to specialty buyers, coffee is usually garbled (hand sorted) by rows of women workers on the factory floor (Figure 3.7). Mixing and bagging machines are then used prior to transport. Conventionally, coffee beans are packed in 60 kg sacks printed with corporate logos, estate identification or other markings based upon the request of the international buyer. In some instances, lower quality coffee is bulk packed into containers without bagging to reduce costs.[29]

Perhaps because of the heightened importance of quality parameters in the global coffee trade, in recent years there has been a trend towards an increasing integration of curers and exporters and, in turn, between exporters and international trading companies. All of the major exporters of Indian coffee (Table 3.8) also operate their own curing operations. In total, there were 74 active coffee exporters in 2006–07, however the trade remains highly concentrated. The top five exporters in 2006–07 accounted for 54 per cent of total exports, and the top 12 were responsible for exporting 85 per cent of India's coffee.[30]

Some of the largest curing works in India are now owned by integrated coffee companies, such as Allana & Sons and Amalgamated Bean Coffee, whose operations we visited in Hassan and Chikmagalur respectively. Allana's, which is a Mumbai-based trading house with diversified agricultural export interests, maintains a network of some 20 collecting agents across the Western Ghats, which operate on direct commissions based on purchases. Operational credit is not supplied to agents due to inherent financial risks, and payment is usually made upon delivery only. Whilst some curers may accept coffee directly from planters, most prefer to mediate their

**Figure 3.7** Hand-sorting (garbling) of coffee green beans. (Tata Coffee curing works, Kushalnagar, Kodagu, Karnataka)

**Table 3.8** Major exporters of Indian coffee, 2006–7

| *Exporter* | *Volume (tonnes)* | *Value (Rs lakh)* |
|---|---|---|
| 1 Allana Sons | 30,864 | 25,982 |
| 2 General Commodities | 28,425 | 21,454 |
| 3 CCL Products (India) | 19,071 | 17,857 |
| 4 Nestle India | 17,668 | 14,025 |
| 5 Olam (India) | 17,116 | 11,846 |
| 6 Tata Coffee | 16,308 | 13,955 |
| 7 Hindustan Lever | 14,550 | 11,928 |
| 8 ITC | 12,970 | 10,770 |
| 9 ABC | 12,664 | 10,681 |
| 10 Ned Commodities | 12,040 | 9,119 |
| 11 Ecom | 8,301 | 5,649 |
| 12 Ramesh Exports | 7,567 | 6,222 |
| Total | 243,059 | 195,832 |

*Source*: Coffee Board of India (2007).

relationship, and the sharing of risk, through agents. Along with the tighter integration of ownership structures at the curing-export stages of the South Indian coffee value chain, there is also a higher level of foreign investment. Large international trading companies such as Ecom Agroindustrial (which acquired Cargill's coffee trading operations in 2000) and Continaf have entered the industry. Ecom established a presence in the South Indian coffee industry in 2000, through a joint venture with Gill & Co, a venerable Indian commodity trading company. The new entity now operates a sales/export office in Bangalore and a curing works near Kushalnagar.

## Grades, standards and market exchange

After the curing stage, coffee green beans are ready for export. The global trade of coffee, unlike that of tea, is organized around a well-established set of grades and standards. The International Standard *Green Coffee – Defect Reference Chart* (ISO 10470) (established in 1993 and revised in 2004) details the main categories of defects considered important for green coffee throughout the world, and individual exchanges have standard contracts with clearly defined levels of acceptable defects. Such definitions are used to specify the buying terms for much of the world's coffee trade. Furthermore, the ICO identifies four categories of coffee for the purposes of tracking global price movements:

- Columbian Milds (*Arabica*: mostly from Colombia, Kenya and Tanzania);
- Other Milds (*Arabica*: from various countries across Africa, Latin America and Asia);
- Brazilian Naturals (Dry-processed *Arabica*: from Brazil, Ethiopia, and Paraguay), and
- Robustas (*Robusta*: from various countries, particularly those in Africa and Asia).[31]

In India, these international grades are complemented by an authoritative and transparent grading system set up by the Coffee Board (Table 3.9).

The relative clarity in grades and standards in the global coffee trade thus provides the basis for a deep and extensive trade in coffee futures, which today act as the primary price-discovery mechanism in the industry. Futures markets act as 'short-term syntheses of market fundamentals (production, consumption and stocks) and technical factors (hedging, trend following, reactions to trigger signals)' (Ponte, 2002a: p. 1104). As a tree crop with a significant (three-to-five-year) lag between planting and bearing, coffee production is poorly responsive to short-term shifts in demand, and this creates endemic volatility within coffee markets. The prospect of

**Table 3.9** Major coffee grades set by the Coffee Board of India

| | |
|---|---|
| **Commercial Arabica grades** | |
| Plantation PB | No sieve requirement, <2% flats, <3% PB triage, clean garbled |
| Plantation A | >90% retained by 6.65 mm sieve, <2% PB, <2% triage, clean garbled |
| Plantation B | >75% retained by 6 mm sieve, <2% PB, <3% triage, clean garbled |
| Plantation C | >75% retained by 5.5 mm sieve, may include some triage, no blacks or damaged beans |
| Plantation Blacks, Bits and Bulk | Various |
| Arabica Cherry PB | No sieve requirement, <2% flats, <3% PB triage, clean garbled |
| Arabica Cherry AB | >90% retained by 6 mm sieve, <2% PB, <3% triage, clean garbled |
| Arabica Cherry C | >75% retained by 5.5 mm sieve, may include some triage, no blacks or damaged beans |
| Arabica Cherry Blacks/Browns, Bits and bulk | Various |
| **Premium Arabica Grades** | |
| Plantation AA | >90% retained by 7.1 mm sieve, <2% PB, clean garbled |
| Plantation PB Bold | 100% retained by 4.75 mm sieve (oblong), <2% AB, <2% PB triage, clean garbled |
| Arabica Cherry AA | >90% retained by 7.1 mm sieve, <2% PB, <1% triage, clean garbled |
| Arabica Cherry A | >90% retained by 6.65 mm sieve, <2% PB, <2% triage, clean garbled |
| Arabica Cherry PB bold | 100% retained by 4.75 mm sieve (oblong), <2% AB, <2% PB triage, clean garbled |
| **Commercial Robusta grades** | |
| Robusta Parchment PB | No sieve requirement, <2% flats, <3% PB triage, clean garbled |
| Robusta Parchment AB | >90% retained by 6 mm sieve, <2% PB, <2% triage, clean garbled |
| Robusta Parchment C | >75% retained by 5.5 mm sieve, may include triage, <2% blacks/browns, or bits |

**Table 3.9** *(cont'd)*

| | |
|---|---|
| Robusta Parchment Blacks, Browns, bits and bulk | Various |
| Robusta Cherry PB | No sieve requirement, <2% flats, <3% PB triage, clean garbled |
| Robusta Cherry AB | >90% retained by 6 mm sieve, <2% PB, <3% triage, clean garbled |
| Robusta Cherry C | >75% retained by 5.5 mm sieve, may include some triage, <2% blacks/browns and bits |
| Robusta Cherry (Blacks, Browns, bits and bulk) | Various |
| **Premium Robusta grades** | |
| Robusta Parchment A | >90% retained by 6.65 mm sieve, <2% PB, clean garbled |
| Robusta Parchment PB Bold | 100% retained by 4.5 mm sieve (oblong), <2% AB, <2%PB triage, clean garbled |
| Robusta Cherry AA | >90% retained by 7.1 mm sieve, <2% PB, <1% triage, clean garbled |
| Robusta Cherry A | >90% retained by 6.65 mm sieve, <2% PB, <2% triage, clean garbled |
| Robusta Cherry PB Bold | 100% retained by 4.5 mm sieve (oblong), <2% AB, <2% PB triage, clean garbled |
| **Specialty grades** | |
| Mysore Nuggets EB (Extra Bold) | Plantation A coffee from Mysore, Coorg, Bababudan, Biligiris and Shevaroys, >90% retained by 7.5 mm sieve, clean garbled, polished |
| Robusta Kaapi Royale | Robusta Parchment AB from Mysore, Coorg, Malabar, Wynaad, Bababudan, Pulneys and Shevaroys, >90% retained by 6.7 mm sieve, clean garbled, polished |
| Monsooned Malabar AA | >90% retained by 7.5 mm sieve, <2% triage, clean garbled, 13–14.5% moisture |
| Monsooned Basanally | >75% retention by 6.5 mm sieve, <3% triage, 13–14.5% moisture |
| Monsooned Robusta AA | >75% retention by 6 mm sieve, <3% triage, 13–14.5% moisture |

*Source*: Coffee Board of India (2004).

production-sapping frosts in Brazil is ever-present in this industry, with 19 frosts during the twentieth century generating sharp and unpredictable price hikes (van Dijk et al., 1998: p. 18). In these contexts, futures enable industry participants to hedge their physical purchases/sales against price fluctuations.

Coffee futures have been traded in New York since the founding of the New York Coffee Exchange in 1882; which subsequently (in 1979) became the New York Coffee, Sugar, and Cocoa Exchange; then (in 1998) a subsidiary of the New York Board of Trade, and finally (in 2007) a subsidiary of InterContinental Exchange. The coffee futures contract traded in New York is called the 'C' contract, based on Central American Arabica coffees. Coffee growers, traders and roasters around the world refer to the New York 'C' for Arabica price developments. With regard to Robusta, futures contracts have been traded on the London International Financial Futures Exchange (LIFFE) since 1975. As discussed by Talbot (2002a), prior to the 1980s the major participants in coffee futures trading were importers and roasters, who hedged against sudden price fluctuations. The relative price stability during the ICA regime meant that financial speculators found coffee futures to be an unattractive market. This situation, however, changed following the 1989 deregulation of international trade, and the earlier (1986) introduction of options trading (where contracts were far cheaper) at New York. Talbot (2002a: p. 231) argues that by the mid-1990s, 'the vast majority of trades made on the coffee futures markets were made for purely speculative purposes, and were not connected to sales of physical coffee', and that the 'combination of pegging the price of coffee to the futures markets and the increased weight of the commodity funds in these markets, has probably *increased the instability* of world market prices' (italics added) (Talbot, 1996: p. 123). This price volatility has occurred to the detriment of producers who, frequently, do not have access to similar price hedging mechanisms.

In India, coffee planters, traders and exporters look to the London and New York futures exchanges for market guidance. During our field research, interview respondents would routinely 'break the ice' with us by discussing the latest market movements in these coffee futures exchanges. Evidently, the centrality of the New York and London markets speaks to the evolution of globalized commodity trade within the coffee sector, and exists in stark contrast to the more nuanced and localized construction of tea markets. Unlike tea, only an estimated 5 per cent of total coffee production is sold through the sole auction in South India, the weekly (Thursday) Bangalore auction organized by the Indian Coffee Trading Association.[32] Although this auction has a well-regarded role in terms of local price-discovery functions (evidenced by the fact that many industry actors attend the auctions without necessarily being involved in actual buying and selling), it remains the case that trends are set by international markets, on the basis of core grades and standards.

Nevertheless, during recent years there have been attempts to further improve the efficiency and transparency of the Indian coffee trade through the establishment of local coffee futures markets. Over the past decade a number of attempts have been made to trade coffee futures within India, starting with the Coffee Futures Exchange of India (COFEI), established in Bangalore in 2000. As a single commodity exchange, COFEI subsequently proved to be unviable when the volumes being traded screeched to a halt in 2002. Then, the National Multi Commodity Exchange (NMCE), with its head office in Ahmedabad, began trading in coffee futures in February 2005. This was followed by coffee futures trading on the National Commodities and Derivative Exchange (NCDEX) in Mumbai in April 2005. However, like COFEI before it, both the NMCE and the NCDEX have not attracted significant interest from traders and speculators. This, apparently, is due to poor contract specifications, lack of speculative activity, and perceptions of excessive government intervention in price regulation hampering the free flow of market forces. Then, most recently, the Multi Commodity Exchange (MCX) in Mumbai (the largest commodity exchange in India, claiming a 72 per cent market share) commenced trading Robusta futures in January 2007. The MCX has developed a strategic alliance with LIFFE, which has significantly improved user confidence in the system. Planters can make deliveries at registered warehouses across the country, with the MCX currently accepting deliveries at Hassan, Kushalnagar and Kalpetta. This has prompted recent attempts to revive the coffee futures contract at both the NMCE and the NCDEX, with three exchanges effectively competing to be recognized as India's leading coffee futures market. Companies such as Geojit Financial Services have established portals in many smaller rural towns which, in theory, allow small traders to trade in futures.

In addition to these coffee futures trading initiatives, online systems have also come to play a keener role in the physical exchange of coffee within South Indian pre-export supply chains. The pivotal actor here has been ITC Ltd, one of India's leading private companies and the sixth largest exporter of green coffee beans in India (Coffee Board of India, 2007). The company has established more than 6,500 *e-Choupal*[33] Internet kiosks across nine states of India to streamline its procuring process for a number of agricultural commodities including soy, tobacco, wheat, shrimp and coffee. The ITC *eChoupal* programme for coffee is 'plantersnet.com', established in December 2000. This system was set up to provide market information and production advice to planters, however, subsequently, it evolved to include 'tradersnet.com', an online marketplace to buy and sell raw and clean coffee. As Anupindi and Sivakumar (2007) have argued, these systems offer the potential to establish within-week price discovery mechanisms, generate liquidity for the 'off-grades' destined for the domestic market, and to move actors beyond existing relationship-based trading systems. Whilst coffee trading through the *eChoupal* model has been modest to date, ITC views the initiative as an investment with long-term benefits as trading systems become increasingly sophisticated.

## Export, roasting and consumption

Green beans are typically exported from South India in 60 kg bags, with between 250 and 400 bags per shipping container, and a more limited use of bulk containerization. The major ports of export for Indian coffee are Cochin and Chennai. As recently as 2003–04, 60 per cent of Indian coffee was exported from Cochin. However, by 2006–07, this share had dropped to only 42 per cent, with Chennai's share increasing (Coffee Board of India, 2007). This appears to reflect the increasing tendency for exporters to utilize the Inland Container Depot (ICD) in Bangalore, a modern facility with cost-effective rail links from Bangalore to Chennai. Bangalore is also the location of the Indian Coffee Board, the headquarters for a number of coffee exporters and diversified plantation companies, and the secretariat for the All India Coffee Exporters' Association (AICEA).

With most of the world's coffee consumed in Europe and North America, a select number of large international ports (Rotterdam, Hamburg, Antwerp and Trieste in Europe; and New Orleans, New York and Oakland in the USA) tend to operate as staging points for the warehousing and distribution of green beans.[34] Once exported, green beans from South India enter a global roasting and retailing industry which has undergone dramatic changes during the past two decades.

To be consumed, coffee needs to be first roasted. Daviron and Ponte (2005: p. 211) characterize the roasting and blending stages of the value chain as 'relatively low-tech', yet it is also this stage that contributes most added value within coffee chains. The ability of roasting firms to maintain the largest profit margins of any actor along the value chain (discussed in greater detail below, in the section on 'governance') resides in the importance of market identities, and the quality information and purported knowledge held by these actors. Roasters are eager to describe their profession as an art and the process itself to be akin to some esoteric form of alchemy (Lingle, 1993), however, in a more prosaic sense, they take rather dull green seeds and transform them into aromatic, chocolate-brown coffee beans ready for consumption.

The roasting process itself can take any time between ten minutes in modern industrial roasters to many hours by traditional pan-frying methods. Any number of variables such as the intensity of heat, type of heat source, duration of roasting, type of cooling and the degree of roast (light or dark) are said to influence the final taste characteristics of coffee. It is the role of the roaster to understand the specific cup properties of each origin and roast the beans in such a way that creates desired characteristics. Carefully selected combinations of particular origins are then blended together to create the desired taste. As is the case with blended tea manufacturers, coffee roasters are able to consistently present consumers with relatively standard taste profile in their trademark blends by continually adjusting the various combinations of each origin based on changing cup characteristics. This blending process is essential to achieve standardization in an otherwise variable agricultural product.

Once roasted and blended, coffee beans can be vacuum packaged (to retain freshness) and sold ready for grinding, or can be further processed and converted to instant coffee. These further processes involve grinding, extraction and drying. As narrated by Pendergrast (1999: p. 147), whilst there are various claimants for the invention of soluble coffee, an American by the name of George Washington invented a soluble coffee in 1906 which was the first to be produced on a large commercial scale and proved to be popular in the US Army during the First World War. The next major technological breakthrough occurred in 1938, when Henri Nestlé introduced a spray-drying technique and subsequently launched the 'Nescafé' brand, which was a considerable taste improvement on Washington's earlier brew (Pendergrast, 1999: p. 213). Whilst spray-drying is still widely used in the industry, it is now generally acknowledged that a more recent freeze-drying technique generates a superior product, albeit at a greater cost. In the instant coffee market, a critically important aspect of value chain dynamics relates to the subordination of origin identities to blenders' proprietary identities. For example, the name of the world's best known instant coffee brand, 'Nescafé', forges an attachment to its corporate parent (Nestlé) rather than informing the consumer of the origins of coffee within the blend.

In general terms, the volume of coffee consumed worldwide has increased very slowly over the last few decades, at just over one per cent per annum. Overall consumption volume trends, however, hide substantial differentiation in terms of geography and product type. In terms of the former, developed countries have been stagnant in terms of growth (Lewin et al., 2004), but this has been offset by emerging markets of Eastern Europe, China and parts of Asia. In terms of product type, there has been an overall increase in consumption of generally higher quality Arabica coffees, reflecting what has been labelled a 'speciality revolution' (Pendergrast, 1999; Fitter and Kaplinsky, 2001; Ponte, 2002a). As suggested by the Specialty Coffee Association of America (SCAA): 'consumers are not drinking *more* coffee, but they are just choosing to drink *better* coffee' (SCAA, 1999: p. 4).

The growth of quality segments within the coffee sector is intimately attached to shifts in points-of-sale. In the USA (the world's largest coffee market), speciality beverage retailers (cafés, carts, roaster-retailers and bars/kiosks) were the most rapidly expanding distribution channel during the late 1990s, and by the end of the decade, half of all coffee expenditure occurred in the out-of-home sector (SCAA, 1999). In the specialty sector, moreover, only 36 per cent was purchased in supermarkets and grocery stores (National Coffee Association, 2002). The turn to speciality coffee in developed country markets, therefore, is entwined with the rise of café and take-out consumption. Weaving these processes together, Ponte (2002a: p. 1112) and Daviron and Ponte (2005: pp. 43–5) argue that the intrinsic quality of coffee beans is only one part of the overall 'flight to quality' in the 'speciality revolution'. Perceived quality within these speciality (out-of-home) consumption spaces is more frequently a product of the way the beverage is brewed and presented, thus pointing to the importance of 'in-person service quality' functions in the addition of value within coffee value chains.

In summary, this delineation of the input–output structures of coffee value chains emphasizes how place-specific upstream activities are articulated with the flexible sourcing capabilities of roaster-blenders. These points also indicate how coffee is a truly global product constructed around highly complex relations of production–exchange–consumption. These arguments lead into the next section of the chapter, which utilizes the analytical dimension of 'territoriality' to locate South Indian production within these global relations.

## The territoriality of South Indian coffee

For the last 150 years, Brazil has been the largest producer and exporter of coffee in the world, although its contribution to global production has shifted markedly during this period. The high point of Brazilian dominance in the world coffee sector was 1900–4, when it accounted for 73 per cent of world production (Daviron and Ponte, 2005: p. 58). By 2006 it accounted for 35 per cent of global production, and 30 per cent of exports (Table 3.10). Traditionally, Brazil's market dominance has meant it has been able to influence global markets. Unilateral Brazilian decisions to retain supply, for example, have had the effect of raising world prices. On the other hand, the persistent risk of frost in the coffee-growing regions of southern Brazil (notably in the states of Parana, Minas Gerais and Sao Paulo) has meant that weather forecasts for South America are the basis for intense price speculation in global markets. Severe Brazilian frosts in 1975 and 1994 led to global panic buying and significant price hikes on world markets. With the expansion into new regions and the implementation of high-input cultivation systems through the 1990s and into the 2000s, Brazilian production

**Table 3.10** The world's major coffee producing countries, 2006

| *Country* | *Production (tonnes)* | *Exports (tonnes)* | *Share of world exports (%)* |
|---|---|---|---|
| 1 Brazil | 2,550,720 | 1,640,299 | 29.7 |
| 2 Viet Nam | 930,000 | 840,058 | 15.2 |
| 3 Colombia | 732,000 | 656,171 | 11.9 |
| 4 Indonesia | 406,200 | 316,826 | 5.7 |
| 5 Ethiopia | 300,000 | 176,134 | 3.2 |
| 6 India | 285,000 | 221,926 | 4.0 |
| 7 Peru | 255,000 | 232,862 | 4.2 |
| 8 Mexico | 252,000 | 154,205 | 2.8 |
| 9 Guatemala | 229,020 | 198,727 | 3.6 |
| 10 Honduras | 162,000 | 173,905 | 3.1 |
| 11 Côte d'Ivoire | 148,920 | 144,123 | 2.6 |
| 12 Uganda | 141,000 | 130,373 | 2.4 |
| 13 Costa Rica | 107,940 | 77,604 | 1.4 |
| 14 Nicaragua | 76,500 | 86,718 | 1.6 |
| 15 Others | 694,920 | 475,723 | 8.6 |
| **World** | **7,271,220** | **5,525,653** | **100.0** |

*Source*: ICO (2007).

volumes have continued to increase. Its share of world production, however, has fallen because of even faster growth in other countries, in particular, Viet Nam. These developments saw coffee prices fall significantly in the late 1990s and, despite a recent recovery, they continue to remain lower than levels of the late 1980s (Figure 3.8).

As indicated in Table 3.10, Viet Nam is now the second-largest exporter of coffee in the world, accounting for 15.2 per cent of world trade. This is remarkable considering that, as recently as 1990, the country was contributing only 1 per cent to world exports. Viet Nam's centre of coffee cultivation is Daklak province, in the Central Highlands northwest of Ho Chi Minh City. This previously heavily forested area has received significant investment in support of high input coffee cultivation, smallholder coffee farmers in Viet Nam now achieve some of the highest yields in the world. For the most part, Viet Nam is producing cheap dry-processed Robusta for the lower end of world markets. Its rise as a major exporter, alongside continued production increases in Brazil, has contributed to conditions of global oversupply and low prices during the period between 2000 and 2004.

Coffee is principally sold into the affluent countries of North America, Western Europe and to Japan (Table 3.11). In terms of per capita

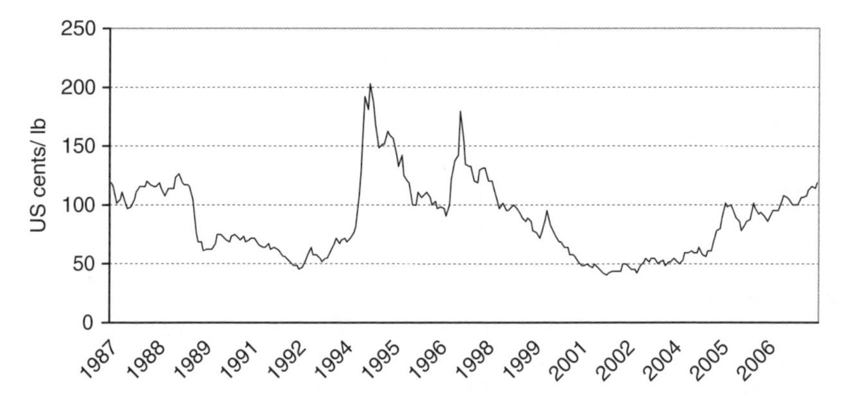


**Figure 3.8** ICO composite price for green coffee (1987–2007).
*Source*: ICO (2007).

**Table 3.11** The world's major coffee importing countries, 2006

| Country | *Imports (tonnes)* | *Re-exports (tonnes)* | *Share of world imports (%)* |
|---|---|---|---|
| 1 USA | 1,382,491 | 147,118 | 19.5 |
| 2 Germany | 1,002,963 | 501,375 | 14.2 |
| 3 Japan | 450,418 | 3,944 | 6.4 |
| 4 Italy | 436,117 | 102,467 | 6.2 |
| 5 France | 342,841 | 61,656 | 4.8 |
| 6 Spain | 261,365 | 87,182 | 3.7 |
| 7 Belgium | 243,794 | 174,311 | 3.4 |
| 8 Canada | 233,753 | 57,070 | 3.3 |
| 9 United Kingdom | 205,998 | 53,113 | 2.9 |
| 10 Russian Federation | 188,354 | 5,958 | 2.7 |
| 11 Netherlands | 179,281 | 63,665 | 2.5 |
| 12 Poland | 167,490 | 31,445 | 2.4 |
| 13 Algeria | 113,962 | – | 1.6 |
| 14 Sweden | 101,580 | 475,723 | 1.4 |
| 15 Other | 1,766,191 | | 25.0 |
| **World** | **7,076,598** | | **100.0** |

*Source*: ICO (2007).

**Table 3.12** The input–output and territoriality dimensions of South Indian tea and coffee

| | *Tea* | *Coffee* |
|---|---|---|
| Input–output structures | | |
| Cultivation | • Weekly plucking cycle;<br>• Plantation and small-holder cultivation;<br>• Importance of permanent labour;<br>• Mono-cultivation on semi-shaded hill slopes;<br>• Threat of pests, disease and frost. | • Annually fruiting tree-crop;<br>• Plantation and small-holder cultivation;<br>• Importance of seasonal labour;<br>• Incorporation of coffee cultivation within multi-storey forest ecosystems;<br>• Significant pest and disease threats. |
| Processing | • Plantation (estate) and 'bought-leaf' factories;<br>• Processing needs to take place immediately after plucking;<br>• CTC versus orthodox methods. | • Diversity of pulping and curing arrangements to prepare green beans, ready for sale/export. |
| Grading and market exchange | • Non-standardization of product defines the institutions of exchange (traditional dominance of local auctions). | • Well-accepted inter-national standards and transparent pricing systems shaped by deep futures trade. |
| Preparation of final product (blending [tea and coffee] and roasting [coffee]) | • Capacity for blenders to flexibly source different teas but create a uniform blended taste. | • Roasting generates large value-addition, through the process itself is relatively simple;<br>• Capacities for roasters to blend and roast different combinations of green beans to attain particular taste qualities |
| Territoriality | • Geographical expansion of tea cultivation into new production countries;<br>• Globally, the liberaliza-tion agenda has encour-aged smallholder production. | • Geographical expansion of coffee cultivation into new production countries; |

*Source*: Own work.

consumption, the Scandinavian countries are the greatest consumers of coffee in the world (International Coffee Organization, 2007). In contrast, China, Russia and Eastern Europe are viewed as important emerging markets, where per capita consumption is well below the global average. Whilst net volume consumption in major markets is not increasing, it is generally the case (as mentioned above) that the value of these markets has experienced tremendous growth due, in part, to a shift to out-of-home consumption associated with espresso-delivered coffee in cafés and other outlets.

## Conclusion

The essential starting point for a Global Value Chain analysis is to document the baseline 'hows' and 'wheres' of the system under focus. Without an appreciation of the agronomic, technological and economic issues discussed here, the analytical intent of later chapters would be impossible to sustain. This chapter has systematically identified and discussed key aspects of the addition of value within the South Indian tea and coffee sectors. Summarizing these factors in Table 3.12, key aspects of the similarities and differences between these two products are apparent. Evidently, the organization of South Indian tea and coffee production occurs in response to the vital biophysical attributes of both crops; and represents the upstream components of export-dependent value chains characterized by considerable further value-addition offshore, and facing intense competitive pressures from producers in other countries.

Traditionally in GVC analyses, these analytical categories are deployed to provide the contexts that frame the more deeply interrogative examinations of 'governance' and 'institutions'. For the purposes of presenting our analysis in an orderly fashion, this book follows this general research tradition; we discuss input–output structures and territoriality as a way of leading into discussions on governance and institutions. However, it is important to recognize that our interest in GVC *geographies* presupposes additional and more complicated consideration to the dynamic interactivity between the various categorical dimensions in chains. Input–output structures and territoriality are not static 'independent variables' that frame systems of governance and institutions, but themselves are co-produced over time by these other categories. Shifts from plantation to smallholder cultivation, for instance, both shape and are shaped by the cross-cutting influences of governance relationships and institutional formations. Nevertheless, as indicated at this chapter's outset, the issues presented here provide a foundational empirical grounding to the basic 'how' and 'where' questions about South India and the international tea and coffee industries. With this task achieved, we now move forward, to address these industries' institutional environments.

# Chapter Four

# The Institutional Environment of the South Indian Tea and Coffee Industries

In Chapter 2, we observed that the GVC literature tends to contextualize the category of 'institutions' in terms of the external framework of laws and rules within which chains are situated. Such conceptualizations, we argue, perpetuate an underdeveloped and undertheorized set of notions about the diversity of institutional forms affecting GVCs. Drawing from recent scholarship associated with the 'institutional turn' in economic geography, we highlighted the scope for a richer elaboration of the institutional dimension within GVC analyses. In the approach we developed in Chapter 2, institutions were understood as multi-scalar structures, with a sphere of influence (an institutional space) ranging from the global to the local. The point is, they interlock and coalesce in places to create a prevailing *institutional environment*: the 'set of fundamental political, social and legal ground rules that establishes the basis for production, exchange and distribution' (Davis and North, 1971: p. 6). The challenge of explaining how and why particular institutional environments evolve in particular places then leads us to the concept of *embeddedness*, referring to the social production of economic activity. In South India, tea and coffee estates, smallholders and workers are embedded within various social–economic–political–cultural formations operating across a range of scales from the local to the global.

Deploying this perspective to the case at hand leads to a quite different 'institutional analysis' to that which is common within GVC studies. Our focus is not simply on the external framework of laws and rules for tea value chains, but a more holistic conception that seeks to identify, firstly, the varying institutional frameworks within which South Indian tea producers are embedded, and, secondly, how this institutional environment interacts with the governance arrangements that connect producers with downstream market interests. Through this approach, we see how the institutional environment of South Indian tea and coffee production shapes a series of struggles over the terms and contexts through which producers engage with other value chain participants, which are elaborated upon in Chapters 5, 6 and 7, respectively.

In the present chapter, we set the scene for those discussions by documenting the recursive relationships between institutional formations and the socio-spatial arenas in which they are embedded. This orientation reflects our emphasis on the notion of institutional *path dependency*. Consistent with the usage of this term by North (1990, 1998) and Williamson (2000) (and see Chapter 2), we understand the current institutional formations in South Indian tea and coffee industries in terms of evolved dynamics. To give these ideas concrete expression, our approach in this chapter is to analyze the institutional environment of South Indian tea and coffee industries through the prism of history. In the following two sections we review these industries during the Colonial and post-Independence periods respectively, and then assess the current (1990s–) institutional environments. It is through these institutional environments that the South Indian plantation complex responds to, informs, and struggles with, the global institutions of tea and coffee, thereby creating the institutional environments of the tea and coffee GVCs emanating from South India. This discussion reveals the importance of path-dependent processes that have created an institutionally thick context for these industries' contemporary engagement with global value chains; the implications of which are assessed in the conclusion section of the chapter.

## Managing agents and plantation life: The insertion of South Indian tea and coffee into Britain's colonial project

The defining starting point in the story of South Indian tea and coffee relates to the insertion of these industries within Britain's Colonial project for the Indian subcontinent. Through the eighteenth and nineteenth centuries, consumers in Europe developed an insatiable appetite for tea and coffee, making these otherwise unassuming products a strategic cog in the politics of Empire. Initially perceived as an exotic stimulant to be drunk in tea and coffee houses, the pot of tea and jug of filtered coffee became essential contributors to lifeworld transformations during the Industrial Revolution. The cup of tea, in particular, provided a short-term physiological 'fix' that helped sustain heightened work intensity (Macfarlane and Macfarlane, 2003).

Control over these commodities, accordingly, became a vital element of Colonial ambition. The politics of tea in British India is a well-trodden historical saga. Until the early 1800s, all of the world's tea was grown in China, and in the course of procuring this vital commodity, British traders racked up significant deficits with Chinese exporters. Aiming to stem this financial outflow, European interests sought to create new import markets in China (most infamously, in opium) and to break the Chinese stranglehold on tea. The first significant step towards this latter ambition was made in 1793 when the Macartney Embassy (see Chapter 1) brought back the first seeds and plants to the Calcutta Botanical Gardens. By the 1830s, after considerable trial and error, the beginnings of a viable industry were underway. Tea was introduced to South India soon after its successes in Calcutta and the Northeast, with a British surgeon by the name of Christie in 1832 successfully cultivated tea at Ootacamund, in the Nilgiris. His initial plantings of Chinese stock were propagated in an experimental farm run by the British Commandant, and the industry took off. In 1854, the first private-owned Nilgiris tea plantation was established; the still-operating Coonoor Tea Estates. By the end of the nineteenth century, some 3,000 ha in the Nilgiris was under tea cultivation, and production was expanding rapidly (Francis, 2001).

Unlike tea, coffee was already being cultivated in India prior to British interests in this crop. Coffee growing had been a widespread, albeit small-scale, economic activity amongst the various agricultural communities of the Western Ghats for perhaps two centuries prior to the first recorded reference to a commercial 'plantation' by the British in 1801 (the Anjarakandi estate near Tellicherry) (Francis, 2001). The seventeenth-century Sufi saint, Baba Budan, is popularly credited with introducing coffee growing to India.[1] According to the *Nilgiris Gazetteer*, first published in 1908 (republished as Francis, 2001), experimental coffee plantations were established by British planters in the Wyanad in 1828 and on the Nilgiris Plateau in 1838. Somewhat of a 'coffee rush' then took place in Kodagu following the first British planting on the Mercara Estate in 1854, such that by 1869, 21,950 ha of coffee was held by Europeans, in addition to 11,780 ha held by 'natives' (Richter, 2002). In 1879 the Coorg Planters' Association (CPA) was established, providing the industry with a lobbying voice and formalizing the industry's presence within Colonial structures.

The conversion of upland forested areas in South India to tea and coffee cultivation was accompanied by the emergence of a specific style of industry governance: dense ensembles of plantation and trading companies that became known as the managing agency system. Managing agents were commission-writing firms which coordinated cultivation, processing and trading activities across a range of commodities, tea and coffee included. This form of business organization rose in the wake of the East India Company's (EIC's) demise (Webster, 2006: p. 743). The Charter Act of 1813 stripped the EIC of its trading monopoly, and subsequent legislation in 1833 effectively terminated the Company's trading activities. As the EIC's stranglehold was relaxed, 'there emerged in the Indian presidencies non-EIC commercial firms known as "agency houses"' (Webster, 2006: p. 746).

Over time, some of these houses branched into the active management of productive enterprises (rather than simply act as trade facilitators) and thereby took control of the operational activities of joint stock companies (Kling, 1966: p. 37). Trading houses moved upstream into plantation management and downstream into the auction and post-auction stages of value chains. An exemplar of these processes in the Indian tea industry is provided by the James Findlay group, which began trading tea from its base in Calcutta (now Kolkata) in the 1880s. By 1901, Findlay operated 108,000 ha of tea through managing agency structures, as well as having investments in warehousing and blending operations in London (Jones and Wale, 1998: pp. 371–2). Thus, different branches of the Findlay group were involved in plantation ownership, plantation management, export–import trade, warehousing, lodging tea at auction, and blending.

Cross-shareholdings then created a dense web of joint interests (Gupta, 1997: p. 159). Describing the structure of the South Indian tea sector at the beginning of the twentieth century, Tharian (1984: p. 38) observes:

> Many of the early companies began to be consolidated under the managing agencies who had access to the developed capital market of London. The marriage of agency house capital with the managerial experience of proprietary planters provided the main foundation for the growth of corporate capital in the South Indian tea industry. The decision making functions were exercised by the managing agencies while the companies were in fact their branches. The structure of the South Indian tea industry during the beginning of the twentieth century had the following characteristics: (i) Large British companies owned and managed most of the estates; (ii) provision of finance by British banks and agency houses; (iii) large-scale operation of the estates using massive forces of labor; (iv) control of import–export trade by the British companies and virtually complete reliance on import supplies of capital equipment, and (v) complete reliance on the British market for the product. The roles of the producer, banker, shipper, broker and distributor were interlocked through the mechanism of interlocking directorships.

In 1910, more than 97 per cent of Indian tea production was exported, with over 60 per cent sold at the London auction (J. Thomas & Co., 2004: p. 33). At this time, 13 managing agents controlled 75 per cent of India's tea production.[2] Amongst tea companies registered in India itself (so-called 'Rupee companies')[3] the level of concentration was even higher, with five managing agencies in India controlling 60 per cent of tea output (Gupta, 1997: pp. 159–60). The dense cross-shareholdings within and between these firms generated a commercial environment in which collusive behavior thrived. Particularly at times of low prices, it is claimed that managing agents would connive to influence supply and demand:

> Despite the large number of plantations [in India], decision making was effectively in the hands of a small number of managing agents, each of whom managed several plantations. This resulted in an oligopolistic structure. Repeated interaction of these managing agents enabled collusion to be sustained ... The managing agents belonged to a small and cohesive social group of British nationals in a colonial environment. (Gupta, 1997: p. 156)

Understood in wider analytical frames, the managing agency system expressed an institutional formation based around the capture and control of profit by elite merchant capital interests. British planters, the actual managers of land under cultivation, were commercially subservient to managing agencies. (A whimsical look into these relations, from a planter's perspective, is provided in Appendix A.) The effect of these arrangements was to construct tea and coffee plantation districts as enclave Colonial economies based around the hierarchy of an exploited native population (see Chapter 5), a social class of planters with the trappings of 'planters' life' (clubs, bungalow housing, etc.) and absent merchant-capital interests whose coordination of trade and finance provided the lifeblood that sustained this regime.

## Institutional reconfigurations after Independence

At Independence, tea and coffee represented potentially important export-earning industries for India's fledgling economy. But with much of these industries under foreign (British) ownership and with the majority of export receipts denominated in Pounds sterling, the industry's institutional structures did not reflect the economic interests and needs of the Republic. Accordingly, the first few decades of Indian Independence saw extensive regulation of these industries by New Delhi. This era can be characterized as one in which state actors exercised key control and coordination functions within these industries.

These processes were more comprehensive in coffee than tea, because of the allocation of wide-ranging economic powers to the Coffee Board. Established as a statutory organization under the (pre-Independence) Ministry of Commerce by the *Coffee Act* of 1942, in the context of wartime austerity and state control, the Board's coordinating role for the industry was embellished further after Independence, particularly from 1962 when India became a signatory to the International Coffee Agreement (ICA), an international regime of managed coffee trade.

As the analysis of the ICA regime for managed trade in coffee has been documented elsewhere in extensive detail (*inter alia*, Bates, 1997; Gilbert, 1997; Maizels, 1997; Robbins, 2003; Talbot, 1995, 1996, 1997a, 1997b, 2002a, 2002b, 2004), there is little need to retrace its detail here. Broadly, however, it owed its origins to international consensus amongst producer and consumer countries that price maintenance and stability in the coffee sector were critical ingredients for development in various Latin American, Asian and African countries. (And perhaps more importantly, by protecting coffee producers from the sharp edges of volatile world markets, a political dividend was paid in the form of forestalling Leftist revolutionary movements which fed off rural discontent.[4]) In 1958, supply retention commitments by Brazil and Columbia led to the establishment of the Latin American Coffee Agreement, a voluntary cartel of 15 Western Hemisphere nations (Fisher, 1972), which was then expanded in 1962 to establish the first ICA. The International Coffee Organization (ICO) was the organizational vehicle through which subsequent intergovernmental discussions and negotiations about coffee took place. Subsequent ICAs were negotiated in 1968, 1976, and 1982, and the system remained in place right up until the end of the Cold War, in 1989. Central to the system in place between 1962 and 1989 was a system of national export quotas to maintain prices within a mutually agreed range. As an ICA signatory, India was bound into this international quota regime, and was thus obligated to closely manage domestic supply and demand.

The Coffee Board was the administering entity for Indian compliance with successive ICAs, and so acted as the *de jure* institutional epicentre of the industry. During these years, all coffee grown in India was pooled at collecting depots in the growing regions, which were operated either by the Board or by approved dealers on behalf of the Board. Growers were paid an amount determined by the Board based on the costs of production plus an amount determined by average export prices. The collecting depots then supplied beans to a limited number (between 43 and 45 during the 1980s) of approved curing works, who cured again according to Coffee Board specifications and stored the beans until requested by the Board. The Board then maintained a monopoly right over exporting coffee from India. The basic premise was that the Board owned all coffee in India and effectively paid different actors (farmers, traders, and curing works) as service providers along the chain.[5] Perhaps not surprisingly, coffee planters we interviewed remembered this time as one of alleged bureaucratic interference and inefficiency, with payments from the Board beset by extensive delays. The Coffee Board paid farmers an advance when they delivered their coffee to the curing works, but the remaining payment would be allocated in installments only after the coffee was eventually sold. Funds were pooled within a Coffee Board account and farmers could expect to wait between one to three years for final payment (Akiyama, 2001: p. 90).

These arrangements stayed in place until the mid-1990s. In 1989, negotiations for a new ICA failed to produce a consensus position on its future, and quotas and controls were suspended. Not coincidentally, the end of the

Cold War saw the USA abandon its erstwhile support for managed trade in the coffee industry. Abandonment of the so-called 'economic clauses' of the ICA had the immediate effect of releasing a significant volume of stocks onto world coffee markets, causing prices to crash (Figure 3.7). The post-1989 role of the ICO has concentrated on such tasks as quality improvement, provision of market information, and administering coffee-related development projects (such as those funded by the Common Fund for Commodities). The most recent ICA, adopted by the International Coffee Council in September 2007, has an overall objective of strengthening the global coffee sector in order to promote sustainability and improve market functions; but includes no provisions for managing world markets.

The collapse of the ICA in 1989 initially did not lead to changed functions for the Coffee Board. Liberalization of the Indian economy during the 1990s, however, saw its erstwhile economic controls deregulated. The pivotal moment here was the 1990–1 balance of payments crisis in the Indian economy, which in January 1991 led to an unavoidable injection of US$1.8 billion from the International Monetary Fund (IMF), and a follow-up structural adjustment loan from the World Bank (in December) of US$500 million (Dash, 1999). The dismantling of government monopolies and state intervention in trade was insisted upon as part and parcel of these bail-outs, and the Coffee Board monopoly came under severe criticism (Akiyama, 2001: p. 88). In practice, liberalization of the coffee sector occurred incrementally over four successive years from 1992 until 1995 (Ravindranath, 1999). An Internal Sale Quota (ISQ) was introduced during the 1992–3 harvest season, which allowed growers to sell 30 per cent of their produce into the domestic market where prices were (at the time) significantly higher than export prices. The following season (1993–4), the Board introduced the Free Sale Quota (FSQ), which allowed growers to sell 50 per cent of production into the free market. The 1994–5 harvest witnessed the FSQ increase to 100 per cent for small growers and 70 per cent for commercial estates. Since 1996, the Coffee Board has not intervened in industry supply and demand. As we elaborate later in this chapter, its role is now limited to R&D, extension services, the implementation of welfare measures for smallholders, provision of quality controls and market information, and generic industry promotion. Although these activities exercise considerable significance in the industry, they do not reflect a coordinating and controlling influence in the industry; the Indian state plays no longer an anchoring role in the coffee sector.

In tea, it was similarly the case that from Independence until the 1990s, the Indian state exercised significant industry coordination and control, albeit without the degree of wholesale bureaucratic economic management that was the case in coffee. The basic reason why the tea sector escaped the economic controls foisted on their compatriots in coffee was that, unlike the latter, there was no comprehensive international commodity agreement for tea in the post-1945 period,[6] and therefore no driving need to enmesh the industry in economic regulation. In tea, the hand of the state mainly took the form of policies to restructure the ownership and operations of plantations.

Upon Indian Independence, many British plantation owners divested their assets, while those who stayed were less inclined to reinvest in a country they no longer controlled. High rates of dividend repatriation back to British owners became a point of political struggle for the Government of India, which initiated a Commission of Inquiry into the Plantation Sector. The Commission concluded its enquiries in 1956. Reporting on that commission, Tharian (1984: pp. 41–2) writes: 'these British companies ... considered their tea estates in India as a wasting asset from which they tried to repatriate maximum possible funds'. Eventually, in the late 1960s, the Government of India took a series of actions to curb the activities of British interests. The managing agency system was abolished in 1970, following the Government of India's *Companies Amendment Act* of 1969, and the Sterling trade cancelled shortly afterwards, through the *Foreign Exchange Regulation Act* of 1973 (requiring that all trade was denominated in Rupees).

These policies ramified onto patterns of trade and domestic systems of market exchange. British tea plantation divestment in the wake of this legislation was accompanied by the decisions of British tea companies to redirect their procurement in favour of Sri Lanka and African countries (Kenya, Tanzania and Malawi). For South Indian tea producers, the loss of key markets in Britain was soon compensated by increasing sales to the Soviet Union and its allies. India was not formally a member of the Eastern bloc, but it dallied with the Soviet Union in order to offset Pakistan's close relations with the USA and China. The 1978 Rupee–Rouble agreement established extensive counter-trade exchange, with Soviet military and industrial hardware bartered for Indian textiles and agro-commodities. Tea and coffee were central components of this exchange. In 1986, when Indo-Soviet trade relations were at their height, the USSR imported 87,362 tonnes of Indian tea. This represented 45 per cent of India's tea exports and 80 per cent of the USSR's tea imports (FAO, 2007). Most of the Soviet buying activity was in South India, where it has been estimated that Soviet purchases accounted for approximately 80 per cent of exports (Ajjan and Raveendran, 2001). In coffee, Soviet bloc members were not signatories to the ICA regime, meaning that India could offload coffee to these customers without contravening its export quota ceiling. According to Akiyama (2001: p. 89), the ICA quota allowed India to export only 50 per cent of its potential exportable production. The barter trade of coffee into the Eastern bloc, therefore, resolved over-supply problems of in the Indian coffee sector. Annual purchase budgets were primarily determined by bilateral diplomacy, as opposed to market signals. In both tea and coffee, therefore, Soviet purchases provided an added fillip to domestic demand, propping up prices at the bottom end of the market.[7]

## Institutions and organizations in the South Indian tea and coffee industries in the contemporary period (1990s–)

The legacies of the Colonial and post-Independence eras have come together to create an intricate institutional–organizational nexus in South India's tea and coffee industries; there are extant vestiges of Colonial era 'plantation life' and its associated social and economic infrastructure (clubs, planters' associations, etc.), and a still-significant presence of bureaucratic agencies (Coffee Board, Tea Board, etc.) associated with the strong regulatory period in post-Independence India. In Chapter 2 we introduced the concept of *institutional thickness* as referring to situations where dense interactivity between organizations helps generate an overarching sense of common purpose (Amin and Thrift, 1994: p. 14). For the most part, this concept has been associated with research on industrial clusters in manufacturing and services sectors, where it is used to explain how organizational interactivity helps to 'hold down the global' and create 'sticky places in slippery space' (Markusen, 1996). However, its applicability can also be seen to extend to regional agri-food complexes, like the plantation industries of South India, where a networked set of relations between various organizations (public agencies, representative trade associations, research organizations, and locally anchored extension agencies) shapes the nature of producer engagement with downstream markets. In the following section we map the vital features of the institutionally thick South Indian tea and coffee sectors, for the overall purpose (brought together in the chapter's conclusion) of establishing a picture of the institutional environment of these industries.

### The institutional environment of the South Indian plantation sector

South Indian tea and coffee production takes place within a complex structure of state, parastatal, industry and civil society organizations (Figure 4.1 illustrates this complex for tea). The web of interactions and interconnections between these entities profoundly shapes both the operating environment of individual economic agents (e.g., estates, smallholders, factories),

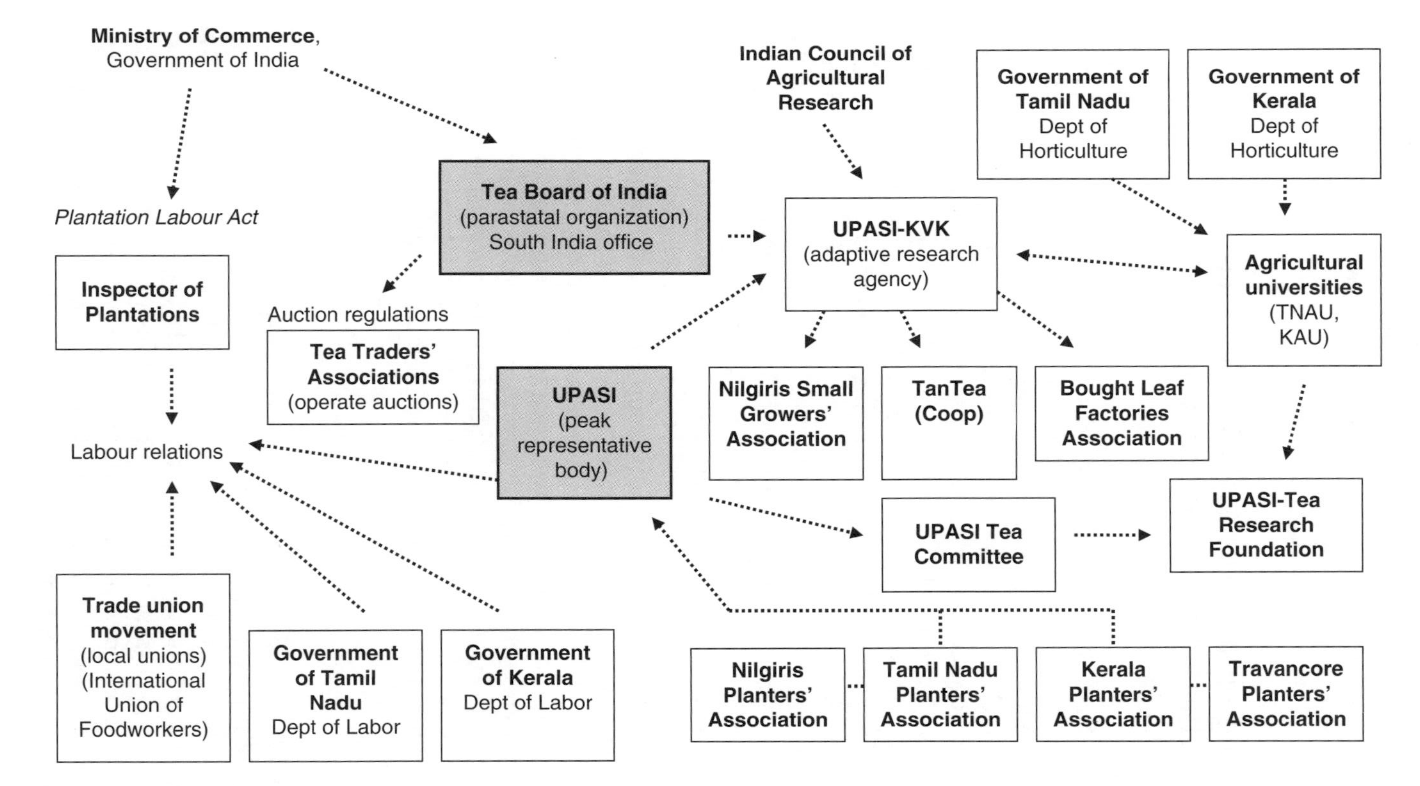


**Figure 4.1** Organizational interconnections in South Indian tea production.
*Note*: Arrows indicate links between organizations; not chains of responsibility.

**Figure 4.2** A typical tea planter's bungalow (Pattumalay Estate, Central Travancore planting district, Kerala)
*Source*: Photo by the authors.

and their capacities to respond and react to external and internal change (ranging from revised export accreditation requirements, to local pest and disease threats).

As mentioned above, the history of tea and coffee as a Colonial plantation crop has led to these crops carrying a particular resonance in the Indian psyche. Despite the departure of the British, the Colonial legacy of planters' clubs and a 'planter's life' persists (Figure 4.2). Indeed, the stark divergence between the hustle and bustle of India's cities and the serenity of bungalow life amongst high-altitude tea gardens provides a neat backdrop to the distinctiveness of the 'planter', in both his British and his Indian forms.[8] In India's national imagination, planters are associated with an 'outdoorsy', hunting culture that works hard and plays hard. This cultural construct works to encourage social capital within the planter community. Although individual plantations may compete with one another in the marketplace, planters themselves readily collaborate with one another on matters of innovation and technology. An expression of the high level of social capital occurred during an interview we conducted with a planter in Central Travancore planting district. Halfway through the interview, his Blackberry 'beeped', and he excited told us that the results of the new 'batting order' had just been posted. What he was referring to was the latest data on yields. Each month, all major estates in the District would submit their yield data and the results would be emailed to all planters. The rank order of estate's performance was colloquially called the 'batting order' (signifying, once again, how cricket analogies seep into almost every part of Indian life!). Such an example illustrates that although estates may be competitors to one another, there is a quotidian aspect to planters' lives based around information sharing and friendly rivalry.[9]

Intimately associated with this plantation culture is the continued existence of planters' associations. Plantations are represented in South India by UPASI, the venerable peak body with origins that date back to the commencement of the British plantation era. UPASI is the recognized representative of the industry by the Ministry of Commerce (the ministry responsible for the plantation sector) and operates out of well-maintained facilities in the township of Coonoor, in the Nilgiris District of Tamil Nadu. With an expansive reach across three states (Kerala, Tamil Nadu and Karnataka), UPASI acts as an umbrella for state-specific planters' associations that, amongst other things, represent management in wage determinations. Additionally, UPASI hosts a set of Commodity Committees representing the major plantation crops of tea, coffee, rubber and spices. In addition to these advocacy functions, UPASI operates the UPASI-Tea Research Foundation (U-TRF) headquartered in the town of Valparai (but with an additional regional presence in the Idduki, Wyanad, Gudalur and Nilgiris districts), and supports UPASI-Krishi Vigyan Kendra (U-KVK), an adaptive rural research and extension agency under the umbrella of the Indian Council of Agricultural Research (ICAR). In addition to their lobbying and research functions, planters' associations are important conduits of innovation and learning. The shared culture of their members facilitates knowledge sharing in informal (for instance, over drinks at the local club) and formal (such as the annual UPASI conference and through association journals) ways.

### *The institutional–organizational nexus in tea*

The tea sector as a whole is beholden to the Tea Board, which under legislation possesses wide powers with respect to authorizing, registering and licensing industry activities. The Tea Board is headquartered in Kolkata, but operates a South Indian office in Coonoor, just a short walk from the UPASI compound. As with state-centred producer marketing boards across the world, endemic within the operation of the Tea Board is a set of fundamental tensions that relate to representation and responsibility. A significant proportion of the Tea Board's finances is contributed by the industry (a Cess for allocation to the Tea Board of Rs 0.30/kg of made tea is levied on every tea factory in India),[10] fermenting producer opinion that the Board should be directly responsible to them. However, in legal terms the Tea Board is a parastatal organization whose office-bearers are appointed by the Government of India, and thus owes it allegiances to these political and legal foundations. Negotiating these tensions remains a vital act for the Board.

The Tea Board's expansive regulatory oversight and its significant financial resources make it a key player in the system. Although in the post-1991 period of market liberalism within India the enactment of many of these powers remains latent, nevertheless the Tea Board has shown a willingness to flex its regulatory muscles when necessary. An illustration of this occurred in 2005 when, following allegations that some South Indian tea producers were failing to comply with the *Prevention of Food Adulteration Act*, the Board independently tested teas from the region, and confiscated and publicly burned some 2,500 kg which did not pass muster. This crackdown, as we report in Chapter 7, had a galvanizing effect within the industry. Systemic problems of non-compliance with mandatory food safety standards have since abated.

As the apex organization for the sector, in practice, much of the Tea Board's influence derives from its extensive collaborations with industry and scientific bodies. In South India, these collaborations include joint activities with UPASI, and joint-funding programmes with the U-TRF and U-KVK. During the past decade, the most prominent collaboration between these organizations has been the *Quality Upgradation Programme,* a joint initiative of the Tea Board, UPASI and UPASI-KVK designed to improve the competitive position of tea smallholders. Given its size and significance an extended discussion of this scheme is warranted, to which we attend in Chapter 7.

Additionally, Tea Board funding has been a critical element in plant-breeding research and linked subsidies for tea bush replanting. The Tea Board provides 50 per cent funding for U-TRF's ongoing research on the development of new clonal varieties suited specifically to the environmental conditions of South India. This funding has culminated in the release of significantly improved clones, the most recent of which (TRF-2) was launched in 2007. The commercialization of TRF-2 may have far-reaching implications for the future of the South Indian tea industry, as it has high resilience to blister blight (a major problem in South India), produces high yields and its plant structure makes it amenable to mechanized harvesting. Dovetailing with the development of improved breeding stock has been a major Tea Board-sponsored push to promote a faster rate of tea bush replanting, known in industry circles as the Tea Board's 'Special Fund'. The origins of this programme lie with research undertaken by the Tea Board in the early 2000s into average bush age across India. In the Nilgiris, it was found that 43 per cent of tea bushes were over 50 years of age (and, therefore, well beyond their prime) yet in the two years from 1998 to 2000, estates had rejuvenated less than 2 per cent of their planted area (Mitra et al., 2003: p. 15). Obviously, estate owners lose their income stream for several years during the replanting process, and with lower tea prices during this time, many plantations were not in a position to afford this. Upon the issue of ageing stock being addressed at a 2004 tea stakeholders' meeting in New Delhi, the Tea Board and the Ministry of Commerce determined that a major strategic intervention was required in order to encourage widespread replanting. This response set out to replace the country's entire aged planting stock with new clonal varieties.[11] To achieve this target, funds of Rs 45 billion (approximately US$1 billion) were appropriated for the duration of the Government of India's Eleventh Five-Year Plan period (2007–12). Under the scheme, estate owners are entitled to have their replanting activities subsidized in the form of a 25 per cent grant from the Tea Board, plus access to loan funding to the maximum equivalent of another 50 per cent (Tea Board of India, undated b: p. 3), with the first tranche of funding disbursed to estates in June 2007.

The Tea Board's influence is further evidenced by its role in providing subsidies for factory upgrading. There are Tea Board programmes which subsidize the costs of ISO 9001 and HACCP accreditation by up to 75 per cent. Furthermore, to also encourage market access to higher-paying markets, the Tea Board has also instituted subsidy programmes for the conversion of factories from CTC to orthodox production (it pays a subsidy of Rs 2/kg for orthodox tea made in converted CTC factories), and for a range of other quality-enhancing factory investments (such as, for example, the implementation of modern stainless steel withering troughs).[12] And, finally, the Tea Board has also funded the ongoing development of proposed new electronic grading mechanisms that potentially could replace cup-tasting as the definitive quality-arbiter in the industry.[13]

The organizational presence of the Tea Board and UPASI have also facilitated widespread participation in various industry events, promotions and cupping competitions in recent years, in both the international and domestic spheres. In the international arena, the South Indian tea (and coffee) industries make sure they are well-represented with the deployment of appropriately credentialed individuals and the distribution of professionally produced promotional materials, which collectively play an anchor role in developing reputational assets. As one small example of this, premium South Indian tea producers organized what was billed (perhaps extravagantly) as the USA's 'first ever tea auction' at the Las Vegas World Tea Expo in 2006, during which Glendale 'SFTGFOP'[14] leaves from the Nilgiris were sold for a record $600/kg. This was followed by a similar specialty auction at the 2007 World Tea Expo held in Atlanta.

In terms of building regional reputation across the board, more significant perhaps is the 'Golden Leaf' competition, the hallmark industry event for South Indian tea producers. The 'Golden Leaf' was first held in September 2005 in Coonoor, shortly after global tea prices reached their lowest point. The 2006 competition was held in Dubai, and in 2007 it was held in Kochi. The Golden Leaf competition is described by its promoters as a 'classic private–public initiative', being organized through UPASI but supported by funding from the Tea Board (UPASI & Tea Board of India, 2007: p. 4). It involves blind cup-tasting of selected teas by an international jury panel, with prizes for each of the grades in Table 3.2 for six local districts, plus additional prizes for the Bought Leaf Factory and organic sectors. A prominent theme within the 'Golden Leaf' competition is how it mounts a discourse of how teas from different growing regions each possess their own distinctive attributes and, moreover, these flavours relate in some way to the natural and social histories of each area (Table 4.1).

A final component of the web of industry organizations is the role played by NGOs and trade unions. The unjust working conditions in the plantation sector during the Colonial era led to progressive unionization of the industry following Independence, and legislative support for the rights of plantation workers through the *Plantation Labor Act* of 1952. This legislation, and the active role played by the union movement, ensured that estate owners provide workers with minimum levels of housing, fuel, community amenities, and local infrastructure. To this extent, it can be said legitimately that the living conditions of plantation workers are superior to those of many other rural labourers in India. Yet at the same time, these gains have been challenged by the harsher economic conditions in the industry since 1998. As detailed in Chapter 5, recent restructuring of the tea industry has been marked by job losses, wage freezes, and a trend of replacing permanent for casual employees. Consequent to these problems, the nadir of the crisis period (circa 2003–04) was featured by high levels of industrial agitation within the sector, notably in Kerala. The various unions across the South Indian plantation sector remain important actors within the industry, not least because of their affiliations with different Indian political parties in the context where recent Indian governments (at both national and state levels) have been comprised of coalitions. The democratic traditions of post-Independence India have also spawned an active culture of critical journalism and NGO activism, which collectively assume an influential role in defining the broader institutional environment of the South Indian tea industry. For example, the Delhi-based NGO, Center for Education and Communication (CEC), which is linked to unions and leftist political interests, was involved in a 2002 fact-finding mission to the abandoned estates of Central Travancore (a concern discussed in greater detail in Chapter 5). The publication of its findings (Chattopadhayay, 2003b) helped raise the profile of the serious

**Table 4.1** South Indian tea growing areas and how they are described in the 2007 'Golden Leaf' awards brochure

| *Growing area* | *Description* | *Tea characteristics* |
|---|---|---|
| Anamallais – The Elephant Hills | 'a magnificent range of hills in the Western Ghats wedged between Tamil Nadu and Kerala …' | 'Medium to high tone fragrance; Biscuity to floral notes, lingering; Biscuity aroma; A great tea to provide the punch and verve to start the day; a morning refresher …' |
| High Ranges: Cloud's Lair | 'Located on the edge of the escarpment facing the Arabian Sea, some of the tea fields in the High Ranges at 2,200 m are among the highest tea growing regions in the world … The Nilgiri Thar or Ibex is seen in large population in the area …' | 'Clean and modern toned fragrance of sweet biscuit in a dip of malt; liquor of golden yellow with an orange depth and a rounded cup; expect the unexpected. The beauty of the hills beckons you to an inspiring morning cup of tea …' |
| Nilgiris – The Blue Mountains | 'Nilgiris, the blue hills or queen of hills, with its beautiful valleys and salubrious climate supports a wide range of flora and fauna … The Neelakurunji, a flower that bathes these hills in lavender, flowers once every twelve years, a natural phenomenon as intriguing as the great wildebeest migration of the Masai Mara …' | 'A beautiful delicious tea; Fragrant and exquisitely aromatic; High tones of delicate floral notes; Golden yellow liquor; Crisp brisk and bright; Lingering notes of dusk flowers with an undercurrent of briskness; Creamy mouth feel; An aromatic bouquet to take you on a high.; Flavourful tea for a stressful day.' |
| Travancore – Tiger Valley | '… The area is surrounded by gorgeous green plantations of tea and other crops besides housing one of the largest wildlife sanctuaries – Periyar Wildlife Sanctuary known for its large tiger population.' | 'Black leaf appearance; Medium fragrance; Reddish liquor with dancing hues of yellow. A fairly balanced tea with body and briskness neck to neck' Good for the elevenses and as your evening companion.' |
| Wyanad – Green for Gold | '… The region with its typical rainforests enjoys tropical to humid sub-tropical climate. The many estates in this area were (literally) erstwhile goldmines.' | 'Black leaf appearance; Clean fragrance; Medium toned biscuit notes; Earthy reddish liquor. Mild and mellow. Full bodied with light briskness; A born soother with flecks of brisk character. All in a gold miner's paradise.' |
| Karnataka – Green Leaves and Red Berries | '… Tea estates are located amongst impressive ecology of coffee farms with gigantic natural shade trees and a large variety of exotic flora and fauna.' | 'Simple and fragrant; Golden ochre liquor. Fair bodied and fairly brisk. Uncomplicated. Balanced. Short and simple finish. A tea you could drink cup after cup throughout the day.' |

labour welfare issues in that district. The international networking of CEC, together with Partners in Change (PIC, another Delhi-based NGO) with the European Union-funded *JustTea* project, then furthered the case for corporate social standards in the tea sector (an issue also discussed in Chapter 5), and dovetailed with the *New Delhi Declaration on the Rights of Tea Workers and Small Growers*, a document (signed by a number of NGOs and unions in December 2005) which sets out an agenda to protect worker interests in the context of the tea crisis.

### *The institutional–organizational nexus in coffee*

Broadly comparable to the situation for tea, South Indian coffee growers are well represented by an institutionally thick network of organizations with dense interconnectivity to state agencies. Various district planters' associations in South India, such as Coorg, Central Travancore, Hassan and Wyanad, were established by coffee planters in the late nineteenth century with the principal aim of promoting a unified approach to pertinent issues affecting the industry and to present a unified lobbying voice to the Colonial government in respect of matters such as taxation and the devastating effects of coffee rust, which decimated the sector in the 1870s. In 1893, UPASI was established to collectively represent the various district planters' associations. The coffee committee of UPASI continues to be an influential lobbying factor affecting national coffee policy. Whilst the larger planters are represented by the planters' associations, smallholders in India are represented by various growers' associations, of which the Karnataka Growers' Federation (KGF), with headquarters at Sakleshpur, is the most significant. This is an umbrella organization with 21 member associations and nine affiliated organizations across Chikmagalur, Hassan and Coorg, with each body consisting of approximately 5,000 members.[15] Many of these smallholder representative associations are closely linked to leftist political elements in India, and are keen to distance themselves from the larger planter class, who they provocatively refer to as 'comprador Imperialist landlords' (Raji, 2004).

The pre-eminent vehicle connecting these various planters' and growers' organizations to the architecture of the Indian state is the Coffee Board of India. This organization has evolved from the period of state industry and has now assumed primary responsibilities for coordinating a raft of extension services, industry promotional activities and regulatory requirements. The Coffee Board undertakes extensive R&D activities; provides assistance to coffee growers on production, processing and quality; supplies market information; implements welfare measures for small growers and plantation workers; and assists with the generic international marketing of Indian coffee. It is headquartered in Bangalore but it also has significant operations in coffee-growing areas that include the Central Coffee Research Institute, at Balehonnur in Chikmagalur District; a coffee research sub-station in Kodagu and four regional research stations in the Eastern Ghats, Assam, the Pulneys, and Wyanad. These research institutes are then linked directly with a network of more than 50 coffee extension offices across the various coffee districts. This Coffee Board infrastructure represents the latest incarnation of a long history of Indian research commitment to the problems of coffee production, such that India continues to be a world leader in coffee science.

The Coffee Board has played a key role in the development of improved and disease-resistant coffee varieties. Even prior to the establishment of the Coffee Board, however, UPASI was responsible for initiating systematic coffee research in South India. UPASI promoted the partially rust-resistant 'Kents' variety over the older 'Chicks' variety from around 1910, and then established the Mysore Experimental Station, established near Balehonnur in 1926 and now under Coffee Board control (Kariappa, 2002). The strength of planters' associations and their affiliated research activities also led to the widespread application of 'Bordeaux' spray (a fungicide made of hydrated lime and copper sulfate) in the 1930s, which has proven effective in controlling leaf rust. The first strains of rust-resistant Arabica (S.288) were released from Balehonnur in 1940, followed later that decade by the release of S.795, a cross between the Kents Variety and S.288 (Eskes and Leroy, 2004). The S.795 variety was a major breakthrough in Arabica breeding and is now probably the most commonly planted Arabica variety in India and Southeast Asia. The Coffee Board continues high-quality research into pest and disease resistance and improved productivity, thus leading to improved efficiencies for estate production.

Improving efficiencies in coffee production also involves developing effective management strategies for the control of pests and disease. The Coffee Board is currently involved in various research activities to address the problem of white stem borer, including an internationally coordinated activity funded by the Common Fund for Commodities (an entity within the UN family of organizations). Research in India was responsible for identifying the presence of sex pheromones released by male borers, which have subsequently been used to develop pheromone traps to lure females (it is the females who are responsible for most plant damage). Indian coffee planters have high expectations of this research system, and look to the Board to develop strategies for their current problems.

Augmenting these Coffee Board R&D activities is a robust extension system. There is a Joint Director of Extension at each of Hassan and Kalpetta (Wyanad), responsible for coordinating the activities of a combined total of 15 senior liaison offices and 28 junior liaison offices across Karnataka, Kerala and Tamil Nadu (as of late 2007). In Wyanad District (where approximately one-quarter of India's Robusta is grown, mainly by smallholders), the local Coffee Board employs eight researchers in a modern office complex, equipped with laboratory and field-testing facilities, who engage in two-way communication with a larger number of extension officers who regularly visit coffee-growing villages. Our visits to the facilities and local coffee-growing communities in Wyanad and Hassan seemed to confirm a high rate of adoption of 'improved' farming techniques (use of chemical fertilizers, pesticides and pruning systems), and a willingness of farmers to consult with extension staff about the latest approaches to pest management for issues of direct concern to them. Further supporting these arrangements is a 'Technical Evaluation Centre' in the town of Kalpetta, which effectively operates as a demonstration farm.[16]

To encourage improved efficiencies and pest/disease resistance within the smallholder sector, the Coffee Board operates a 'Support to Small Grower Sector Scheme'. This scheme empowers the Board to allocate financial incentives for small growers (defined as owning less than 10 ha) for a range of productivity-enhancing activities, including: (i) re-planting and water augmentation; (ii) quality upgradation; and (iii) implementation of pollution abatement measures (Coffee Board of India, 2003). A variety of activities are included under the rubric of 'quality upgradation', including purchase of pulping equipment, construction works, establishing drying yards and warehouses. This support includes a direct 20 per cent subsidy for approved activities, which is generally linked to credit provision from a financial institution and, like comparable programmes within other arms of the Government of India, requires proof of land ownership and a feasibility study prior to allocations being approved. Somewhat inevitably, these requirements have meant that the larger, more established 'smallholders' have been more able to obtain support, whilst many of the lower acreage smallholders (<2 ha) that we visited in Kodagu, Wyanad and Hassan claimed that these requirements were too burdensome and may involve indirect or hidden costs exceeding the actual amount of the subsidy. Prior access to bank credit is a prerequisite for this assistance, and tends therefore not to benefit the poorest coffee growers. For marginal smallholders, the Coffee Board facilitates the creation of Self-Help Groups (SHGs), which are used as conduits for the provision of research and extension services. Small growers are encouraged to initiate SHGs, of between 15 and 50 members, through cooperative village-level arrangements. Through a targeted policy to encourage the functioning of SHGs, the Coffee Board provides grants of Rs 200,000 (approximately US$4,600) to SHGs to 'find a solution to common problems through a participatory group approach' (Coffee Board of India, 2008: p. 1). We were told that 400 small growers in Kodagu had benefited from the scheme between 2002 and 2005, including three SHGs which had collectively purchased pulping machines and, in one instance, established a coffee roasting business.[17]

These state-funded activities interact with growers' tacit knowledge about the crop. The shaded, multi-crop ecosystems in which coffee is grown in South India (see Chapter 6) produce complex biophysical interactivity, to which South Indian coffee cultivators (especially the mid-sized 'planters' class', described in Chapter 6) are closely attuned. Our interviews with these industry participants elicited a wealth of well-informed, geographically contextualized knowledge about the specificities of coffee growing on their estates. For example, understanding the life cycle of pests such as white stem borer and coffee berry borer has enabled planter strategies to manage these pests effectively, including tracing for stem borer and the use of alcohol-based berry borer traps near pulping stations. Ongoing planter experimentation is shared through informal knowledge networks based around planters' club culture and associated organizations. The large corporate-owned plantations also exhibit commitment to R&D, with the Tata Coffee research laboratory at Pollibetta being a noted industry leader. Via routine leaf sampling and absorbed atomic spectrometry analysis, Tata Coffee can identify micronutrient deficiencies in the soil and address these through precision-based fertilizer application. In conjunction with the development and application of biological pest controls, such as *Beauveria bassiana* fungus for berry borer control, *trichoderma* for pepper wilt and pheromones to attract stem borer, these techniques have contributed to very high yields on Tata Coffee estates, averaging 1,300 kg/ha for Arabica and 1,800 kg/ha for Robusta.[18] Tata Coffee's competitor, Amalgamated Bean Coffee (ABC), also maintains an R&D centre at its curing works in Chikmagalur where, amongst other things, it invests significantly in the development of coffee extracts and decoctions for the domestic market.[19]

The initiatives by the Coffee Board and private sector interests to improve industry productivity and pest/disease resistance are complemented by commitments to enshrine reputable grades and standards, and to cooperate on generic promotion. With respect to the former, the issue of grades and standards plays a critical role in the coffee sector in terms of building industry capacities to ensure buyers' minimum requirements are met. Notably, the Coffee Board has invested considerable effort into assuring an authoritative and consistent system of grading and standards. The Board offers a fee-for-service quality analysis for planters, curing works and exporters at a cost of Rs 100/sample, which includes a cup-testing component. A similar quality analysis service is provided by Coffeelab, a private sector entity in Bangalore. The 'Quality Certification' provided by Coffeelab covers various factors, including estate name, variety of coffee, use of shade trees, altitude, environmental and social conditions, biodiversity on the estate, water use, wastewater management, processing methods, defect count, cup characteristics, suggested use in blends and recommended roasting styles. Such quality certificates have no statutory basis; rather they are intended as a means to add value to product. The professionalism of the certificate and the rigour and independence of Coffeelab is seen as a basis for ensuring a positive perception amongst buyers of the producers' coffee. Tata Coffee also operates an accredited quality analysis laboratory, similar to the Coffee Board and Coffeelab.

In terms of collective marketing and promotion, the Coffee Board runs an active agenda promoting the quality attributes of Indian coffee to the mainly tea-drinking Indian public. The annual India International Coffee Festival was first organized in Bangalore in 2002, with joint funding from the Coffee Board, UPASI and corporate actors, in an attempt to showcase Indian coffee to domestic and international audiences. This inaugural festival was also accompanied by the publication of the attractive coffee table book, *The Connoisseur's Book of Indian Coffee* (Kariappa, 2002). Since 2003, the Coffee Board has organized the annual *Flavor of India – Fine Cup Awards*, which involves an international jury of experts from major coffee-consuming countries selecting the best coffees from the various coffee-growing regions across India. The 2003 awards were announced by the Coffee Board at a ceremony held at the Specialty Coffee Association of America's annual convention. These activities are important inasmuch as they provide launch pads for developing and honing a set of discourses that spell out a message about the distinctive attributes of a region's product. Journalists and trade magazines act as carriers of these discourses, which are often lent authority by expert cuppers. The publication by the Coffee Board of the *Kenneth Davids Portfolio: Ten Best Indian Coffees of 2004* provides such an example of how a high-profile personage in the industry is utilized to help construct associations of quality that are attached to specific production regions.[20]

## Conclusion

Quite intentionally, this chapter has taken pains to emphasize the rich detail of the institutional–organizational nexus in South Indian tea and coffee sectors. We have done this because, firstly, it sheds light on the importance of path dependence in shaping contemporary industry arrangements and, secondly, it illustrates the importance of the institutional environment in shaping engagement with global value chains.

Considering the first of these issues, it is clearly apparent that the contemporary status of these industries cannot be understood without recourse to a historical perspective. Of key relevance to the arguments of this book, in South India important aspects of tea and coffee production in the Colonial and post-Independence eras live on. Although the British are long gone, an Indian planter class has appropriated vital aspects of that legacy, including the interactive network of clubs and associations. The way that the planter community is embedded in socio-spatial articulations of trust, collaboration and shared circumstance provides a vitally important ingredient in the construction of the institutional environment of these industries. It provides the plantation sector with robust vehicles to foster managerial innovation and technological change. The interventionism of the Government of India in the post-Independence period provides a second leg to this story. The heavy bureaucratic hand of the Indian state in the decades after Independence was reflected in the development of a raft of programmes, policies and administering agencies. During this period, the Tea Board and the Coffee Board gained extensive legislative powers to manage these industries in accordance with national priorities. As discussed, these functions were more extensively applied in coffee than tea, because of India's signatory status within the ICAs.

Beyond these economic regulatory functions, however, these parastatals also developed significant institutional capacities in the form of statistical data collection, pest and disease monitoring, and agronomic research. Although the government no longer actively exercises supply and demand management in these sectors, the parastatal commodity boards that were built up to serve those purposes continue to operate. Their institutional capacity provides a strong foundational element in these sectors. Via their implementation of subsidy programmes, investment in R&D and industry promotional activities, the Indian tea and coffee sectors are provided with a level of support that not all their international competitors can boast.

These aspects of the industries' institutional environments have a crucially determining influence in terms of sectoral engagement with global value chains. In later chapters, we discuss how lead firms in the international tea and coffee industries have set down increasingly stringent compliance requirements with respect to the terms through which they make purchases from upstream suppliers. These include a more rigorous articulation of quality parameters, stricter rules on pesticide residues and other aspects of hygiene and food safety, and direct monitoring of (at least some) labour and environmental issues. For the broader concerns of this book, the themes addressed in this chapter point to the fact that South India's tea and coffee sectors are embedded within an institutional environment with rich capacity to engage with the demands and requirements of these (external) global value chain actors. The networked institutional capacity of the Tea Board, UPASI, trade unions, NGOs and other related agencies needs to be understood as an element of considerable strategic importance to understanding the fortune and fate of South Indian tea producers. To the extent to which this institutional environment provides an enabling context for producers to meet these requirements, a *strategic coupling* (Coe et al., 2004) of institutional-governance formations can be said to exist.

Recognition of these points feeds into the following three chapters of this book, which identify and assess various political-economic struggles that have come to the fore in South India as the institutional environments of tea and coffee production interact with restructured systems of chain governance. Those analyses are framed by the material presented in this chapter, where we have shown the importance of *path dependence* in terms of shaping the institutional environment.

# Chapter Five

# Struggles over Labour and Livelihoods

Issues relating to labour and livelihoods have obvious significance for economic geographers researching global value chains (GVC) and/or Global Production Networks (GPN). It is through labour that economic value is created. And in an increasingly interconnected global economy, differences in the cost of labour between different production sites provide sharper motivations for companies to reinvigorate profits through relocating various activities to lower cost sites; a *spatial fix* to the problem of capitalist over-accumulation (Harvey, 2001: pp. 284–311). Yet labour geographies have not always found a prominent place within GVC/GPN studies (Coe et al., 2008b: p. 284). This chapter goes towards redressing the relative neglect of labour issues in the GVC/GPN literature. We contend that attempts by lead firms in western countries to regulate labour relations in upstream production sites (what Hughes et al. [2008: p. 350] label 'caring at a distance') generate struggles and silences which are often occluded in top-down analyses of these schemes.

For reasons of clarity of focus, we examine these issues with reference to the tea industry of South India. Despite some recent trends towards mechanization, tea production in India remains a highly labour-intensive activity, with plantation-based labour costs (mainly plucking) contributing an estimated 60 per cent to the cost of production. Tea plucking is physically harsh and repetitive, and is performed almost exclusively by women. Pluckers are exposed to insect and snake bites and, in the rainy months, spend long days in highly uncomfortable weather draped by plastic rain sheets. Furthermore, the history of tea production is implicated in extensive abuses. Being located in remote upland areas with no proximate labour force, the colonial origins of tea cultivation in India are connected to bonded workforces who were often forcibly relocated to tea estates. Debt bondage practices were frequently inter-generational, with sons and daughters responsible for the unpaid debts of their parents (Bales, 1999: p. 19).

It is in this light that this chapter focuses on recent initiatives emanating from affluent countries (mainly Western Europe) to introduce ethical sourcing arrangements within the global tea industry. Multinational corporations are conscious of the need to dissociate their lucrative brands from allegations of unethical sourcing and, to this end, have authored a range of schemes seeking to enshrine ethical principles within their upstream supply arrangements. However, these schemes are being applied in a climate of depressed international prices and intense global competition where plantation managers have little choice but to reduce production costs, including those relating to labour. The clash between the *race-to-the-bottom* tendencies of international commerce and attempts by multinational firms and NGOs to assert and codify *ethical accountability* within the industry provides an ongoing source of struggle and contest. In South India, the backdrop to these discussions is a regional industry which in 2002 was directly employing 359,200 people (Tea Board of India, 2003: p. 4), and indirectly supporting perhaps four times that number.

We analyze these issues of labour and livelihoods by maintaining a focus on ways globally coordinated ethical sourcing practices intersect with embedded institutions of labour regulation. Building on the conceptual framework of this book, this analysis demonstrates the mutual structuring of value chain governance with the institutional environments associated with particular places. The ethical accountability agenda represents a new set of conventions unleashed in the contexts of captive supplier forms of value chain governance; but it is played out in an institutional environment of pre-existing socio-political regulation of labour and livelihoods.

This focus represents a substantive variation to how these issues are usually considered. In the main, the ballooning research literature on the politics of global labour regulation – the so-called 'moral turn' in geography – constructs and casts this issue solely within the prism of the value chain. Key concerns include the issues of why lead firms seek to institute selected labour standards, who determines and arbitrates vital benchmarks, and how claims are brokered, monitored and reported on (Hughes, 2005a, 2005b; Christopherson and Lillie, 2005; O'Rourke, 2006). Contrastingly, all too rarely do we hear about how these agendas are enmeshed within regional producer communities; the bottom-up perspective of how these agendas are made manifest in economic landscapes. As illustrated in Figure 5.1, our interest is not just on tea produced within systems of ethical accountability (the vertical downwards arrow on the left-hand side of the diagram) but how these flows intersect with the wider regional economy of South Indian tea production and trade.

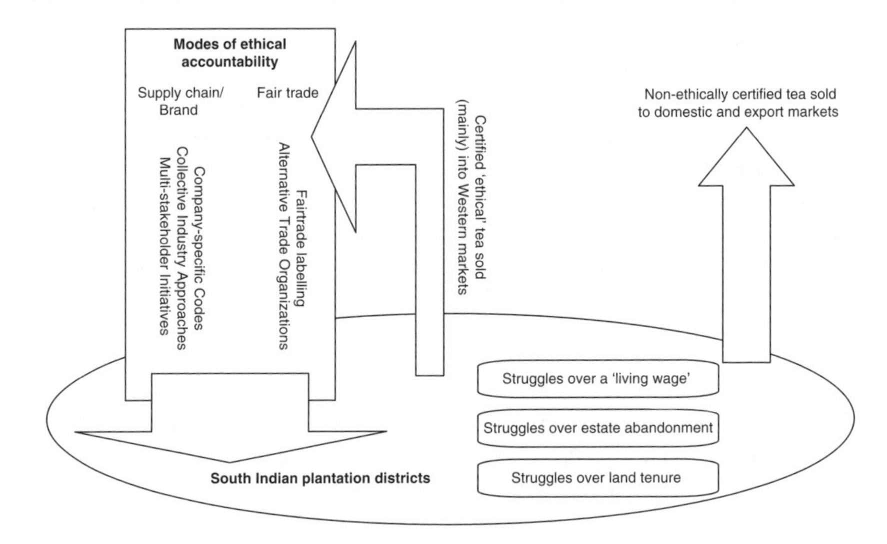


**Figure 5.1** The ethical landscape of the South Indian tea sector.
*Source*: Own work.

Viewing the ethical accountability agenda within the broader regional context of tea production reveals its interstices and points of struggle. Central to these considerations is the fact, discussed in Chapter 3, that much of South India's tea production is destined for markets where issues relating to the ethicality of production currently exercise little significance, if any at all. No more than 5 per cent of South Indian tea production ends up in affluent markets where the ethical accountability agenda resonates. Yet notwithstanding the small proportion of South Indian tea potentially subject to ethical accountability, this issue is not insignificant for producers; crucially, ethical compliance is seen as a precursor to accessing these more lucrative markets. Consequently, this issue has key strategic importance for the industry as a whole. Consistent with this argument, industry leaders take pains to emphasize the ethical credentials of South Indian tea production. In the brochure for the 2007 'Golden Leaf' industry awards, for example, the organizers write: 'Possessing arguably the best Human Capital Index, for any community, this industry can easily stake a claim to be the most ethical source for world teas' (UPASI & Tea Board of India, 2007: p. 5).

In this chapter we seek to widen the terms under which the politics of ethical accountability is debated and understood. Ethical initiatives can make a positive contribution in engendering an improved moral basis for international trade, but it is equally important to recognize that such arrangements represent partial solutions only in terms of contributing to social justice and pro-poor development in industries such as tea.

This argument is developed by, firstly, providing a brief introduction to key concepts we use in this chapter. This provides a conceptual framework for the second section of the chapter, which investigates how recent developments in ethical accountability are being played out in the South Indian tea sector. Whilst the various initiatives described here all encapsulate a wider regime of global private regulation of labour, there is a set of contests relating to authorship and control of these agendas which influence livelihood outcomes for intended beneficiaries. We then focus on the specific circumstances of estate labour in South India, with an initial overview of the institutional settings that have hitherto regulated labour welfare. This discussion adds a crucial dimension to the debate on ethical trade because, when viewed with the knowledge of the institutional environment of South India, various 'silences' emerge in the way the ethical accountability agenda intersects with this regional production system. Ethical initiatives tend to regulate benchmarked behaviour among certain industry participants (mainly, the larger plantation estate sector), whilst other participants are either beyond its purview, or problematic to include within its existing structural forms. We crystallize this argument using examples of three issues that the ethical accountability agenda (at least in its present form) is not addressing:

- The moral authority of ethical accountability presumes the notion of a 'living wage'; however most schemes have little to say with respect to wage-setting regimes on estates, and livelihood supports for smallholders;
- The most pressing immediate threat to tea-worker livelihoods in South India is estate abandonment and the devastating neglect of worker communities, yet these concerns are wholly outside the remit of value chain accountability;
- Contests over rights of access to productive land continue to generate conflicts in the form of smallholder encroachment of estates, yet again, these issues do not figure in any systematic way within ethical agendas.

## What is Ethical Accountability?

The concept of ethical accountability refers to an array of schemes which purport to regulate or coordinate upstream production systems in ways that enshrine (variously defined) ethical benchmarks and principles. Inescapably, these activities and agendas encapsulate a complex terrain of debate which an extensive recent literature has addressed (*inter alia*, Barrientos and Dolan, 2006; Brown and Fraser, 2006; Neilson and Pritchard, forthcoming; Raynolds et al., 2007a; Tallontire, 2006). Our discussion here focuses on the evolution of these agendas and the key differences between core concepts.

Barrientos and Dolan (2006: p. 5) underline the important distinction between 'ethical trade', which involve schemes which regulate labour conditions 'through the implementation of codes of labour practice in the supply chains of large food corporations and retailers' and 'fair trade', which focuses on 'equity in trading relations, and particularly for small producers and farmers'. Essentially, the broad church of 'ethical trade' grew out of the moral arguments articulated by smaller-scale fair trade initiatives, and so it is here that we begin our analysis.[1] Fair trade's origins were with Alternative Trade Organizations (ATOs) in the 1950s and 1960s, which purchased and marketed goods on behalf of disadvantaged producers in developing countries. The development of fair trade emerged from the aid and charity agendas of ATOs such as Oxfam Trading, Traidcraft and SOS Wereldhandel (Tallontire, 2006) who were motivated to provide market access and remunerative prices to marginalized small producers. 'Fairtrade' labeling represented an extension of such ATO initiatives, with guarantees of fairness being encapsulated in the form of an intangible trademark, as opposed to being defined by the ATO itself. The advent of Fairtrade labelling, with the establishment of the Max Havelaar Foundation in the Netherlands, took place in 1988, and by 1997, the three major international Fairtrade labels (Max Havelaar, Fairtrade Mark and Transfair) were harmonized under the Fairtrade Labeling Organization (FLO). As of December 2007, there were 20 national labeling initiatives coordinated under FLO (FLO, 2007).

Fair trade's evolution corresponded to increasing disquiet amongst consumer segments in the West regarding the social and environmental implications of globalization. Encouraged by a succession of media exposés and targeted campaigns by NGOs, by the late 1990s consumer perceptions in affluent western countries had began to question whether the 'race-to-the-bottom' principles of global competition were offering an inadequate moral basis for international commerce. In response, international firms with lucrative brand assets to protect have sought to reconfigure their value chains to meet consumer demands for *ethical accountability*: Consequently, issues of labour welfare and worker livelihoods have interceded forthrightly in the normative lexicon of global trade and investment. Consumer brands that are exposed as sourcing product from sites with unacceptably poor labour conditions can face a pillory of public opprobrium and collapse in sales.

As the influence of these ideas has spread, it has spurred questions of who should be responsible for defining and certifying 'ethical conduct' (Raynolds et al., 2007a). Issues of authorship and control circumscribe the ways that schemes define and deal with social impacts at sites of production. With regards to labour issues, the baseline for all credible ethical accountability schemes rests on four core International Labour Organization principles: (i) freedom from forced labour; (ii) no child labour; (iii) freedom of association and the right to collective bargaining; and (iv) freedom from discrimination on grounds of gender, race, etc. In addition to these four principles, the Ethical Trade Initiative (ETI) base code further stipulates that: (i) working conditions must be safe and hygienic; (ii) living wages are paid; (iii) working hours are not excessive; (iv) regular employment is provided; and (v) no harsh or inhumane treatment is evident (Ethical Trade Initiative, 2007). However, the extent to which schemes move beyond these basic principles is contingent on their origins, motivations, scope and transparency (Hughes, 2005b). This diversity reflects the inherent contestability of the question of 'what is ethical?' The entirety of this discursive field is a shade of greys, as political interests seek to define and/or challenge notions of ethicality to suit their own interests. As a result, the area of ethical accountability represents a continuum of evolving institutional forms which have had the intention of codifying varied notions of 'acceptable' ethical labour practice. Once defined, these 'acceptable' standards are monitored and 'signed off' by global outsiders who are often obliged for reasons of cost and time to rely on intermittent and fleeting audits (Hughes, 2005a).

Accordingly, 'ethical trade' now encompasses a variety of organizational modes, including company-specific codes, sector-specific codes, multistakeholder initiatives and international umbrella organizations, such as Social Accountability International (SAI) and the Global Reporting Initiative (Arthurs, 2001; Graham and Woods, 2006; Neilson and Pritchard, forthcoming). Committing value chain participants to ethically accountable behaviour ensures that the (mainly) western consumers who buy their products are not end-points in the perpetuation of flagrant social injustices but, clearly, it is also big business. Generally speaking, whilst the NGO roots of fair trade were generally motivated by conventions and discourses that shape a common understanding of purpose linking producers with cause-conscious consumers; notions of 'partnership, alliance, responsibility and fairness' (Whatmore and Thorne, 1997: p. 295), corporate-initiated schemes have more complex origins and motivations connected to the maintenance of corporate reputation and the ensuing protection of intangible assets, notably brand names (Hughes, 2005b: Brown and Fraser, 2006).

## Ethical Accountability in Tea and its Influence in South India

Comparable to the case of coffee production and trade (Neilson and Pritchard, 2007b), the gradual decline of commodity prices in tea has produced a galvanizing effect in terms of raising public awareness in the affluent West to the social hardships associated with tea cultivation. The cycle of public awareness and corporate response regarding social conditions in the tea industry has probably been played out more prominently in the UK than anywhere else. This has resulted in the establishment of various non-government attempts to infuse global tea value chains with some kind of ethical accountability. This accountability agenda can be typecast as being constructed through various organizational vehicles: (i) fair trade tea; (ii) organic certification; (iii) company-specific Codes of Conduct; (iv) collective private associations from the West; (v) collective industry standards developed locally; and (vi) multi-stakeholder initiatives. In the sections below, we address key elements of these schemes. We elaborate upon them further in Neilson and Pritchard (forthcoming).

## Fair trade tea

As of January 2008, tea was one of 18 commodities for which FLO has published specific guidelines. The FLO *Generic Fairtrade Standards*[2] provide minimum requirements for social development (including democratic organizational structures and appropriate use of revenue), economic development (such as a fair-trade premium and economic strengthening of the organization), environmental development, and conditions for hired labour. In South India, Fairtrade minimum prices are set at US$1.40/kg (except for the Nilgiris, which is set at US$1.75/kg).

Whilst Fairtrade certification was initially only available to cooperatives of small producers, FLO now also routinely certifies larger estates.[3] The criteria for certification requires that, in addition to following various minimum labour standards, estates must paid a 'premium' on top of the minimum market price (from February 2008, this is set at US$0.50/kg for CTC tea and US$1.10/kg for orthodox teas, with some flexibility for orthodox fannings and dust to be set at the CTC level). This premium is paid into a development fund jointly managed by management and workers, and must be used to support improvements in working conditions. This premium is not to be used to fund infrastructure projects which the company is otherwise required to provide through the *Plantation Labour Act.*

At 18 December 2007, 17 of the 18 tea producers in India which had obtained Fairtrade certification were large estates.[4] Generally speaking, the estates applying for Fairtrade certification in South India are those which are already committed to high labour standards.[5] Fairtrade is seen as a tool (alongside organic certification) to improve the marketability of their tea in a crowded marketplace, where product differentiation offers some potential for accessing quality-conscious markets. In contrast, only one smallholder organization has obtained Fairtrade tea certification in India. The development and management of the 1,200-member Sahayadri Tea Farmers' Consortium, located in Central Travancore, was facilitated by the Peermade Development Society, a church-based philanthropic organization, which employs ten extension officers to serve approximately 1,000 farmers. However, establishing this scheme required substantial donor funding (from the European Union) which was put to effect in the construction of a state-of-the-art tea-processing factory. Our two separate field visits to this organization led us to conclude that the considerable costs in terms of member organization, certification and extension would likely make a similar activity unviable without similar donor support.

## Organic certification of tea

Whilst not specifically addressing issues of labour welfare, the promotion of organic tea has been an important component within broader trends towards ethical production systems. Specific organic conversion of tea in India first occurred back in 1987, by the Bombay Burmah Trading Company at its Singampatti estate. Since then, only a few estates (and even a lesser volume of smallholder production) have been certified organic. The slow take-up of organic methods by South Indian tea producers relates closely to the costs and uncertainties associated with conversion. Tea is generally grown as a mono-crop and, as such, is beset with pest and disease problems. The undulating topography in which tea is cultivated leads to soil and nutrient loss which pose particular challenges for organic cultivation. Standard organic strategies used in other agricultural sectors (for example, use of legume cultivation to replace nitrogen content in soils) are not really possible in tea because of the close planting of bushes (Naqvi, 2002: p. 4). Further to this, the 'Soviet period' of the 1980s instilled a culture of coarse plucking techniques and yield optimization which was consistent with chemical over-use (Jothikumar, 2002: p. 1). Low tea prices obviously raise the risk threshold for companies seeking organic conversion. Organic conversion seems to offer most economic potential for high-cost, high-return enterprises that are already tapping into lucrative niche markets, with certification costs estimated at Rs 100,000 (US$2,175) per estate (Damodaran, 2002b: p. 16).[6] Hence, organic conversion tends to be associated with high-quality estates, or in contexts attached to donor support.[7]

## Company-specific Codes of Conduct

The third ethical scheme is discussed here is the company-specific Code of Conduct. The first company to address ethical trade concerns through this avenue was Premier Brands (at the time the owner of the Typhoo Tea brand in the UK), which in 1992 introduced a Quality Assurance Program including internal social audits of its suppliers (Typhoo Tea, 2007). Notwithstanding these vanguard efforts, it was the involvement of Unilever, the world's largest tea company, which gave the corporate-led accountability agenda a substantial kick-along within the tea industry. Unilever's 'Growing for the Future' programme, developed in 1998, was initially implemented for five crops which the company had globally significant purchases: processing tomatoes, green peas, spinach, palm oil and tea. For each crop, the company proposed extensive environmental and social sustainability benchmarking, with the view to implementing a change in its sourcing practices. Unilever initially ensured its own plantations complied with variously defined sustainability indicators, however its purchases of teas from third party sources were not explicitly incorporated within its own sustainability requirements until May 2007, when the company committed to buying 100 per cent of its tea from sustainable, ethical sources which are to be externally certified by the international environmental NGO, the Rainforest Alliance (Unilever, 2007). This certification will include components for fair treatment and good working conditions for workers. Significantly, Unilever's lead in this area encouraged other agri-food transnationals to follow suit, and has evolved into a broader consortia of the world's largest food companies (under a banner known as the Sustainable Agriculture Initiative [SAI]), which provides a broad forum for member firms to collaborate on issues relating to sustainable sourcing (Alison, 2006).[8]

Other leading global brands, including Tata Tea, operate their own codes of ethical conduct, which govern the corporate behaviour of their owned businesses (but are not generally extended to incorporate supplier behaviour). The Tata Group of companies has its own highly evolved notion of ethical responsibility engendered by the culture of its eponymous family owners, the Parsee Tatas. The family bring to their operations a strong moral position regarding workplace care and a reputation, well known within India, which has became the basis for a stronger articulation of ethical claims within Tetley.

As recently as 2005, the leading gourmet tea brand, Twinings, stated on its website: 'We buy on the open market and have no direct control over how tea is grown' (Twinings, 2005). This was in response to concerns over possible pesticide residues in tea. However, such a position is becoming increasingly difficult to maintain in the current global context, where the private regulation of value chains is commonplace. According to an updated version of the Twinings website: 'Twinings is committed to the ethical sourcing of tea and we do this through our membership of a growing international organisation called the Ethical Tea Partnership' (Twinings, 2007). This brings us to the next of our ethical categories, collective industry standards.

## Collective industry standards: western initiatives

Collective industry authorship of ethical tea standards originated as a response to individualized corporate engagement with the ethical agenda and is premised on the advantages that can accrue to corporate participants from adopting a common platform that defines ethically acceptable behaviour across an entire industry. The exemplar here is Global-GAP, known, prior to September 2007 as Eurep-GAP, a prior name which reflected its organizational origins as representing the interests of a consortium of European supermarket chains. Global-GAP is now an international organization representing such retail giants as Ahold, Aldi, Metro and Tesco, which has the function of setting on-farm standards for Good Agricultural Practices. Global-GAP's approach is premised on a control point, check-list method of vindicating compliance. Certified producers are expected to gain preferential access to the major retail markets that Global-GAP's membership base represents. In March 2006, Eurep-GAP issued a module of critical points and compliance criteria for tea producers, and this subsequently was updated in September 2007.[9] Reflecting the origins of the Global-GAP model in quality assurance and food safety, however there are relatively few requirements in terms of 'Worker health, safety and welfare' in the September 2007 version. Presumably, therefore, whilst Global-GAP certification increasingly will be a requirement for all suppliers of tea into the branded retail sector, in its current form at least, such certification imposes relatively few requirements for producers in terms of labour and workplace obligations.

However, the most influential collective ethical standard now operating in the tea industry is the Ethical Tea Partnership (ETP). A non-profit unincorporated alliance of tea-packing companies, the initiative was initially conceived as the Tea Sourcing Partnership (TSP) in 1997, subsequently evolving into the ETP in 2004. Its purpose is to promote social responsibility by monitoring the conditions of tea production around the world and, with a belief in reinforcing cycles of continual change, to encourage gradual improvement where needed. At the time of writing, the ETP has membership from 24 of Western Europe's and North America's major tea-packing companies,[10] covering more than 60 brands. By September 2007, the ETP had monitored 671 production sites, making up over half of its members' supplier base across the world's 12 major growing regions (Ethical Tea Partnership, 2007). The ETP coordinates and maintains a list of 'approved suppliers' for its membership. These suppliers represent tea estates that have been graded by an audit (paid for by the ETP) that is undertaken by the Partnership's selected auditor (to date, these audits have been carried out exclusively by the multinational accountancy firm, PwC). The ETP standards are consistent with the ETI base code, incorporating the nine

ETI principles mentioned above, supplemented with compliance to national laws and collective bargaining agreements for each country. The ETP submits annual reports to ETI outlining producers' compliance with the base code.

The monitoring process involves a producer questionnaire, followed up by an arranged site visit by monitors, and the subsequent issuing of certificates with grades according to the level of producer conformance. The graded certification system provides a basis for companies to be notified of non-conformance problems, and take action to rectify these. The ETP is committed to engagement with producers within an ethos of 'continual improvement', and only when an estate receives the lowest grade (a 'Fail') is it removed (as a 'last resort') from the ETP's approved supplier list. Unwillingness to participate in the programme also results in removal from the list. Because there is no legally binding mandatory obligation for ETP members to source their tea only from this list, its existence needs to be understood as ultimately providing a guiding recommendation to direct members' purchasing decisions.

This sense of the audit as managerial prerogative flows into the subsequent life of the audit as a prepared document. As an organization owned and operated by major tea companies, the ETP does not share audit reports with other stakeholders and, although the Partnership has demonstrated a commitment to engage with civil society actors through various forums, its structure does not allow for stakeholder participation in any formal manner. Without belittling the personal commitments to ethical principles embodied in individual ETP staff and relevant managerial actors in its member companies, ultimately it is an obvious and important point that the ETP is designed to act on behalf of its membership.

## Collective industry standards: local initiatives

The schemes described above represent arrangements authored outside the region. As a point of contrast, in 2006, the Nilgiris Planters' Association (NPA), representing 36 member estates in the Nilgiris (with a total production of some 16,000 tonnes), registered a trademark for quality tea and ethical compliance. In addition to minimum quality standards, the NPA scheme would include an ethical assurance regarding labour practices on the estates and specifications regarding pesticide use. Use of the logo is conditioned on estates adhering to the required standards of the *Plantation Labor Act*, guarantees of appropriate agro-chemical use, and minimum standards in terms of leaf quality.[11] The NPA's actions were prompted by disquiet amongst the planter community in relation to the alleged 'ethical imperialism' associated with schemes such as the ETP. For Nilgiris planters, establishing their own ethical scheme represented a means of taking matters into their own hands. The NPA views this initiative as a means of creating market recognition which encapsulates both quality and ethics, although it is unclear (at this stage) whether the scheme will be able to successfully gain credibility in the international market. In light of the fact that the majority of tea production in the Nilgiris derives from smallholders, the NPA collective trademark can be interpreted as a means of segmenting and privatizing this geographical identity. By definition, the only Nilgiris tea producers having access to this trademark are the Association's own members, who constitute the higher-quality estate sector. Moreover, its development can also be interpreted as part of a wider strategy to re-localize control over the ethical trade agenda, vis-à-vis the other externally authored arrangements described in this chapter.

## Multi-stakeholder initiatives

Multi-stakeholder initiatives are those involving conjoined industry and NGO participants. In the South Indian tea sector, the *JustTea* project represents the most advanced example of such an arrangement, commencing in 2004 on the basis of European Union funding. Its objective was to establish a business-case argument for corporate environmental and social standards in the tea sector. The programme was implemented through the German-based NGO *FAKT Consult*, in cooperation with the Centre for Education and Communication (CEC), the Delhi-based NGO. The efforts of *JustTea* then dovetailed with the *New Delhi Declaration on the Rights of Tea Workers and Small Growers*, a document (signed by a number of NGOs and unions in December 2005) which sets out an agenda to protect worker interests in the context of the tea crisis. CEC was also a key instigator of this declaration in India. Evidently, the forthright stakeholder incursion of NGOs is having major implications for the politics of the sector. Under *JustTea*, working groups comprising regional stakeholders, planters, factory owners, trade unions, small growers and brokers, were established in each of the major tea-producing regions of India. However, the initiative was unable to garner industry support in India due to a perceived confrontationist approach, and attempts to establish a multi-stakeholder code through this mechanism lapsed with the expiry of funding in 2006 (CEC, FAKT and Traidcraft Exchange, 2006). More recently, building on the experience of *JustTea*, the Center for Research on Multinational Corporations (SOMO, a Dutch-based NGO) in collaboration with Partners in Change (New Delhi) have reinvigorated NGO attempts to drive a multi-stakeholder code in the tea sector. To the time of writing, however, these efforts have not born fruit. In September 2006, we attended a workshop in

Darjeeling as part of this initiative which, despite its multi-stakeholder approach, was characterized by planter resistance to perceived foreign imposition of ethical standards.

Notwithstanding the limited success of these activities, the presence of international NGOs marks a significant development in the institutional configuration of tea chains. As a mode of global private regulation, the ethical accountability agenda gives space to the participation (requested or not) to international civil society actors. Potentially, it also gives space to enhanced international networking by trade unions; however, the outputs from such efforts have been mixed. Consequently, global private agendas of ethical accountability potentially can be seen as inflecting a change in orientation for the ways that civil society issues are understood, considered and articulated; from the (horizontal) sphere of a sovereign jurisdiction with territorial responsibilities for its citizenry, to the (vertical) framework of regulating particular (export-oriented) value chains. The multi-stakeholder nature of meetings, such as that in Darjeeling mentioned above, thereby brings to the fore a set of tensions in terms of worker representation. The shifting trend towards worker representation by international actors such as NGOs and consumer associations can cut across the traditional domain of how these issues are framed by actors embedded in national spaces (trade unions, planters' associations, etc.).

## Divergence and commonalties: implications of an audit regime

Underpinning the ethical accountability schemes described above is product traceability. Consequently, discussion of ethical accountability necessarily must revolve around the praxis and politics of the two critical notions of auditing/monitoring and accreditation. Audits of producers have the function of establishing compliance with set standards. Ongoing monitoring ensures that these are upheld. Accreditation then formalizes and acknowledges compliance in codified form. Whilst there is clear divergence in the origins and authorship of the various modes of ethical accountability, for industry participants in South India these various schemes follow a similar operational pattern: (i) a set of standards are established (commonly elsewhere and with little local consultation); (ii) a brief visit to the estate is made by an auditor; and then (iii) compliance to the audit is somehow linked to market access.

The legitimacy of the system is dependent, of course, on the veracity of the audit regime itself. Over time, ethical or social audits have evolved from first party (where a company audits itself) to second party (a company audits its suppliers) through to third party (an external auditor is appointed).

Alternatively, O'Rourke (2006) categorizes regulatory strategies as either 'internal monitoring', 'external monitoring' or as 'verification' (where independent evaluations are not paid for by those being monitored). Indeed, the growth of global codes of conduct and third party audits has engendered a corresponding growth in an audit industry comprised of various organizations and consultancies. Tendering, appointment and payment structures within this audit industry vary considerably. Key questions about process, moreover, shape the politics of audit. In most cases, audits generally work in such a way that visits are announced in advance (potentially enabling management to 'clean up' any problems) and take place within relatively short timeframes. In this environment, issues of non-compliance can readily escape view. Moreover, although the protocols for interviewing workers theoretically create the preconditions for free information exchange, despite the best will in the world it may be difficult to disabuse vulnerable workers of the notion that unknown outsiders conducting the audit are, in fact, independent from management.

The accepted limitations, therefore, of snap-shot audits are driving more recent demands for a new auditing model, such as Auret and Barrientos' (2006) call for 'participatory social auditing'. Such an approach involves the use of local auditors who are intimately familiar with the language and culture of workers and who are gender-sensitive. For corporate lead firms, of course, the introduction of such arrangements would carry obvious risks as the politics of control within audit processes shifts from the situation where audits verify compliance, to situations where they potentially document and expose non-compliance. These issues, thus, remain a field of struggle and contest.

## Labour Welfare on South Indian Tea Estates

In the words of one analyst, 'tea plantations are enclaves, alien and inward looking and cut off from the outside world' (Chattopadhayay, 2003b: p. 6). Worker populations are frequently wholly reliant on company masters for the provision of the basic necessities of life, and generations of workers can live and die in the enclave economy of a single estate. Plantation companies are responsible by law for providing the schools, health clinics and housing, which elsewhere are generally considered to be the responsibility of government. Children of tea pluckers can literally grow up in the gardens, and in the past, the fine line defining what constitutes 'child labour' was readily breeched:

> Given few child care alternatives, women who pluck tea often bring their young children with them into the fields. Young children may begin by helping

> their mothers for recreation, but in countries where schooling is not compulsory, it is a short step to carrying their own baskets for pay. (United States Department of Labor/Bureau of International Labor Affairs, 1997: not paginated)

The use of child labour was historically endemic in Indian tea estates, particularly in the northern states of Assam and West Bengal. A research report supported by Friedrich Ebert Stiftung (Bhowmik et al., 1996: pp. 20–1) found that in 1990 there was an estimated 97,000 children working on tea estates in Assam alone. By 1994, however, adverse publicity and the threat of trade sanctions had caused the plantation sector to embark on steps to curtail this practice. These days, as a general rule, estate owners actively discourage the use of child labour, and a recent Dutch NGO report asserted that child labour had been eliminated from the Indian tea plantation sector (van der Wal, 2008: p. 36). This is despite the fact that amendments to the *Plantation Labor Act* specifying a minimum age of employment of 15 years on tea estates were introduced to the Lok Sabha (National Legislature) in 1990 (Bhowmik, 2003: p. 32), but have not been voted on. Of course, in the smallholder sector, child labour remains part and parcel of the employment landscape, with families relying on the work of their children in situations of dire poverty.

Plantation workers' interests in India are defended by a prevalence of trade unions, although there is widespread disagreement on their effectiveness. Trade union agendas and representational strategies are contextualized by the often-fractious and self-absorbed character of Indian domestic party politics. For trade unions, worker populations in tea estates generally serve as vote-banks for party affiliates. Internecine struggles for trade union membership (and, therefore, the spoils of party politics) are an endemic feature of Indian democracy. Indeed, it is not insignificant that a Dutch NGO, representing European trade unions and other civil society interests, sums up the union presence on tea estates as follows:

> Although the plantation workers in the Indian tea sector are highly organized, there may be strong rivalry between the various unions on the tea plantations, which is not always to the advantage of the workers. Labor unrest is not always sustained and often turns out to benefit the management. The 'elected' union leaders do not always represent the interests of the plantation's workers. Plantation workers generally belong to the lowest socio-economic groups, while union leaders are often 'outsiders' from the middle class. (SOMO et al., 2006: p. 41)

There is a readily observable social and economic gulf separating manual workers from estate management. Plantation bungalows may be only a few hundred metres away from tea pluckers' 'housing lines', but there is a veritable universe of difference between these classes in terms of economic, social and educational status. These structures provide both a ready target for labour agitation, and an arena in which ensuing struggles are played out.

The *Plantation Labor Act* provides the foundational basis for workers' rights in the sector. As discussed briefly in Chapter 1, this legislation was enacted with the intent to recompose the politics of the plantation industry in the wake of the British departure from India. In effect, it formalized, standardized and extended the *dastoor* (standing practices or customs) of plantations. Under the Raj these differed from estate to estate, largely owing their origins to the varied conditions that colonial masters initially provided to their migrant workforces in these remote localities. By way of contrast, the *Plantation Labor Act* promulgated the obligations of owners in black-letter terms. Under the Act, the plantation company is obliged to construct and service housing, as well as provide water supply, sanitation, electricity, local roads, firewood and kerosene, statutory staple food items, blankets, overcoats, space for a kitchen garden, and community infrastructure such as primary health stations, community hospitals and crèches, as well as the operation of local buses and primary schools. The office of the Inspector of Plantations has responsibility to monitor compliance with these requirements, and is legally required to visit all plantations at least once every four months (Bhowmik et al., 1996: p. 97). Our interviews detailed a range of views on the effectiveness of this system, however, it should be noted that a comparative study of the Act's operation suggests that South Indian estates have a high rate of compliance, particularly compared to North India (Bhowmik et al., 1996: pp. 110–14).

The operation of the *Plantation Labor Act* also provides an ongoing sore point within the planter community. Plantation sector representatives point to the fact that employers in other sectors are not required to bear the cost of social welfare provision and, perhaps more tellingly, neither (so they claim) are many of their international competitors. According to one estimate, the cost of labour overheads in South India is four times higher per kilogramme of made tea than that incurred by tea estates in Kenya, a major competitor (Accenture, 2002: 81). As management was keen to inform us in our many visits to tea estates, the extreme isolation which marked the plantation sector during Colonial times (and which formed the justificatory basis for these obligations) is now much less an issue, and hence, in their view, the legislation should be amended.

In this light, estimating the cost of the *Plantation Labor Act* has become a point of contention within recent political debate over the plantation sector. Perhaps the most authoritative estimate comes from a study commissioned in 2005, by an interministerial committee set up by the national government, which concluded that the Act's provisions in South India were costing the industry Rs 3 per kilogramme of made tea (*The Hindu*, 2005a). Estimating an average daily plucking rate of 20kg per worker, the Ministry's figures suggest that the Act's requirements add up to Rs 12 to the daily cost of labour.[12] Our own data gleaned from field interviews broadly support this estimate. For example, the Kanan Devan Hills Plantation Company at Munnar (Kerala) estimated that their 2006 expenditures for complying with the *Plantation Labor Act* would amount to at least Rs 30 million (approximately US$680,000), not including capital costs relating to schools, health care facilities etc.[13] With a workforce of approximately 11,000, this translates to approximately Rs 9 per worker per working day. Around one-third of this expenditure was accounted for by immediate direct annual expenditures on housing, health, water and sanitation, with the remainder accruing to schooling, transport, firewood, the salary costs of doctors, and energy provision.[14] In some cases, moreover, local social circumstances may dictate a need for social welfare expenditures in excess of *Plantation Labor Act* provisions. In one plantation we visited in Wyanad, the estate's health care facilities were burdened by additional demand from significant population growth in adjacent communities. With state government health services not keeping pace with population growth, the existing plantation facilities inevitably took up the shortfall.[15]

Management frequently iterated to us that, in addition to these expenditures, employers are also obliged to bear a range of salary on-costs, including bonus payments, holiday pay, annual leave, provident fund contributions, and gratuity (payment on retirement). Again, estimates varied as to the size of these payments, but based on financial data sighted for one Kerala estate these on-costs can be assumed to represent approximately 36 per cent of monetary wages.[16] To put these figures in context, if we assume a tea plucker earns Rs 100/day in wages then the total daily cost to the employer will be Rs 148 (Rs 12 in *Plantation Labor Act* requirements and Rs 36 in on-cost obligations).

Whilst it is quite understandable that estate managers point out how their international competitiveness is impaired by the cost of labour, it is sobering to remember that tea plucking is arduous and repetitive work. For their part, tea workers are required to work from 7.30 a.m. to 4.30 p.m. from Mondays to Fridays, and a half-day on Saturday. Workers are entitled to an hour break for lunch, and breaks for morning and afternoon tea. In Tamil Nadu, tea pluckers are entitled to receive the minimum wage once they pluck 25kg in a day. In Kerala, the benchmark rate is 16kg, or 12kg in the low-growth months of January and February and the rainy month of August. Whilst it is true to say that tea pluckers are not at the absolute bottom rung of India's agricultural workforce, and via the *Plantation Labor Act* have access to social benefits that many other rural workers do not, it is equally the case that they remain significant bearers of disadvantage, inequality and poverty.

## Struggles and Silences: The Interstices of the Ethical Accountability Agenda in South India

So far, this chapter has discussed the rise of the ethical accountability agenda in the tea industry, and the general circumstances relating to the ethical debates over labour in the South Indian context. Attention now turns to the vexed questions of 'struggles and silences': the unresolved contests over how the ethical accountability agenda defines what it means to 'be ethical' in South Indian tea production. Coming to grips with these questions requires detailed assessment of place-bound processes of industry restructuring, which are often outside the domain of much that is written about ethical accountability. To reiterate points made earlier, consideration of these themes brings to the surface *geographical* issues of how industries are institutionally embedded within localities, regions and nations. For the case at hand, this problematic set of questions is telescoped by considering four issues that exercise considerable concern and debate within South India, yet rarely rate a mention in the body of reports and analysis that is produced by stakeholders within the field of ethical accountability: how do ethical accountability agendas intersect with wider struggles over a living wage; how do systems of ethical accountability cope with the specific problems of smallholders and informal workers; how does the ethical accountability agenda deal with situations of estate closure and abandonment; and how does ethical accountability relate to disputes about land tenure?

### Struggles over wages and smallholder incomes

In a climate of weak commodity prices and intense global competition, an ongoing struggle exists between the implied presumption in ethical accountability that workers should be paid a 'living wage', and the imperatives of management to keep a lid on their wages bill. Yet, with the notable exception of fair trade, the ethical accountability agenda is largely non-interventionist with regards to income-setting. As far as most schemes are concerned, the issue at stake is whether employers comply with the relevant legal obligations to pay workers their due; not the question of whether wage regimes are appropriately consistent with broader aims to adequately sustain rural livelihoods. And, moreover, most schemes have blind spots when it comes to worker livelihood issues outside of the formal, permanent sector. As argued by O'Rourke (2006, p. 907): 'even when monitoring is effective, some of the most hazardous jobs may be shifted further down the supply chain or into the informal sector to avoid the selective gaze of non-governmental regulation'. When implemented in contexts of the deregulation

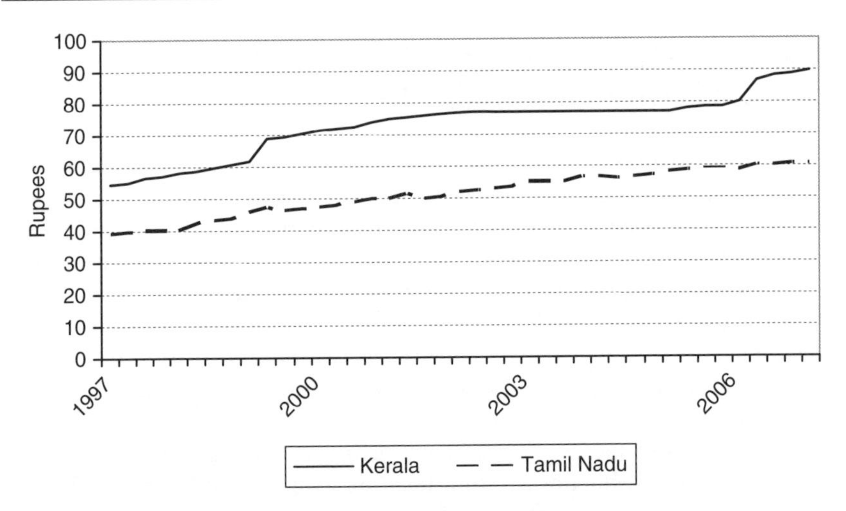


**Figure 5.2** Daily minimum adult wages in tea plantations, 1997–2007.
*Source*: Documents provided to the authors by UPASI, Coonoor, 18 June 2007.

and de-collectivization of formal workforces, and their progressive replacement by casual workers and/or product supplied from smallholders, the silences of ethical schemes on this issue gain pressing significance. In this section we open up this often-occluded debate.

The system for establishing workers' incomes in South Indian tea plantations is bureaucratically cumbersome and enacted in an often politically charged environment which sets up workplace relations in party-political shadows. India's national legislature passed a minimum wages law soon after Independence, but the principles upon which minimum wages are set remains the responsibility of individual states. In the South Indian tea sector, the Government of Kerala (with a legacy of political domination by Communist administrations) has historically instituted minimum wage levels that are considerably higher than those across the border in Tamil Nadu (Figure 5.2). Debate on the differences between daily wage minimums in Kerala and Tamil Nadu was rekindled in early 2006 when the Kerala Minister for Labour instituted a Minimum Wage Rate Notification that raised daily minimum tea pluckers' wages by 12.9 per cent (from Rs 78.6 to Rs 88.8) within the space of a year. This was done in the context where the state-wide Plantation Labour Committee, the tripartite body representing owners, unions and the state government, had failed to come to agreement on wages and a state election was looming. Upon the Minister for Labour's Notification, the Kerala Planters' Association mounted a legal challenge on the grounds that the decision would be ruinous to the industry, but at the time of writing (more than one year after the Notification was lodged) the case remains unresolved in the Kerala Supreme Court and so Kerala plantations have become accustomed to paying the new rate.

During the tea crisis, larger corporate groups generally had the financial strength to maintain regular fortnightly payments to their workers throughout this period, but many smaller entities were forced into the situation where they had to cut or delay wage payments. One estate we interviewed in Tamil Nadu told us how it cut a deal with the union to reduce daily wages from Rs 74 to Rs 72 during the nadir of the crisis period in 2002–3.[17] However, not all such deals were negotiated with common consent. Delayed or reduced wages created industrial turmoil, especially in Kerala, where strikes and agitations increased in intensity during these years.

These contexts have encouraged plantation management across the region to instigate historical breeches to the traditional use of permanent workforces, paid at standardized rates. Given the inherent difficulties in winding back the social provisions of the *Plantation Labor Act*, management has instead begun reducing the number of permanent workers and is increasingly relying on cheaper, flexible, seasonal and contract labourers. According to a report by ActionAid (2005: p. 6) 'the percentage of casual labourers on tea plantations rose from 13 per cent to 24 per cent between 1997 and 2000'. No recent official data exists on the prevalence of this trend (and it is difficult to ascertain the authority of the ActionAid data), but our visits to tea estates across South India certainly support this notion, albeit in the context where permanent workers still provide the overwhelming majority of estate labour.

For smallholders the situation has also been dire, and again their circumstances generally fall outside the overall gaze of the ethical accountability agenda. The question of whether or not smallholders are paid a 'living price' for their product is a question that is extremely difficult to resolve and for interested parties to monitor. In recognition of the importance and complexity of applying codes to smallholders, the ETI has produced specific *Smallholder Guidelines* (ETI, 2005). The ETP's coverage, however, remains anchored on plantation estates, despite efforts from 2005 onwards to expand its reporting scope into the smallholder sector. As of late 2007, the ETP was still struggling to incorporate smallholder monitoring into its framework. Moreover, despite its historical origins in supporting smallholder production, the fair trade movement faces similar problems of certifying and monitoring smallholder producer systems. As noted earlier in this chapter, it is somewhat ironic that the number of Fairtrade-certified *estates* in India (17) far outnumbers the number of Fairtrade-certified smallholder production systems (only one) (FLO, 2007). The costs of establishing and maintaining smallholder organizations, which facilitate the documentation and certification required to satisfy fair trade specifications, may be prohibitive for smallholder inclusion.[18] As argued elsewhere by Smith and Dolan (2006), ethical codes tend to neglect the more complex (and often gendered) needs of the informal and smallholder workforce.

Turning to wage-setting processes, wage levels in the industry are shaped not only by planters' arguments about their abilities to pay workers a given rate, but also by the supply-side conditions of labour market availability. Critical in this regard are tendencies towards labour shortages in the plantation sector. As long ago as 1996, an International Labour Organization report suggested: 'Looking ahead, a shortage of plucking labour is clearly foreseeable' (Sivaram, 1996: not paginated [Part 4]). Tellingly, plantation managers interviewed in the course of our research reported that the next generation of tea estate workers is 'going missing', as the daughters and sons of estate workforces leave the 'lines' for greener employment or educational pastures elsewhere. India's booming economy has generated more new jobs each year during the 2000s – an average of 11.3 million – than any other country, including China (OECD, 2007). A significant proportion of this new employment is taking the form of low- and semi-skilled labouring positions potentially available to the children of plantation workers. Also relevant is the fact that India's 'education revolution' is also leaving its mark on the plantations. Traditionally high birth rates within plantation workforces have come down rapidly, and our interviews with plantation workers revealed a thrifty intent to use hard-earned savings to fund their children's education.[19] As this next generation obtains diplomas and degrees in fields like IT, engineering, hospitality and health care, the future of labour-intensive plantations becomes increasingly problematic.

Returning to the central issue, then, there is a regulatory disjuncture between the rather imprecise language of ethical accountability schemes regarding the need to pay workers 'a living wage', and the reality of how those wages are determined. With the exception of fair trade schemes, which are premised on the creation of price premiums that are repatriated to the benefit of workers, the ethical accountability agenda authored by companies has generally little to say on this account.

## Estate closure and abandonment

*If the crisis continues for another year, all the planters will abandon their tea fields and become owners of BLFs. A. E. Joseph, Chairman, the Association of Planters of Kerala, Peermade (Joseph, 2003: p. 66)*

During the period of low tea prices from the late 1990s, the planting district of Central Travancore, in Kerala, was, without doubt, the site of greatest human misery and suffering across the entire South Indian tea industry. It has been buffeted by a spate of estate closures and abandonment which has left workers without wages and facilities. Between 1997 and 2006, tea production (ex-factory) fell from 23 million kg to 8 million kg. In February 2006, 20 of the district's 39 estates had been abandoned by management (Table 5.1). On estates which have been either wholly or partly abandoned, there lived a permanent workforce of 12,500 (Chattopadhayay, undated: pp. 11–13) with at least the same number again of temporary workers and dependents (International Union of Foodworkers, 2003).

**Table 5.1** Status of tea estates in Central Travancore, 2006

| *Owner* | *Estate name* | *Size (ha)* | *Status* |
|---|---|---|---|
| AVT | Arnakal | 517 | Operating |
| | Caradygoody | 295 | Operating |
| | Pasuparai | 183 | Operating |
| HML | Pattumalay | 232 | Operating |
| | Moongalaar | 739 | Operating |
| | Wallardie | 518 | Operating |
| Malankara | Penshurst | 112 | Operating |
| | Karimtharuvi | 170 | Operating |
| Tyford | Tyford | n.a. | Operating |
| | Rohit | n.a. | Operating |
| Tropical Products | Connemara | 201 | Operating |
| Owner-manager | Alampally | 144 | Operating |
| Owner-manager | Semnivalley | 372 | Operating |
| Owner-manager | Chidambaran | 45 | Operating |
| Owner-manager | Haileyburia | 252 | Operating |
| Owner-manager | Kudukarnam | 305 | Operating |
| Owner-manager | Churakulam | 195 | Operating |
| Owner-manager | Aban-loyd | 761 | Operating |
| Owner-manager | Mlamallay | 174 | Operating |
| A.V. George | Stagbrook | 184 | Non-functioning (a) |
| | Ashley | 160 | Non-functioning (a) |
| Hope Estates | Glenmary | 377 | Non-functioning (a) |
| | Kudukarnum | 305 | Non-functioning (a) |
| | Ladrum | 289 | Non-functioning (a) |
| MMJ | Vaghamon | 330 | Non-functioning (b) |
| | Kottamulai | 322 | Non-functioning (b) |
| | Bon Ami | 243 | Non-functioning (b) |
| Peermade Tea | Lone Tree | 304 | Non-functioning (b) |
| Company | Pirmed | 332 | Non-functioning (b) |

**Table 5.1** *(cont'd)*

| *Owner* | *Estate name* | *Size (ha)* | *Status* |
|---|---|---|---|
| Ram Bahadur | Munjamullay | 279 | Non-functioning (b) |
| Thakur (1) | Thengakal | 229 | Non-functioning (b) |
| | Nellikai | 173 | Non-functioning (b) |
| | Pambanar | 243 | Non-functioning (b) |
| Ram Bahadur | Tungamullai | 342 | Non-functioning (b) |
| Thakur (2) | Passumullai | 193 | Non-functioning (b) |
| | Koliekanam | 299 | Non-functioning (b) |
| | Granby | 216 | Non-functioning (b) |
| | Mount | 214 | Non-functioning (b) |
| | Pullikanam | 250 | Non-functioning (c) |

*Key*: Non-functioning (a) = Factory closed and workers plucking leaf sold to management.
Non-functioning (b) = Abandoned estate and workers plucking leaf and selling it via unions.
Non-function (c) = Abandoned estate and no workers present.
*Source*: various documents obtained during field research.

The point we make here is that this dire situation has fallen completely outside the ambit of the ethical accountability agenda. Because ethical accountability has concern only for product that ultimately finds its way to affluent consumer markets, it has no capability, interest or mandate to address ethical issues associated with the closing down of production facilities. In making these points, we underline the way that these schemes, as modes of social regulation, are partial and selective as to whose ethical interests they attend.

The human suffering in Central Travancore deserves focus. One tea worker from an abandoned estate told a Fact Finding Mission in January 2003:

> Our wages have not been paid for the last 20 months. The Provident Fund has not been deposited for the last four years. The Bonus has not been given for the last two years. The electricity has been cut in the plantation and in the worker lines for non-payment of dues. Drinking water supply to the worker lines in the plantation has been stopped completely. The estate hospital is non-functional with no medicines, doctor or nurse. (Chattopadhayay, 2003b: p. 10)

According to testimony from another worker on an abandoned estate:

> Getting water is the most tedious work. One has to walk kilometres and wait relentlessly to get potable water ... We could have operated the pump to collect

> water in the tank, but there is no electricity [so] the pumps cannot be operated ... Our life was already in darkness and the cut in electricity has made it all the more dark ... When there was work in the estate, we used to buy firewood. But now we go to the far-off woods to collect fuel ... We live in dilapidated houses. We have been staying in our house for many years ... During the rainy season it is as good as staying outside. Water pours heavily inside the house. (Menon, 2003a: p. 54)

In 2003, at the height of the crisis, there were 12 deaths from malnutrition (Jacob, 2003), and uncounted levels of hunger. Following on from the problems of hunger, malnutrition and despair, was a spate of suicides. In 2003 there were eight documented suicides (Jacob, 2003), one of which gained wide media attention because it concerned a 14-year-old girl called Velankanni. According to the Fact Finding Mission's account of her circumstances:

> [she was studying] in the Panchayat High Scholl in Vandiperiyer. She was a bright student who went to school walking the 18 km every day.... Velankanni did not even have even one pair of school uniform [sic]. Every day for two years she wore a borrowed, tattered uniform. The estate management had stopped giving wages for months and Velankanni's father ... had to manage the home with the Rs 200 he got every week. The teachers in Velankanni's school asked her to deposit Rs 35 for a youth festival and a stamp collection fund. Unable to get the money, she stopped going to school. One day, [her mother] found her daughter dead, hanging from the ceiling of their home. (Chattopadhayay, 2003b: pp. 13–14)

Not surprisingly, these conditions have provoked large-scale agitation. In March 2003, unions staged a 300 km protest march from Central Travancore to Kerala's capital, Thiruvananthapuram (Trivandrum). Before and after these events, workers and unions had, in any case, taken matters into their own hands. From 2003 onwards, the unions had essentially taken over from management and allotted permanent workers specific areas in the fields and were paying them Rs 100 to Rs 150 per week (approximately US$2.30 to US$3.40) for plucking (Menon, 2003b: p. 69). Amounting to roughly one-fifth (or less) what these same workers would have earned per week prior to abandonment, these monies were well below basic survival levels. Evidently, the ad hoc and haphazard nature of plucking leaf from appropriated estates provided only a temporary livelihood fix for these workers. Appropriated fields were not nourished with crop and soil maintenance and, on our observation, vigorous and/or desperate plucking was causing lasting damage to bushes (Figure 5.3; Figure 5.4). Aggravating this situation was workers' vulnerability once outside the protective umbrella of plantation life. In the course of our fieldwork in Central Travancore, interviewed informants from different walks of life alleged that union leaders collecting leaf from pluckers

**Figure 5.3** Degraded tea landscape in Central Travancore, 2006.
*Source*: Photo by the authors. Note the extensive patches of bare earth visible on the far hillside.

were skimming profits for their own individual and party-political purposes. (One estimate we heard was that union leaders were charging pluckers a dubious 'commission' of Rs 2/kg for green leaf, in the context where pluckers would have been lucky to receive a gross price of Rs 6/kg.) Living in the grey economy outside of the formal rule of law, former estate workers had little recourse for any such abuses, except migrating out of the district; which, in fact, large numbers did.

The severity of the crisis in Central Travancore underlines the general effects of low international tea prices on the sector's profitability and viability. Further to these considerations, however, a number of locally specific circumstances point to why Central Travancore in particular has been the focal point for social and economic crisis in the South Indian tea sector. What seems to have occurred in Travancore is a 'perfect storm' of geographical vulnerability and estate mismanagement by some interests in the district. Travancore is an old plantation district with many estates dating to the early years of British pioneering efforts in tea. Its aged bushes now deliver lower productivity and are particularly susceptible to pest and disease threat. Red spider mite and leaf blister blight, two banes of the South Indian tea industry, have had wider purchase in Travancore compared to other districts.

**Figure 5.4** Abandoned tea factory, Hope Estates, Central Travancore, 2006.
*Source*: Photo by the authors.

According to the local representative of the UPASI Tea Research Foundation, in the 1950s and 1960s, large amounts of DDT and other persistent organochlorines and organophosphates were used on the district's tea estates which upset the ecological balance and led to large insect populations including thrips, red spider mite, short stem borer and helopeltis (tea mosquito).[20] Dealing with these problems has required astute managerial expertise. While some estates in the districts have weathered the storm (notably, estates owned by well-established plantation companies such as Harrison Malayalam Ltd and AVT Thomas Ltd), others have not been up to the task. According to Babu Divakaran, Kerala's Minister for Labour:

> In Peermade [Central Travancore] the companies in their good times never thought about the lean period. Most of the companies have diverted their

> funds to many other areas. In the planters' association general body meeting I had told them about this. The reserve funds [were] not kept for plantations. Mismanagement is one of the root causes of the problem. (John, 2003b: p. 71)[21]

The collapse and abandonment of estates in Central Travancore emphasizes two key aspects of the politics of labour and livelihoods in the South Indian tea plantation sector. Firstly, the management of a number of estates lacked the inclination, imagination or expertise to effectively reposition their operations in light of changed global market circumstances. The plantation system being what it is, workers and their families bore significant hardship once companies abandoned their estates. Secondly, estate abandonment focuses attention on the importance of the regional institutional context. For several years following the abandonment of various estates, the apparatus of the Indian state struggled to mount an appropriate response to the dire human suffering in the district. Some emergency relief was provided, but this was neither sufficient nor enduring. Following the April 2006 Kerala election, however, the newly elected Left Democratic Front government (led by the Communist Party of India [Marxist]), in consultation with the Tea Board of India, announced a progressive reopening of abandoned estates in accordance with recommendations made by a Parliamentary Standing Committee in the Lok Sabha, headed by Kailash Joshi. As of January 2008, nine tea estates had been re-opened under this scheme (Asia Pulse, 2008). However, despite these initiatives, the future of the abandoned estates in Central Travancore remains uncertain.

At a broader scale, this case exemplifies the central argument of this chapter; namely, that as a mode of social regulation for the tea industry, ethical accountability schemes privilege not the most needy, but the needs of western consumers to drink their tea in the assumed knowledge that it has been produced ethically. There is nothing wrong in this, of course. It is right and proper that western consumers should consider the ethicality that underpins the production of the goods and services they consume. However, when ethical accountability is considered from a *geographical perspective* as a mode of social regulation authored by outside interests and intervening within production regions, its unevenness and, indeed, capriciousness, becomes starkly apparent.

## Encroachment: ethical accountability and the complexities of land tenure

The last of our three examples of the struggles and silences that are implicated within the ethical accountability agenda concerns the highly fraught issue of land tenure. In the context of its rising population and limited land resources, encroachment is an increasingly problematic issue in India. Encroachment refers to the, technically speaking, illegal expansion of smallholder cultivation into government land and privately owned estates. Focusing on how this problem manifests in Gudalur Taluk, a sub-district of the Nilgiris in Tamil Nadu, we make the point that the concern within ethical accountability schemes for how tea estates are internally organized (i.e. whether they comply with relevant legislative requirements about their use of labour, etc.) can have the effect of treating these entities as 'islands' of production without due consideration to how they are embedded within their local societies, economies and environments. Gudalur possesses a particular politico-administrative context which lends currency to the consideration of these issues. Tea production there takes place in a situation where the social and economic landscape is driven by unresolved land tenure disputes, which have created significant political friction. The vignette we presented at the outset of Chapter 1, of smallholders up in arms at meeting convened by the Tea Board of India, was one episode in this long-running dispute.

The background to these problems lies in the arcanae of Indian land law.[22] Traditionally forested with a sizeable tribal population including the Paniyans and Kurumbas, the first sedentary agriculturists in the area were scattered communities of Malabar Chettiars. However, in the late 1790s a wave of settler migration occurred in Gudalur in the context of Tipu Sultan's challenge to the authority of the Mysore Maharajah.[23] Gudalur was then ceded to the British (Government of Madras) under the 1799 Seringapatam Treaty with Tipu Sultan, however this was initially resisted by the forces of Kerala Varma, a king from (what is now) Kottayam District who became known as the 'Pychy Rebel'. British settlement of Gudalur occurred later following the first survey of the District performed by Col. John Ouchterlony in 1847. Large plantation leases were issued (at first) for coffee cultivation with labour brought in primarily from Mysore. This was then followed by another wave of migration during the period 1879–82 in response to (highly exaggerated) claims of gold in the region. By 1908, Francis (2001: p. 366) describes the district as a 'melancholy tract … having seen better days', and it was not until a concerted government effort to eradicate malaria in the late 1940s and mid 1950s led to a population explosion (Adams, 1989).

It was around the time of the 'Pychy Rebel', and Muslim incursions under Tipu Sultan, that the Keralan nobility (the so-called *Janmi* or 64 noble families of Kerala) began to assert claims to Gudalur as *Janmam* land (Kjosavik and Shanmuganratnam, 2007). A uniquely Keralan arrangement, *Janmam* was based on the feudalist principle that the *Janmi* held exclusive rights to land, and cultivators were no more than rent-paying lessees on those lands. In Wyanad-Gudalur, the colonial government even created a new class of land tenure, *Government Janmam*, based on lands of the 'Pychy rebel', where indigenous cultivators would pay rent to the government (ibid.).[24]

As with the comparable *Zamindar* feudalist system which operated through much of the rest of India, *Janmam* tenure were outlawed in the Indian Constitution of 1947, with the states given charge of enacting legislation that would free cultivators ('*ryots*') from their subjugated relationship with feudalist intermediaries. In Tamil Nadu, the *Zamindari* land tenure system was abolished with the *Tamil Nadu Estates (Abolition and Conversion into Ryotwari) Act*, 1948.[25] At this time, however, Gudalur was still a district of Malabar (now Kerala) and so Keralan *Janmam* land tenure systems prevailed. However, in 1956, Gudalur's historic ties to Kerala were cut because in-migration of Tamil-speaking peoples made Tamil the (slightly) majority language, and under the language-based redrawing of state boundaries, Gudalur became incorporated within Tamil Nadu state. Upon being transferred to Tamil Nadu in 1956 it was initially assumed that the 1948 legislation would act to abolish the *Janmam* system and thus give cultivators their freehold ('ryot') entitlements to land. However, subsequent litigation proved that the 1948 Tamil Nadu legislation did not apply to the *Janmam* system in Gudalur, and so legal rights over land remained in limbo.[26] To address this flaw in the execution of agrarian reform, the Indian Constitution was amended in 1964 to clarify the meaning of the word 'estate'. Under the status quo, the *Janmi* were effectively the owners of 'estates' (though tilled by rent-paying tenants) and were therefore unaffected by the agrarian reforms. The Indian Constitutional Amendment of 1964 thus read: '*Estate* will include in the States of Tamil Nadu and Kerala under *Janmam* tenure' (italics in original). This Constitutional amendment was enshrined legislatively in Tamil Nadu through the *Tamil Nadu Gudalur Janmam Estates (Abolition and Conversion into Ryotwari) Act*, 1969. Effectively, the relevant provisions of this legislation mirrored those of the earlier 1948 Act with the similar name, although expressly including the *Janmam* situation in Gudalur. Similarly to that earlier Act, the spirit of this legislation was not to interrupt the ongoing land tenure of cultivators, but to remove their exposure to the exploitative social context of their tenure. Conversion of their status to *ryotwari* vested ownership of the land in the state but gave these cultivators, like their counterparts elsewhere, almost all the trappings of full proprietary ownership.

Yet although seeming to resolve the inadequacies of the 1948 Act, the 1969 legislation opened more areas of legal dispute. Central in this regard was the status of plantations. The original British pioneers into Gudalur during the nineteenth century took ownership of their lands via the *Janmam* system, as lessees to the local *Janmi*. The land tenure fate of these plantations rested with the *Tamil Nadu Gudalur Janmam Estates Act*, and legal dispute over the interpretation of different provisions of the legislation has created uncertainty over their title ever since. On the one hand, one part of the legislation (sections 8, 9 and 10) seemed to indicate that all tenants under the *Janmam* system were entitled to conversion of their land to *ryotwari* status. On the other, another part (section 17) specified that in the case of plantations operating under the *Janmam* system, the state may terminate all rights over land by giving three month's notice, subject to the lessee being compensated and the state acting in 'the public interest'.

In litigation that began in 1970, the planters argued that as existing tenants operating within the *Janmam* system their whole properties should be converted to *ryotwari* tenure and that, accordingly, the provisions of section 17 were redundant and discriminatory. This position was opposed by the government of Tamil Nadu, which supported the differential treatment of plantations vis-à-vis other (smaller) cultivators. Importantly for the central theme of this book, this dispute between the government and the planters took place in the context of considerable population growth in Gudalur from the 1970s fuelled by the migration of landless populations. These pressures led to the landless encroaching upon the non-cultivated (forested) lands of plantation estates. Despite calls by planters for these smallholders to be evicted, the government of Tamil Nadu did not act, evidently not wanting to be party to the forced eviction of otherwise landless small cultivators. The issue eventually reached the Supreme Court in New Delhi in 1999, but to this day remains unresolved. The Supreme Court supported the planters' rights to lodge applications for *ryotwari patta* over their whole lands, but the government of Tamil Nadu has failed to implement the procedures to process such applications.

The upshot of this uncertainty over land tenure has been to create a large population of smallholders who are denied access to a raft of government and commercial services. To gain access to subsidies and programmes operated by the Tea Board and related instrumentalities, cultivators need to show proof of land holding. But with these issues in dispute, government authorities have not issued these documents.

Seen more broadly, this dispute underlines the socially and politically precarious pressures facing the plantation sector. Compared to elsewhere in the world, it can reasonably be said that the South Indian plantation sector operates on the basis of robust (albeit required by legislation) levels of corporate social responsibility. Yet these practices can exist nevertheless within social, economic and environmental contexts of landlessness, protest and poverty. The purpose of the preceding discussion of the legal fissures surrounding the conversion of properties from *Janmam* into *ryotwari* tenure in Gudalur is to highlight the complexities of geographically specific issues affecting livelihoods in the upstream segments of global value chains. In Gudalur, any serious questioning of the ethical basis of production needs to confront unresolved land tenure issues and their ensuing effects on smallholder marginalization and impoverishment. However, the top-down, audit-centric approach of many ethical accountability schemes structurally omits consideration of such deeper questions relating to how and why particular production arrangements come to evolve the ways they do in particular places. To chime again the central argument of this book, these issues point to the relevance of a geographical perspective within value chain analysis, that jointly considers not just the vertical arrangements of production systems, but the institutional arrangements that connect them to specific spaces, places and territories.

## Conclusion

Bringing together the issues raised in this chapter, we contend that the challenge of understanding labour and livelihood issues in the South Indian tea plantation sector cannot rest wholly on upbeat accounts of how ethical accountability schemes are supposedly raising the global safety net, or on despairing ('race-to-the-bottom') accounts of how neo-liberalism and intensified global competition are leading to wage freezes, work intensification and estate abandonment. What needs to be appreciated is the bigger picture of how global processes are simultaneously restructuring geographies of opportunity and marginality within this regional production system.

In the South Indian tea sector, a profound reorganization of production spaces is occurring. The ethical accountability agenda is helping to create a highly variegated production system marked by a dichotomy between compliant estate companies able to meet new standards imposed along global value chains, and a vulnerable mass of others in various states of crisis-ridden restructuring. In other words, the additional layer of ethical accountability may be enshrining further advantage to those producers already better enabled to compete in the global marketplace (as either well-managed corporate plantations or donor-assisted smallholder organizations). In general, these producers can more easily fulfil the obligatory 'ticks' that international audit systems require and thus achieve ethical certification. In turn, this helps generate new geographies of inclusion and exclusion amongst regional producers. The ethical accountability agenda scrutinizes and gives voice to workers engaged in productive enterprises that supply affluent western consumers, whilst remaining essentially uninterested in the fate of those who lie outside its umbrella of concern. In this vein, the pursuit of ethical accountability in South Indian tea districts accords to what Mutersbaugh (2005b) terms 'just-in space' production, namely, islands of global engagement and relative privilege set amidst the restructuring and crisis of 'non-linked' agrarian participants. Reflecting on comparable issues and themes in the case of Ghanaian perennial crops, Fold (2008) paints a similar picture of highly uneven ethical accountabilities.

Recalling Chapter 3, the developed country markets of Western Europe, Japan and North America are the destinations for less than one-quarter of the world's internationally traded tea. In South India, the proportion of production destined for these markets is less than 5 per cent. If we accept that the ethical accountability agenda has minimal purchase in markets outside of these developed country contexts, then its limitations become clear. It needs to be recognized that ethical accountability agendas are frequently developed with the sole and directed purpose to ensure consumers in affluent western countries may rest easy with the knowledge that the tea or coffee they drink is produced in accordance with defined ethical benchmarks. This is wholly appropriate but, as also noted earlier, the agenda has little to say with regards to production arenas not directly upstream from the cups, mouths and stomachs of western consumers. To link social justice to the moral priorities of western consumers is to forget or ignore these already marginalized producers. This is the case whether we are talking about uncertified smallholders, people living on abandoned estates in Central Travancore, or encroachers in Gudalur.

Whilst the goal of enshrining ethical values in global value chains is certainly worth pursuing, a geographical perspective on this issue requires that we do not lose sight of how such initiatives are situated in their broader industry and regional contexts. From a development agency perspective, the arguments of this chapter suggest that 'jumping on the ethical bandwagon' will generate patchy outcomes at a regional level. Benefits may flow to some participants, but with no guarantee that this will include those most in need. In short, the goal of ethicality in value chains should not be confused with the goal of social justice within regional production systems.

# Chapter Six

# Struggles over Environmental Governance in the Coffee Forests of Kodagu

The coffee forests of Kodagu are unique. Not only is the district India's leading producer of coffee, it is also a storehouse of threatened biodiversity, providing habitat for such flagship species as tigers, elephants, bison and leopards, as well as home to abundant birdlife. The relationship between coffee production and environmental management in Kodagu, however, is not at all straightforward. Since 1977, the total area of land planted with coffee has expanded significantly, mostly at the expense of medium-elevation wet evergreen forest on private holdings (Garcia et al., 2007 and Figure 6.1). Notwithstanding changes in local landuse patterns, coffee planters in Kodagu have traditionally retained native trees for shade on their plantations, such that these cultivated areas, together with formal protected areas and community-managed sacred groves, constitute a mostly contiguous forested landscape, which in total covers nearly three-quarters of the district. The integrity of the biodiversity values of Kodagu is thus dependent on the enrolment of coffee planters as environmental stewards. Evidently, however, coffee planters' capacities and attitudes to respond to conservation imperatives are intimately linked to global value chain dynamics. Positioned at the upstream end of the global value chain for coffee, planters are beholden to many of the governance structures we described in Chapter 1. Economic and regulatory signals transmitted from lead firms upstream to coffee planters affect conservation incentives on estates. In this chapter we review how these processes are presently unfolding in Kodagu. This discussion has the dual purpose of informing: (i) contemporary policy debates on conservation strategies for the district; and (ii) the more general set of arguments about the implications for environmental governance structures within the changing global value chain for coffee. In doing so, we highlight the complex – and context-specific – set of relationships and institutions

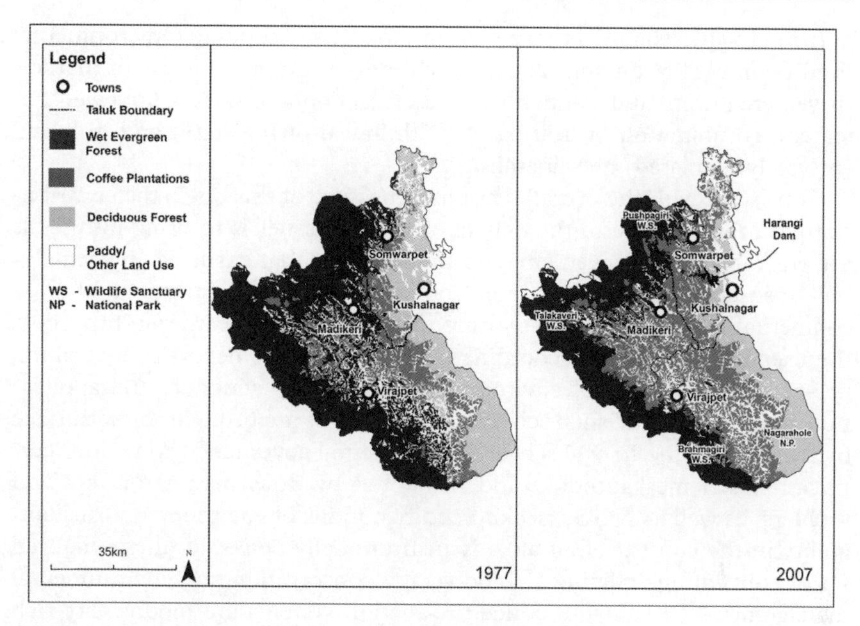


**Figure 6.1** Landuse map and coffee expansion in Kodagu, 1977–2007.
*Source*: Modified by the authors from maps provided by K.M. Nanaya and the French Institute of Pondicherry.

that bind together concerns over brand management, global certification schemes and environmental outcomes.

The background for our discussion of these issues is provided by the recent development of a plethora of *global private regulation* schemes seeking to document and authenticate the environmental and social bases for coffee production (Giovannucci and Ponte, 2005; Kolk, 2005; Macdonald, 2007; Muradian and Pelupessy, 2005; Neilson and Pritchard, 2007b; Raynolds et al., 2007b; Renard, 2005). As introduced in Chapter 1, the concept of global private regulation generally relates to the development and enforcement of rules and standards on upstream producers by downstream private sector actors. Whereas Chapter 5 focused on trends towards social and ethical accountability in the global value chain for tea, this chapter takes a sharper look at environmental governance. In coffee, economically dominant retailers and roasters have aggressively pursued audit-based environmental accountability schemes during the past decade. Frequently, these schemes have been developed and enacted through an entanglement of corporate and NGO interests, reflective of the notion of the 'NGO–industrial complex' described by Gereffi et al. (2001). In a broader sense, moreover, the leading agri-food researcher Harriet Friedmann describes the convergence of environmental and social politics, and the retail-led reorganization of agri-food value chains, in terms of an emergent 'corporate-environmental food regime' (Friedmann, 2005). Such ideas presuppose a rise of market-driven environmental standards and certification alongside a hollowed-out state: a combination of 'roll-back' (withdrawal of the state) and 'roll-out' (privately regulated) neo-liberalism.

Consistent with the overall tenor and message of this book, the contribution we seek to make to the debate over these issues is to bring into focus the *geographies* of engagement, and the struggles that ensue when such globally coordinated initiatives intersect with embedded institutions of environmental management. Reiterating themes expressed earlier in this book, the execution of globally coordinated schemes never occurs in a vacuum; pre-existing institutional environments necessarily configure the application and outcomes of such schemes. These issues are brought to the surface by focusing on the struggles over environmental governance, as various certification schemes authored and promoted by downstream roasters and retailers, as well as NGOs, seek to codify notions of environmental sustainability in the context of an already institutionally dense, locally prefigured set of conventions relating to biodiversity conservation and environmental management. In Kodagu, coffee production systems are biodiversity-rich and institutionally complex. As with our parallel discussion of ethical accountability in South Indian tea production (Chapter 5), what interests us in this chapter is not just the dynamics of how environmental accountability is implemented *within* coffee value chains, but the broader question of how this agenda plays out at a regional scale: what its significance is in terms of shaping overall environmental outcomes in Kodagu.

We do this as follows. In the first major section of the chapter, we introduce and critically review product certification systems in the global coffee industry. This discussion explains the wider international context for contemporary struggles over environmental governance in Kodagu. Then, we focus in on the situated realities of environmental management in Kodagu. This ensuing analysis sets out the complex relations through which coffee production is inserted within systems of land tenure and environmental conservation in Kodagu, and then evaluates the issues confronting the region with regard to the incipient incursions of roaster/retailer-led environmental certification schemes. As such, the analysis positions these schemes within an appreciation of wider conservation strategies for the district. This approach reveals the limited capacity for private environmental governance schemes orchestrated by downstream retailers and roasters to induce positive environmental change in Kodagu. They are both limited in their reach within Kodagu and, in their present guise at least, ill-suited to the grounded realities of the district's conservation needs. We make the point that such schemes initially emerged from an attempt to further a particular conservation agenda in Latin America by providing financial incentives to farmers to maintain the habitat services provided by their plantations. This very context-specific mechanism has since been scaled up, adopted by corporate coffee companies, expanded and adjusted to incorporate a variety of coffee-growing environments around the world. The problem, as we argue, is that their roll out has not always been sufficiently encompassing to the specific conservation issues facing individual coffee-growing regions. We conclude, therefore, by asserting that the potential of global private regulation to effect positive environmental change in regional production complexes such as Kodagu is ultimately limited. Such an outcome is a function of the struggles between local environmental needs and governance systems and the needs of lead firms elsewhere in the value chain.

## Product Certification in the Coffee Industry

The growth in global coffee production over the last 30 years has inevitably occurred through either an expanded area under cultivation or the intensification of production on existing plantations. In a general sense, this has either implied expansion of coffee cultivation into new areas (often into previously uncultivated tropical forests) or the adoption of high-input, intensive coffee production (commonly with a corresponding loss of on-farm shade cover). Consequently, coffee-related deforestation has occurred in all of the major producing regions across Central and South America, Asia and Africa. Originally an understorey species from the East African highlands, coffee is now cultivated within a great diversity of agro-ecological systems. These range from multi-strata systems where coffee has been introduced as an understorey plant beneath an otherwise 'natural' forest canopy; to coffee planted amongst sparse scatterings of mono-shade (single species of shade tree); to mixed cropping systems where coffee is grown alongside other crops such as cocoa, citrus or pepper, and to systems with a total absence of shade cover (so-called 'sun coffee'). In many instances, coffee agro-forests function as buffer zones around protected forest areas, performing important hydrological functions, and sometimes the coffee plantations themselves constitute important wildlife habitat (as is the case in India).

Growing consumer awareness of the role played by multi-strata coffee agroforests in biodiversity conservation was motivated by research performed at the Smithsonian Institution in Washington DC (refer to Rice and Ward, 1996), which specifically linked the intensification of coffee production in Latin America with habitat loss for migratory bird species in the USA.[1] The habitat value of shade-grown coffee plantations can be considerable in parts of Central America, such as in El Salvador, where the area planted with shade coffee is ten times larger than all protected areas (Varangis et al., 2003). The intensification, or 'technification', of coffee production in Central America is generally associated with gradual removal of shade cover and an increase in fertilizer application and sometimes irrigation. Recognition of the habitat value of traditional coffee systems, and the threat imposed by their conversion to intensified systems, led to efforts to promote environmentally-friendly coffee production through value chain mechanisms providing economic incentives for coffee farmers. In essence, such mechanisms attempt to provide market-driven Payments for Environmental Services (or PES) to coffee farmers for providing an ecosystem service (in this case, habitat protection[2]) valued by coffee consumers in key markets.[3] What is a PES? Sven Wunder (2007) defines PES schemes as: (i) being a voluntary transaction; (ii) which provides a well-defined environmental service; (iii) which is bought by a buyer; (iv) from a provider; and (v) is conditional on the continuous provision of that service. The supply chain schemes to promote habitat protection in Central America would appear to satisfy this definition.

The implementation of PES schemes was performed initially through certification arrangements such as Smithsonian Bird-Friendly coffee (shade-grown) and Eco-OK (later the Rainforest Alliance), which had very specific environmental objectives in the context of coffee 'technification' (Varangis et al., 2003). The background, and therefore motivations, of the conservation organizations involved in such schemes provides important insights into the rationale for their promotion of coffee certification. For instance, the New York-based Rainforest Alliance was established in 1987, and has been principally involved in the Sustainable Agriculture Network (SAN) programme (formerly known as ECO-OK). Its principal mission is to: '... work to conserve biodiversity and ensure sustainable livelihoods by transforming land-use practices, business practices and consumer behaviour' (Rainforest Alliance, 2007a: not paginated).

With conservation activities focused in Central America, it is perhaps not surprising that this region continues to dominate the list of Rainforest Alliance-certified coffee producers (although this is changing rapidly as the programme expands its activities globally). The first shipment of coffee certified by the Rainforest Alliance to the USA took place in 1996 from a farm south of Guatemala City (Rainforest Alliance, 2007a). Of the 274 producers certified under the Rainforest Alliance in late 2007, only six are located outside the Americas (three producers in Ethiopia, two in Indonesia and one in Tanzania), with a large concentration found within Colombia and El Salvador (Rainforest Alliance, 2007a). It is important to point out that virtually all of these certified producers are Arabica growers, as there is apparently little current market interest in certified Robusta coffee. By 2007, 160,000 hectares of coffee-growing areas were certified for sustainable farming by the Rainforest Alliance, representing a massive 90 per cent increase in the certified area in 2006 (Rainforest Alliance, 2007a). From small beginnings, the Rainforest Alliance has scaled up to be one of the dominant labels of sustainable coffee certification in the industry. As Raynolds et al. (2007b) suggest, this is on account, in no small measure, to the fact that the organization is increasingly viewed as delivering a 'business-friendly' mode of certification. Indeed, the commitment of Kraft foods to purchase Rainforest Alliance certified coffee for one of its leading US brands, Yuban, has now brought the seal squarely into the coffee mainstream (Rainforest Alliance, 2006).

A smaller 'eco-friendly' coffee programme is that administered by the Smithsonian Institution, with 6,525 hectares of 'Bird-Friendly' coffee certified in November 2007, all in Latin America (and 85 per cent in Mexico and Peru alone). The Smithsonian Institution was established in 1846, following a bequest to the US government from the British scientist James Smithson to be used 'for the increase and diffusion of knowledge among men'. The institution is now the world's largest museum complex comprised of various research centres, natural history collections and libraries and effectively operates as a quasi-government agency. One such research centre within the Institution is the Smithsonian Migratory Bird Center, which has developed a certification scheme for 'Bird-Friendly' coffee from farms in Latin America that 'provide good, forest-like habitat for birds rather than being grown on land that has been cleared of all other vegetation' (SMBC, 2007). Unlike Rainforest Alliance specifications, Smithsonian bird-friendly coffee must also be organic-certified. The establishment of criteria for 'bird-friendly' coffee was developed by the Institution following a period of intensive ecological research and is highly specific to the conservation requirements of coffee systems in the Western hemisphere's tropics and sub-tropics.

Both Rainforest Alliance and Smithsonian Bird-friendly coffee were initially established with the specific objective of encouraging habitat protection in Central America through a certification-based PES scheme linking producers and consumers. To reiterate, the key point of these schemes was that coffee growers would be rewarded financially for their role in habitat conservation. Whilst initially, biodiversity-friendly coffee was generally associated with isolated niche markets, mainstream coffee companies such as Kraft, McDonalds, UCC and Tchibo are now offering Rainforest Alliance certified coffees, thereby significantly widening the potential market for coffee PES schemes. The success of eco-friendly coffees is premised on a market demand for habitat protection amongst coffee consumers, a clear environmental service provided by coffee producers, and a reliable mechanism (certification) which ensures that price premiums are conditional upon service provision. Yet as PES coffees have become more mainstream, the field of 'sustainable coffees' as a whole has become more crowded, due to the arrival of an array of corporate sustainability codes. Unlike PES arrangements, these codes have the prime function of providing defensive brand management for corporate interests. We now turn to these schemes.

## The rise of corporate sustainability codes in the coffee sector

During the past decade, PES-like schemes coalesced with calls for social equity and fair trade in the global coffee sector, and have been increasingly subsumed within broader doctrines of corporate social responsibility (CSR). Generally speaking, there now exists a strong business case for the corporate self-regulation of environmental performance, built around an argument of enhanced shareholder value through effective risk management (Brown and Fraser, 2006, provide an overview of the Business Studies literature on this subject). Reflecting this business case for CSR, the second half of the 1990s saw the dramatic rise of a CSR agenda across the corporate world, as companies actively sought to engage NGOs, consultants and civil society organizations to improve their ethical credentials. For many observers, the coalition of corporate business interests, non-government organizations and private certifying agencies has been accepted as a cost-effective means for reining in unacceptable behaviour and improving social and environmental conditions amongst workers and supplier communities. In the case of coffee, exposure of the environmental implications of intensified production, deforestation, and the very promotion of eco-friendly coffee (through the PES schemes described above) left major corporate brands exposed to allegations of neglect.

There are clearly considerable corporate benefits to be gained through pre-empting formal regulation and freeing business from government intervention. For multinational corporations with activities extending deep into developing countries, private regulation also provides a valuable defence against allegations of social and environmental abuses in regions of the world where the regulatory capacity of states can be limited. Even in developed countries, it is not uncommon for private sector standards to now exceed in strictness those being enforced by the state, as evidenced in sectoral issues such as food safety (Konefal et al., 2005; OECD, 2006), thereby challenging established state-led regulatory structures.

The expanding influence and reach of 'voluntary' corporate codes reflect strategic business responses to a growing concern over the need to protect a firm's 'reputational capital' and operational legitimacy (Angel and Rock, 2005). Graham and Woods (2006) identify changes in corporate behaviour towards an embrace of CSR as being driven by diverse pressures, including risk management, ethical investment pressures, consumers and activists, and also by the desire to retain and attract quality employees. Consequently,

Werther and Chandler (2005) argue that the risk to reputation is strongest when companies rely heavily on the value of their brand. For many corporate leaders, the potential risk to brand reputation is the leading driver of upstream supply self-regulation. The confluence of high brand stakes and the exposure of 'sweatshop' conditions among upstream suppliers of major fashion labels led to the early emergence of CSR in the global apparel industry. Initial public outcry and campaigns against Nike, The Gap and other leading fashion labels, encouraged the promulgation of corporate responses pegged to a belief in self-regulation through Codes of Conduct (O'Rourke, 2005). Clearly, in the eyes of a discerning public, outsourcing did not absolve companies from meeting social and worker obligations amongst their production base, resulting in the, then, new concept of supply chain accountability. The importance of brand assets in the coffee industry has seen a similar rush, by global roasting firms and retailers, to establish ethical credibility in the marketplace. As described by Macdonald (2007: p. 807), the Oxfam coffee campaign specifically targeted the brand exposure of major roasting firms:

> Activists have also attempted to transform the supply chain practices of major roasters by building non-state mechanisms of accountability, in which market-based sanctioning mechanisms are deployed as means of enforcing activist demands for increased corporate responsibility within global supply chains.

In this regard, the emergence of global private regulation (stimulated by corporate strategies of defensive brand management) in the coffee industry can be viewed as a direct consequence of NGO pressures. By encouraging agendas of corporate social responsibility, the Oxfam coffee campaign both prompted and legitimized corporate regulation of the supply chain. As a consequence, the original conceptualization of product certification (that is, a system to generate payments for environmental services undertaken by growers) has been besieged subsequently by a corporate agenda of defensive brand management. The end-result has been a series of baseline standards for acceptable practices related to coffee production. In a previous paper (Neilson and Pritchard, 2007b), we described the mechanics of corporate engagement with sustainable coffee agendas as occurring via three overlapping sets of processes: (i) the corporate adoption of NGO-certified 'niche' product lines; (ii) the development of firm-specific corporate codes of conduct; and (iii) an agenda to work towards collective baseline standards for the industry as a whole. According to this framework, the initial adoption of organic, fair-trade and biodiversity-friendly product lines has gradually given way to the development of firm-specific and industry-wide agreements. In the coffee sector, key industry initiatives now include: the

**Table 6.1** Stated objectives of various 'Sustainability' codes in the coffee sector

| *Label* | *Geographical range* | *Year established* | *Stated objective* |
|---|---|---|---|
| Rainforest Alliance/ Eco-OK | Primarily Latin America | 1996 | 'Certification is a conservation approach that the Rainforest Alliance uses to award growers who meet our guidelines for sustainability.' (www.rainforest-alliance.org) |
| Utz Kapeh | Global | 1997 | 'With UTZ CERTIFIED, your favourite coffee brand knows exactly where their coffee comes from and that it was grown responsibly.' |
| Smithsonian 'Bird Friendly®' coffee program | Latin America | 2000 | '"Bird Friendly®" coffee is coffee that comes from farms in Latin America that provide good, forest-like habitat for birds.' (www.nationalzoo.si.edu) |
| Starbucks CAFÉ Practices | Global | Guidelines introduced in 2001 (operational in 2004) | 'These guidelines are designed to help us work with coffee farmers to ensure high-quality coffee and promote equitable relationships with farmers, workers and communities, as well as protect the environment.' |
| Global-gap Code for Green Coffee | Global | 2003 (Eurep-GAP) | 'The Global-GAP standard is primarily designed to reassure consumers about how food is produced on the farm.' (www.globalgap.org) |
| Common Code for the Coffee Community (4C) | Global | 2005 | 'The objective is to achieve a baseline level of social, ecological and economic sustainability for all types of coffee, all coffee production systems and all coffee producers.' |

*Sources*: As specified.

Eurep protocols for Good Agricultural Practice (Eurep-GAP, known since September 2007 as Global-GAP); Utz Kapeh certification; the 'Common Code for the Coffee Community' (widely known as '4C'); and Starbucks' CAFÉ Practices (Coffee And Farmer Equity) Program. All of these codes attempt, in some way, to address the environmental conditions of coffee production. The volume of such coffee has rapidly increased in recent years. In 2007, for example, the volume of Utz Kapeh certified coffee on the market reached 750,000 bags (45,000 tonnes), mostly from Brazil, Columbia and Vietnam (Utz Kapeh, 2007), reinforcing its image as a mainstream, and easily accessible, mode of coffee certification.

As summarized in Table 6.1, the stated objectives of the major coffee codes suggest a broad, but often loosely defined, notion of sustainability. Indeed, it could be argued that the timorous ways that 'sustainability' is enshrined within codes suggest it has as much to do with the sustained supply of coffee beans to their roasting plants, than the sustenance of complex coffee ecosystems in producing countries. Reflecting the rising importance of brand management, following the release of its code for green coffee code in 2002, a press release for Eurep-GAP claimed that: 'The Eurep-GAP Coffee Code responds to two key needs for coffee brands: first, a baseline for defining good agricultural practices and for providing recognition to good producers, and second, a trustworthy system for farm-to-fork accountability' (Eurep-GAP, 2003).

Furthermore, the motivations for the code were explained as:

> In 1998, Ahold Coffee Company in the Netherlands looked to Eurep-GAP as a way to address assurance issues in origin. As the supplier of coffee to the Ahold retail companies in Europe, Ahold Coffee Company saw that it was only a matter of time before the attention of retailers, consumers and advocacy groups turned to coffee.

The Global-GAP code maintains a checklist and control-point approach, with many similarities to food safety systems such as HACCP. Clearly, a driving objective of the Global-GAP code for coffee (and allied systems such as Utz Kapeh) is to protect branded coffee from consumer allegations of environmental and social negligence. As this occurs, the links between certification schemes and incentives for farmers to contribute to biodiversity conservation are becoming increasingly tenuous. Instead, certification is driven by a motivation to appease the ethical consciences of affluent consumers and maintain brand credibility. As explained by Raynolds et al. (2007b: p. 149): 'Given recent NGO and media reports of severe social and environmental abuses involved in production in the global South, wary Northern consumers rely increasingly on new certification initiatives to ensure they are not unwittingly supporting such abuses.' Similarly, the 4C initiative: '... aims at preparing the entire coffee sector for the increasing demand from consumers worldwide for an all-round good coffee quality. This quality refers not only to the product itself, but also to sustainable production and trading methods' (CCCCA, 2008).

The compliance requirements of such certification are deliberately set to flexibly accommodate various production contexts. The 4C code, for example, states:

> The set of practices which are used in the coffee sector guides participants on the way towards a more sustainable production, post-harvest processing and trading of coffee. Through its flexible design it has a moderate entry level and combines this with the commitment to continuous improvement. Like this it is easier accessible for all actors but makes sure that the road towards sustainability is entered. (CCCCA, 2008)

The very inclusiveness of these schemes has important implications for their ability to generate positive environmental outcomes in producing regions. Whilst it is not necessarily the intention of corporate-driven codes to operate as a PES scheme in coffee-producing communities, the way these schemes are packaged and presented to consumers certainly mimics the PES schemes which preceded them. However, their aspirations and motivations are distinctly different. The broad nature of compliance criteria means that such schemes are more concerned with preventing gross environmental abuses than with improving the habitat value of coffee agro-forests. Notwithstanding claims by the authors of corporate sustainability codes that their codes augment (rather than directly compete with) pre-existing certification mechanisms such as fair-trade or eco-friendly coffee, this distinction for consumers is often lost. Moreover, Raynolds et al. (2007b: p. 147) argue that: 'certifications that seek to raise ecological and social expectations are likely to be increasingly challenged by those that seek to simply uphold current standards'.

Corporate sustainability codes thus offer a win–win outcome for their authors. They muddy the waters among environmental claims, thus diminishing the moral high-ground initially captured by PES schemes. At the same time, however, they represent a backstop risk management tool enabling corporations to protect themselves against claims of gross environmental negligence.

Further, as schemes are rolled out, coffee producers are now finding that conformance to the requirements of Utz Kapeh, Global-GAP, CAFÉ Practices or (in the imminent future) 4C is becoming a mandatory requirement to sell green beans to certain buyers and markets. The extent to which producer compliance is voluntary is a highly contested, and sometimes bitterly fought, aspect within the emerging arena of global private regulation.

The ability of branded manufacturers and retailers to enforce producer compliance to externally authored codes of conduct and environmental standards is built upon the increasing 'buyer-drivenness' of the global value chain for coffee (Chapter 4). As a result, then, compliance is by no means voluntary, but has instead become a precondition for involvement within global value chains (remembering also that the voluntary nature of a scheme was one of five criteria identified by Wunder, 2007, to be the principles of PES). The remainder of this chapter addresses the implications of buyer-driven environmental regulation in the global value chain for coffee on environmental governance systems within the district of Kodagu.

## The Coffee Forests of Kodagu

Some time after its initial introduction to India in the seventeenth century, coffee cultivation was embraced by the Kodavas, also known as Coorgs, who are the dominant community within the modern-day district of Kodagu in Western Karnataka. Today, Kodagu is the single most important site for coffee production in India, contributing 28 per cent of national production in 2006–07 (Table 3.7). Kodagu is highly rural, with 86 per cent of its total population (estimated at 548,561 in the 2001 Census) living outside Madikeri, the major town in the district (Directorate of Census Operations, undated). Settlements in Kodagu are not generally comprised of condensed village units, as is common in many other parts of rural India (such as the tea-growing Badagas of the Nilgiris, described in the following chapter), but, instead, *Okka* family groups are scattered across agricultural and forested holdings, where traditional *ainemane* houses form focal meeting points within the forest landscape (Figure 6.2). The district is thickly forested, with an estimated 46 per cent of its area covered by natural forest (encompassing reserve forest, conservation areas and privately managed natural forest) and an additional 29 per cent under shade-grown coffee plantations (Moppert, 2000). Figure 6.1 presents the landscape mosaic of Kodagu, showing the expanding areas under coffee cultivation (which has increased substantially since 1977), amidst forests and protected areas. Whilst there are a number of large corporate coffee estates in Kodagu with holding size greater than 100 ha (such as the extensive holdings of Tata Coffee based in South Kodagu), the vast majority of land under coffee is cultivated by smallholders (with less than 10 ha) (Table 6.2).

The Kodavas are the culturally dominant group and principal landowners within Kodagu district, despite only constituting a minority of the district's total population. As a result, Kodava customs and traditions have historically shaped, and continue to shape, landscape evolution in the district. The Kodavas consider themselves 'sons of the soil', along with the various

**Figure 6.2** Ainemane house, Kodagu.
*Source*: Photo by the authors.

**Table 6.2** Distribution of coffee farm holdings in India

| *Farm size* | *Number of farms* | *Area (%)* | *Production (%)* |
|---|---|---|---|
| <4 ha | 164,758 | 53 | 60 |
| 4–10 ha | 10,717 | 18 | |
| 10–60 ha | 2,605 | 16 | 40 |
| >60 ha | 354 | 13 | |

*Source*: Coffee Board of India (2006).

scheduled tribes, such as the *Jenu Kurubas*, *Betta Kurubas*, and *Yeravas*, that make up 8.4 per cent of the Kodagu population (Directorate of Census Operations, undated). According to Srinivas (1965), the Kodava generally consider themselves *Kshatriya* (warrior caste) according to the system of Hindu *Varnas*, and yet their elaborate rituals are closely aligned to forms of nature and ancestor worship, rather than following strict *Vedic* principles. For example, sacred groves known as *devarakadu* (*devara* = God's and *kadu* = forest) are maintained in their natural state amongst the coffee plantations (Chandrakanth et al., 2004; Garcia and Pascal, 2005). Each 'village' usually has at least one *devarakadu,* resulting in approximately one in every 300 ha of land in Kodagu being under this traditional management institution (Kushalappa and Bhagwat, 2001). These forests have considerable spiritual value, as well as moderate ecological and social value, providing minor forest products for the poorer members of Kodagu society (Chandrakanth et al., 2004; Garcia and Pascal, 2005; Bhagwat et al., 2005a). Originally rice growers, coffee cultivation now defines Kodava identity, along with dress and traditional customs. For the most part, modern Kodagu coffee planters still consider themselves to be environmental stewards[4] and, according to a study by Ninan and Sathyapalan (2005) and Ninan (2006), coffee planters in Kodagu show an overall positive attitude to biodiversity conservation on their estates.

The British Colonial government annexed Kodagu in 1834, arresting control from Viraraja Vodeya, the last 'Rajah of Coorg', whose family (the *Lingayat Rajas*) had ruled Kodagu since 1600.[5] The Rajah's deposition in 1834 was justified, by the British, on the grounds of alleged oppression and violence enacted by the Rajahs upon the population of the district. Whatever the case, soon after the annexation large numbers of European planters began settling in the forested mountains to cultivate coffee, dramatically changing the economic and social structures of Kodava society. By 1870, 85,780 acres of Kodagu land was planted with coffee, 65 per cent of which was held by Europeans (Richter, 1870). The 1834 annexation allowed the Colonial government to assume control of the so-called 'waste' lands from the Rajas and, despite the establishment of a Forest Conservancy Department in Kodagu in 1864, much forest land had already been cleared for coffee. Whilst legal forest protection in Kodagu preceded a national law (the *Indian Forest Act* of 1878), its implementation was consistent with the broader tradition of Indian forestry law where the forests of Kodagu were eventually designated as either 'reserve' forest, so-called as they were to be reserved for cultivation, or protected forest (Ribbentrop, 1900). The colonial government reserved the exclusive legal right to release these lands for cultivation by planters.

The cultural dominance of the Kodavas in Kodagu is reflected in patterns of land ownership, as the group is easily the most important landowning community in the district. The system of land tenure in Kodagu is particularly complex, and rights over land and trees continue to be hotly contested. A system of land tenure, known as *Jamma*, was formerly instituted in Kodagu during the times of the *Lingayat Rajas* (Srinivas, 1965). Based on this system, *Jamma* agricultural lands in Kodagu (generally reserved for wet-rice cultivation) were held exclusively by Kodavas as a hereditary right granted by the *Rajas*, and were both indivisible and inalienable (as well as being only lightly taxed). In return, the Kodavas would perform military service as warriors to the *Rajas*. Especially relevant for our discussion here, rights over forest lands adjacent to the cultivated fields were also attached to *Jamma* tenure, and were known as *bane*. *Bane* lands were set aside for the primary purpose of generating a source of green manure for the rice fields and firewood for fuel. As a result, relatively expansive agricultural-forestry estates (the *Jamma* estates) have remained intact across Kodagu, and have not generally been exposed to an open real estate market. By the late nineteenth century, however, non-Kodavas (such as Holeyas and Gaudas) also owned some agricultural land in Kodagu, although rarely as *Jamma* tenure, which remained predominantly a Kodavan institution (Richter, 1870). The complexity of land tenure in Kodagu is such, however, that tree rights on *bane* lands, even those converted to coffee plantations (and unredeemed) remain in the hands of the government (Uthappa, 2004). The same is true for land held as *Jamma Malais* cardamom forests. Whilst these restrictions on tree ownership are a continuing source of tension in the community (*The Hindu*, 2006), they do appear to have helped protect the biodiversity of trees, birds and other biota on the plantations, and have slowed down the replacement of native trees by exotic fast growing shade trees (Satish et al., 2007).[6]

The land on which coffee was cultivated in Kodagu, then, was an amalgam of: (i) large forest tracts, released by the colonial government for European planters; (ii) individually owned *ryots* (known in Kodagu as *Sagu* tenure) cultivated mostly by Gauda and Holeyas; and (iii) the *Jamma* agro-forestry estates (included the accompanying *bane* forests) institutionalized mainly for the Kodavas under the *Lingayat Rajas*. In a very general sense, forested *Jamma* land implied unredeemed tenure, and tree rights on that land passed from the Rajas to the colonial administration to the post-Independence government of India through the Forestry Service of India.

## Conservation and Coffee in Kodagu

Since the early days of coffee planting in India, the role of shade trees on coffee estates has been hotly debated (see, for example, early writings from Kariappa et al., 2004, in their compilation, *Planting Times*) and the merits of natural shade continue to be discussed within the industry today (Jayarama and D'Souza, 2006). As discussed in Chapter 3, Indian coffee is traditionally planted as part of a three-tiered canopy, including a mid-level of nitrogen-fixing *Erythrina* or *Glyricidia*, sometimes intermixed with a secondary canopy of *Grevillia*, and then with a layer of remnant rainforest trees, including species of *Ficus*, *Artocarpus*, and *Terminalia* (Jayarama and D'Souza, 2006; Titus and Pereira, 2006). Such shade is considered an appropriate response to local climatic factors (an extended dry season and strong prevailing winds) with advantages such as improving soil fertility, reducing pest incidence (especially white stem borer), suppressing weed growth and preventing overbearing.

An inherited planter culture of shade maintenance, combined with the restricted rights of land tenure for many *bane*-based coffee plantations and strong cultural beliefs (including maintenance of sacred groves), has resulted in an immense wealth of biodiversity across the coffee forests of Kodagu. Conservation International recognises the Western Ghats of India as one of only 34 global biodiversity 'hotspots' (Conservation International, 2008), based on the definition of an ecological hotspot provided by ecologist Norman Myers (Myers et al., 2000). A survey by Bhagwat et al. (2005b) found that tree, bird and fungal diversity in Kodagu was comparable between coffee plantations and adjacent protected forest and sacred groves. An estimated two-thirds of Kodagu District is covered by forest (Chandrakanth et al., 2004), which includes the coffee plantations, *devarakadu* groves, three Wildlife Sanctuaries (Brahmagiri, Talakaveri and Pushpagiri), and one National Park (Nagarahole). The formal protected areas further constitute site elements within the Talacauvery sub-cluster nomination for World Heritage recognition (UNESCO, 2007). Bhagwat et al. (2005a) call for a landscape approach to conservation in Kodagu, describing the importance of heavily shaded coffee plantations in connecting remnants of native forest in the Western Ghats, and the need to conserve such habitats adjoining protected areas to increase the 'landscape-level connectivity' of patches. Maintaining the ecological integrity of coffee plantations, within a broader landscape of formal protected areas and *devarakadu* groves, is therefore a vital component of wider biodiversity conservation efforts in the region.

An essential part of the ecosystem value of the coffee estates rests with the provision of habitat to India's large flagship wildlife species and, perhaps not surprisingly, human–wildlife conflict is a serious social and economic issue in the district. In densely populated India, tensions between the resource needs of local communities and wildlife objectives are widespread. There can be fundamental conflicts between local communities and the organization entrusted with formal wildlife management in India, the Indian Forest Service. These tensions severely complicate the conventional, exclusionary approach to conservation in India (Mahanty, 2002; Arjunan et al., 2006), highlighting the reality that formal protection (and command-and-control approaches) is likely to remain only one element, albeit an integral one, within a broader approach to conservation in strategic ecological sites such as Kodagu. Conflict between humans and wildlife (notably elephants and, to a lesser extent, tigers) outside the formal protected areas is also considerable. A recent report funded by the Karnataka State Forest Department (Kulkarni et al., 2007) reported 51 human deaths and 51 serious injuries caused by encounters with elephants between 1995 and 2005 in Kodagu, in addition to an average 600 reported cases of elephant-related crop damage each year. The considerable costs incurred due to crop damage are further exacerbated by the increased costs of establishing and maintaining deterrent systems such as electric fences and canal barriers (which, for the most part, remain disappointingly ineffective). Whilst the government operates a compensation scheme for crop damage, many farmers believe these payments are insufficient (Kulkarni et al., 2007). Elephants enter the coffee plantations on a seasonal basis in search of food and water (the availability of which is improved by irrigation and water retention systems on the estates) in response to the ever-shrinking area of remaining natural forest. Due to these concerns, there is currently a strong disincentive against planters maintaining forest-type habitat on their estates if this encourages elephant migrations, crop damage and safety concerns.

Further to the significant habitat function provided by coffee plantations in Kodagu is a critical hydrological function. The headwaters of the Cauvery River, sometimes referred to as the 'Ganges of the South', are located in Kodagu at Talacauvery (Figure 6.1 and Figure 6.3). From Talacauvery, the sacred river flows east through the mid-section of Kodagu, south of Bangalore,

**Figure 6.3** Tala Kauvery, Kodagu.
*Source*: Photo by the authors.

into Tamil Nadu and finally into the Bay of Bengal. The river supplies a hydroelectric plant at Srirangapatnam, near Mysore, and the Thippagondanahalli Reservoir, east of Bangalore, which is the primary source of drinking water for that rapidly growing metropolis. The ecosystem services provided by the coffee forest landscape of Kodagu, therefore, are both varied and of crucial significance for many living outside the district.

These ecological and hydrological functions, however, are currently under threat. Damodaran (2002a) explains how the diversity of shade cover on Indian coffee estates has been significantly reduced since the 1970s as a result of planter attempts to intensify production. The introduction of irrigation technologies on many coffee estates has also reduced the reliance on a dense shade canopy and is associated with declining tree diversity and reduced shade. (Traditionally, Kodagu has stood out amongst the coffee plantation districts of the Western Ghats as exhibiting considerable shade diversity due, in part, to the cultural-historic factors related to tenure systems.) Figure 6.1 also shows how the expansion of coffee cultivation since 1977 in Kodagu has occurred largely at the expense of wet evergreen forest (Garcia et al., 2007). Importantly, this forest conversion has generally taken place on privately owned *bane* lands (*Jamma* estates) and not on land administered by the Indian Forest Service. The historical importance of *bane* land as a source of green manure and fuelwood has declined over time with the introduction of green revolution technologies to rice cultivation and the availability of alternative energy sources. The ecosystem services of the coffee forests of Kodagu, then, are being threatened by the twin tendencies of: (i) expansion of coffee cultivation at the expense of forested areas, albeit on privately held *bane* land; and (ii) the loss of a diversified shade cover due to the intensification of production systems.

## Conservation management through Payments for Ecosystem Services (PES)

A study in South Kodagu by Ninan and Sathyapalan (2005) demonstrates that the opportunity costs of biodiversity conservation on the estates, in terms of coffee benefits foregone, can be significant. Yet presently, there are no widely implemented incentive mechanisms that recompense Kodagu coffee planters and smallholders for the provision of their ecosystem services. Realizing the limitations of formal protected area management in the context of India, conservationists are now looking towards value chain mechanisms and PES schemes to create behavioural incentives for land managers to protect India's threatened wildlife. In Kodagu, a recent initiative funded by the European Commission, known as CAFNET,[7] seeks to redress this gap. The logic behind CAFNET draws a stark distinction with the kind of corporate sustainability codes described earlier in this chapter. CAFNET aims to link the sustainable management of coffee forests with appropriate remuneration for producers through: (i) better access to environmentally conscious product markets; and (ii) direct payments for environmental services. Importantly, CAFNET is investigating PES mechanisms appropriate to specific conservation requirements in Kodagu rather than attempting to import a certification system from elsewhere. A rationale for this activity is the increasing interest in environmental sustainability that now characterizes the global coffee sector.

With regards to the first of CAFNET's aspirations, in recent years, schemes that seek to give producers access to premium product markets, have been heavily promoted by agencies such as the World Bank as a win–win scenario for the environment and poverty reduction.[8] Nevertheless, the capacity of such schemes to effect environmental improvements is still widely debated. Rappole et al. (2003) have argued that any economic incentives through shade certification are generally insufficient to actually encourage farmers to convert existing sun-grown coffee to shade grown. This is due to the productivity losses associated with shade-grown conversion (or losses due to maintaining shade cover). A recent paper by researchers from the Smithsonian Institution (Philpott et al., 2007) similarly suggests that the financial gains of certification (organic and fair-trade) in Chiapas, Mexico are not actually enough to outweigh the costs. The environmental benefits of market-based price premiums for eco-friendly production practices in low-income countries have also been questioned in a paper by Ferraro et al. (2005). Through economic modelling, they show that such indirect premiums are less effective than direct payments made to land managers, thereby questioning whether market incentives are the most efficient use of funding by conservation agencies (due to the relatively high transaction costs associated with administering such programmes).

Furthermore, some authors (Rappole et al., 2003) have argued that endorsement of shade coffee itself represents a 'lowering of the bar' in terms of conservation goals, as shade coffee cannot provide ecosystem services comparable to native forests. The debate here, of course, revolves around what the ecological value of the coffee system is being assessed against. This, however, is much more than simply an academic question and is central to the effectiveness of PES schemes. The environmental services must somehow be measured against a counterfactual baseline, and the difference between this baseline and the measured environmental outcomes is sometimes referred to as the 'additionality' of the scheme (Wunder, 2007). The 'additionality' of introducing a certification program or corporate sustainability code to a coffee-producing region is highly variable depending on the geographical and institutional context of production. Drawing on research in Lampung, Indonesia, O'Brien and Kinnaird (2003) have argued for environmental governance in value chains to be more sensitive to the varied contexts within which coffee is grown globally. For example, the primary conservation threat in Lampung is Robusta expansion into Bukit Barisan National Park, and therefore there is a need to develop mechanisms which encourage intensification of existing low productivity plantations, rather than extending production into the rainforest frontier. In Viet Nam, by contrast, the key environmental issue on the coffee plantations is over-extraction of groundwater (Cheesman et al., 2007; D'haeze et al., 2005 ), such that conformance to responsible extraction practices would constitute an environmental improvement (thus generating environmental additionality). Theoretically, focusing on ensuring financial incentives for environmental improvements in these contexts may entail the certification of sun coffee, which would be abhorrent to advocates of shade cover in Central America. The key message here is that the specific conservation requirements in one coffee region (Central America, for example) cannot be extrapolated to meet the requirements of coffee-related environmental issues elsewhere. As noted by Dankers (2003: p. 89): 'Standards developed in one particular country or geographical area may discriminate against producers of other countries or areas if they do not take into account different local conditions.' Understanding this notion of additionality is central to assessing environmental benefits and the varying counterfactual baselines across the coffee-growing world presents a strong argument against the standardization of sustainability schemes. Indeed, it is also a central problematic in the current push to develop market mechanisms for the reduction of carbon emissions, especially in respect to payments for foregone opportunities, or payments not to clear a tract a forest.

In this conceptual context, the unique institutional context of conservation management issues within Kodagu, including *Jamma* tenure arrangements, cultural attitudes towards biodiversity conservation, an inherited planter culture of shade management, forest area management, and the specific issues of human–animal conflict (and associated compensation schemes) all emphasize the importance of place-based conservation strategies. It is highly unlikely that, from a conservation perspective, the promotion of a certification scheme designed elsewhere will be able to respond effectively to the specific requirements of this institutional environment. The ability of a baseline corporate sustainability code to generate additionality is even less likely. In respect to shade cover, for example, the requirements of certification schemes such as Utz Kapeh and 4C are certainly not particularly onerous. According to Control Point IIb.7 of the Utz Kapeh code, 'If the producer uses shade trees on his coffee planting he should preferably use native tree species', and as a recommendation, Point 11.B.8 suggests 'the producer preferably uses native tree species to plant within and around his coffee plantings to provide fruit, timber and wildlife habitat'.

Similarly, to achieve a green light ranking in the biodiversity category of the 4C code, it is sufficient to ensure that 'native flora including watersheds and biodiversity habitats are protected and enhanced'. Details of how this should be implemented are not provided. This contrasts with the strict shade specifications (relevant to the Central American context) of systems such as Smithsonian Bird-friendly coffee which requires a minimum 40 per cent canopy cover, including less than 5 per cent of the canopy consisting of 'unacceptable' species such as *Erythrina*, *Grevillea* and *Gliricidia*; at least ten different tree species in addition to the backbone species, and the presence of epiphytes (SMBC, undated). Similarly, Rainforest Alliance certification requires more than 12 different species (and a total of at least 70 trees) per hectare, as well as 40 per cent minimum canopy cover (Philpott et al., 2007).

The basis for a more geographically sensitive approach to conservation issues is, however, emerging in Kodagu. In 2005, Kodagu joined the International Model Forest Network (IMFN), a global partnership (with headquarters in Ottawa, Canada) aimed at developing a participatory approach to sustainable forest-based landscape management. The Kodagu Model Forest Trust (KMFT) has been established to facilitate a voluntary, partnership-based approach to managing the entire Kodagu landscape, encompassing protected areas, coffee plantations, wet-rice agriculture and riverine ecosystems. The active involvement of coffee planters will be integral to the success of the KMFT. Attempts are being made to address issues related to land tenure, shade tree management, human–wildlife conflict, appropriate remuneration for producers, and payments for environmental services. The clear point to be made here about initiatives such as the KMFT (and those of other local NGOs such as the Coorg Wildlife Society) is their explicitly landscape approach to sustainability, which identifies priorities and strategic action plans against geographically informed criteria specific to the Kodagu context (as opposed to the generic checklist approach of corporate sustainability codes instigated by external authors). In any case, the imposition of an externally dictated management regime on the planter is likely to receive only cool support. The perceived loss of local planter control to global environmental dictates may, counterproductively, undermine local social institutions leading to declining stewardship sentiments.

If value chain certification mechanisms are to promote improved environmental practices in Kodagu, then a specific locally contextualized, producer-informed *modus operandi* is required. It should be anticipated, however, that the establishment of such a producer-driven mechanism is likely to be resisted by downstream lead firms. Furthermore, PES certification schemes require a consumer willingness to pay for the environmental services being provided by the coffee estates. Globally, it appears that consumers are generally willing to pay a biodiversity premium for speciality coffees, but not necessarily for commercial grades. As a result, the market for biodiversity-friendly Arabica coffee is greater than that for Robusta coffee. (Although it is acknowledged that India is one of the few producers of speciality Robustas and so certified Indian Robusta may have a greater market potential than many other producing regions.) Furthermore, the biological requirements of Robusta coffee are such that it is commonly grown under a more open shade canopy than Arabica. With 73 per cent of Kodagu coffee production comprising Robusta coffee, and with less than 4 per cent of Indian exports considered 'Speciality Coffees', the market potential of certified eco-friendly coffee from Kodagu certainly needs to be assessed in greater detail. The current concentration of biodiversity-friendly certified coffees in Latin America reflects its proximity, and relationship, to the North American market. In contrast, Indian coffee is currently sold predominately into the markets of southern Europe and the former Soviet bloc (as presented in Chapter 4), where consumer interest in paying for ecosystem services along with their coffee is currently limited. Similarly, the considerable domestic consumer market within India is not, as yet, overtly interested in the environmental attributes of coffee products.

## Conclusion

This chapter has presented the emergence of two distinct modes of environmental regulation along the value chain for coffee. The first mode is the introduction of PES schemes which provide payments for coffee farmers to maintain the habitat value of their estates. This is a relationship between consumers and producers mediated by a certification mechanism. The second mode of certification is that promoted by branded coffee interests, which enables value chain regulation and accountability. Conformance to corporate sustainability codes is evolving as a requirement for market access and entails a substantial shift in governance patterns. Due to their inherently global scope, initiatives such as Utz Kapeh, 4C, and CAFÉ Practices, rely on a checklist approach to social and environmental responsibility. Suppliers (producers) are audited against compliance to specific criteria in an attempt to encourage improved environmental performance. The explicit aim of 4C, for example, is to eliminate unacceptable practices through the establishment of baseline social and ecological standards. In this way, corporate Codes of Conduct may help address the worst forms of environmental abuse, but in places such as Kodagu, where production practices are likely to already meet these minimum standards, their capacity to induce positive environmental change is limited Whereas the primary aim of earlier 'bird-friendly', 'shade-grown' and otherwise ecologically sound coffees was to reward producers for environmental services, the recent wave of corporate sustainability coffee codes may actually threaten the ability of producers to generate economic rents through product differentiation (Mutersbaugh, 2005a). The more accessible standards present consumers with an assurance of acceptable environmental performance and, in doing so, threaten to undermine the market for eco-friendly coffees attempting to deliver PES schemes. Most existing Kodagu plantations (even those with a relative mono-shade of silver-oak) would probably satisfy the environmental requirements of those generic codes of conduct designed to ensure the protection of brand assets, such as Utz Kapeh and 4C. As a result, such certification would be unlikely to result in any real benefits for biodiversity conservation in Kodagu; and in fact, might be counter-productive if these externally-authored schemes earn the ire of the Kodagu planter class.

Generally speaking, the Indian coffee industry has been a vocal critic of global attempts to link certification and traceability with market access (Menon, 2005; Venkatachalam, 2005; Neilson and Pritchard, 2007b). The fundamental objection is to the imposition of environmental and social standards by outsiders, often framed as neo-colonial imperialism. This highlights a complex set of issues regarding environmental and social governance within a regime of global private regulation. Global regulatory shifts in agricultural trade are fundamentally changing the way environmental issues are being addressed by farmers, NGOs, governments and other stakeholders. Concerns over 'green protectionism' as a Technical Barrier to Trade (TBT) have hitherto limited the ability of the World Trade Organization (WTO) to include environmental standards within bilateral and multilateral trade agreements. The awkward position of environmental standards within international trade negotiations, however, is increasingly being resolved through the rise of corporate self-regulation motivated by brand management concerns in consuming markets.

The implications of increasing global value chain regulation of environmental governance are profound, and parallel Swyngedouw's (2000) exploration of the disempowering effects of 'glocalization' through a rescaling of governance structures and the 'hollowing out' of the state. Indeed, the emergence of 'post-sovereign' environmental governance (Karkkainen, 2004) has been presented as arising from the limitations of top-down, territorially defined state structures. However, in many instances, market-based environmental instruments have been added to, rather than supplanted, traditional governance structures, leading to complex and sometimes conflicting practices 'on the ground' (Meadowcroft, 2002). The present situation of environmental governance in the global coffee industry can be conceptualized as a scattered assortment of partial regulatory forms, articulated through both state and non-state actors. As argued by Macdonald (2007), control over key outcomes has become highly fragmented with a 'structural disconnect' between distributions of responsibility and control.

Similarly, Raynolds et al. (2007b: p. 147) reminds us that: 'The vulnerability of these initiatives to market pressures highlights the need for private regulation to work in tandem with public regulation in enhancing social and environmental sustainability.'

It would seem that an extra-territorial system of environmental governance, orchestrated by downstream coffee roasters under the umbrella of schemes like 4C, further risks divorcing environmental management from the place-specific contexts of local agro-ecological problems. As argued by Lockie and Goodman (2006), 'duty of care' approaches towards environmental management by landholders in Australia are in sharp contrast to the value chain monitoring/audit approaches employed (for instance) by the Fairtrade movement. The rise of corporate self-regulation has far-reaching implications for global environmental governance more broadly. Associated with the 'hollowing out' of the state, CSR initiatives are increasingly defining the boundaries of acceptable and unacceptable social and environmental performance in a way which both pre-empts and displaces state intervention. This implies a shift in environmental governance away from the local scale towards globally defined systems and structures. The purpose of this chapter has been to use the grounded contexts of coffee cultivation in Kodagu to underline the struggles and contradictions associated with constructing environmental management systems at a global scale. The environmental outcomes in sites such as Kodagu cannot be understood by reference to global regulatory structures alone, but need to be examined through the interplay of such schemes with the embedded realities of coffee production at the headwaters of the global value chain. In many cases, those global systems cannot compensate adequately for the diversity of institutional forms found in local geographical settings. When the standards are being set by organizations with commercial brand-oriented objectives, corporate self-regulation further removes the capacity for producer-driven product differentiation, thereby affecting incentive structures and potentially undermining the market potential of PES-linked conservation objectives.

# Chapter Seven

# Smallholder Engagement in Global Value Chains: Initiatives in the Nilgiris

The problematic status of smallholders is one of the most persistent and important debates within the field of Global Value Chain analysis. Their apparent marginalization in the face of shifts towards buyer-centric forms of chain governance is a pressing development issue that has gained considerable attention within major multilateral agencies (see FAO, 2004; UNCTAD, 2005, 2006, 2007; OECD, 2006; WTO, 2007), and has spurred a veritable avalanche of case studied interventions (see the edited collections by Vellema and Boselie, 2003; Batt, 2006; Ton et al., 2007). Symptomatic of the central relevance of this issue for debates on global poverty, The World Bank's *World Development Report* for 2008 argues:

> Structural adjustment in the 1980s dismantled the elaborate system of public agencies that provided farmers with access to land, credit, insurance, inputs, and cooperative organizations. The expectation was that removing the state would free the market for private actors to take over these functions – reducing their costs, improving their quality, and eliminating their regressive bias. Too often, this didn't happen. [This failure left]... many smallholders exposed to extensive market failures, high transaction costs and risks, and service gaps. Incomplete markets and institutional gaps impose huge costs in forgone growth and welfare losses for smallholders, threatening their competitiveness and, in some cases, their survival. (World Bank, 2007b: p. 138)

Such analyses paint grim forecasts for the millions of smallholders across the developing world, whose livelihoods are frequently hinged to issues of market access. In terms of the ambit of this book, these issues could have been examined with respect to both tea and coffee producers; however, for reasons of clarity, we focus on the specific situation facing tea smallholders in the Nilgiri Hills of Tamil Nadu. Investigating these processes in this place-specific context, this chapter: (i) identifies and discusses the position of these participants in tea value chains; (ii) reviews attempts to engage these producers with market requirements; and (iii) critically examines these interventions in terms of how they intersect with smallholder livelihood agendas at the regional scale. Through these foci, the chapter mirrors the frame of reference in Chapters 5 and 6, where ethical and environmental accountability agendas respectively were placed within regional socio-environmental perspectives. Such a framework, moreover, reiterates central arguments in this book relating to the co-production of governance arrangements and the institutional environment.

## Smallholders and Global Value Chains: The General Argument

The power of downstream lead firms to coordinate and control upstream suppliers was influentially brought to the attention of agri-food researchers in work on the export fresh fruit, vegetables and floriculture sectors of East Africa (Barrett et al., 1999; Dolan and Humphrey, 2000, 2004; Hughes, 2001; Humphrey et al., 2004). Linking together the research on retail restructuring in Western Europe with the development studies literature on export agriculture, these researchers identified and elaborated how consumer demands for increasingly sophisticated products (including those possessing certain environmental or social attributes) and government regulations requiring due diligence and enforcement of traceability systems, critically affected the industry arrangements for the international supply of fresh fruit and vegetables. According to Dolan and Humphrey (2000: p. 157):

> As the supermarkets began to require increasing product variety, customer-specific processing and greater knowledge of production and processing systems (in order to reduce the risk of exposure to failings by their suppliers), they greatly increased the degree of control exercised in the chain by reducing the number of suppliers and tightening linkages in the chain.

More or less contemporaneously, Thomas Reardon and associates began to document in painstaking detail and with global scope the upstream implications of rapid supermarket diffusion (Reardon et al., 2002; Reardon and Berdegué, 2002; Weatherspoon and Reardon, 2003; Neven and Reardon, 2004; Reardon et al., 2004). Coming out of Reardon's work is an appreciation of how 'waves' of supermarket diffusion wreak monumental change on upstream procurement via four 'pillars': (i) centralized procurement by lead firms; (ii) use of specialized wholesalers and logistics firms for supplier selection; (iii) listing of 'preferred suppliers' who satisfy downstream actors' demands through contracted arrangements; and (iv) development of private standards that define chain relations (Reardon et al., 2004).

These research agendas were then pulled together in the *Regoverning Markets* project coordinated by the International Institute of Environment and Development (IIED), commencing in 2003. Involving dozens of researchers from around the world, the project sought to develop the major lines of argument about the various impacts of GVC-induced restructuring on smallholders (see Vorley et al., 2007). Through evidence gathered from 16 country studies, the researchers point to how 'farmers require technology, financial capital, human capital and organization to avoid being excluded from this rapidly growing sector and take advantage of the opportunities for growth that clearly exist', yet '[t]he high capital requirements for entering buyer-driven chains mean that the higher land and labor efficiency of small producer production may no longer be a comparative advantage' (Biénabe et al., 2007: p. 3).

In these circumstances, the demands by downstream lead firms for suppliers to meet more stringently defined grades and standards has emerged as centrally relevant to the problem of smallholder marginalization. Grades and standards have become more rigorous and more embracing in scope. Their more extensive usage has encouraged an overall shift to a 'specification regime', in which downstream lead firms define the criteria and set the bar for products to be admitted into their supply channels. Grades and standards increasingly encompass both 'performance attributes', which can be loosely defined as the tangible/measurable properties of products (e.g., colour, taste, presentation, existence of chemical residues, etc.) and 'process attributes'; the bundle of information which relates to a product's life (its credence, the nature of the farmer/farming system which produced it, etc.). The role of 'process attributes' has mushroomed over recent years, as lead firms increasingly have demanded upstream suppliers gain accreditation as an basic requirement for chain participation.

The UK *Food Safety Act* (1990) was an important milestone in these developments, as it responded to food safety concerns by effectively establishing food retailers as guardians of public health. Retailers were required by law to demonstrate due diligence in regard to the sourcing of food products (Marsden and Wrigley, 1996). The private sector, and supermarkets in particular, then commenced the task of establishing a parallel set of food safety-related grades and standards which complemented, augmented and in the end, effectively replaced public systems. As discussed in Chapters 3 and 4 respectively, tea and coffee manufacturers have assumed a key role in standard setting within these industries. This has significantly altered the role played by the existing infrastructure of public grades and standards, authored by national governments and managed internationally through such organizations as Codex Alimentarius (Barling and Lang, 2005). In developing countries, existing laws have not always been monitored adequately (or at all), such that private standards employed by lead firms have quickly taken on roles as *de facto* enforcement vehicles. The general pattern has been for downstream lead firms to pitch their standards above mandatory requirements, and in so doing, take 'regulatory leadership' in developing countries with respect to quality and food safety (Reardon and Farina, 2002).

These trajectories of change have caused serious problems for smallholders in the tropical products sector (Neilson and Pritchard, 2006; Lewin et al., 2004; Humphrey, 2006). With this in mind, we now turn to the precise manifestations of the 'smallholder problem' in the tea industry.

## Tea Smallholders in the Nilgiris

The problematic position of smallholders in global value chains is exemplified by the case of Nilgiris tea growers. The Nilgiris is an undulating plateau mostly over 1,000 m altitude and rising to 2,637 m (at Doddabetta peak). Prior to the first incursions by the colonial British in 1819, it was blanketed in forest, though few remnants of this primary ecology still remain. The 2001 Census of India recorded a district population of 762,141, some 74.6 per cent of whom lived outside its four major towns (Directorate of Census Operations, Tamil Nadu, undated). The district capital is officially called Udhagamandalam (2001 population 93,987), but is known universally as 'Ooty', a shortened version of its British-era name, Ootacamund. As a former British hill station with an elevation of 2,200 m above sea level, Ooty is a cool climate tourist and service centre. The second major town in the Nilgiris is Coonoor (2001 population 50,196), located some 30 km to Ooty's south and fulfilling a quite different service function than its larger neighbour. Coonoor's tourist and education sectors are less advanced than Ooty's, but it is the place where plantation industries are administratively headquartered, being the home for UPASI and the Tea Board's South India office. The only other major towns in the Nilgiris are Kotagiri (2001 population 29,192), which acts as a minor service centre for the local population; and Wellington (2001 population 20,217), which is a major Indian military barracks and officer training centre.

The rural-based cultural landscape of the Nilgiris takes the form of a chequerboard of tribal and ethnic distinctions that pre-date the British. Four major groups inhabited the Nilgiris with little contact from the outside world until the arrival of the first British settlers: the Badagas (agriculturists), the Todas (pastoralists), the Kotas (craft-oriented) and the Kurumbas and Irulas (hunter-gatherers). For decades, the Nilgiris has had the reputation as being somewhat of an anthropologist's paradise, because the relationships between these groups has been argued to present key clues into the evolution of human society (see Hockings, 1997). As the reviewer for the Hockings' anthology on 'Nilgiris studies' has asserted: 'the regional scientific and historical literature on the Nilgiris is so extensive that one could almost literally paper the Nilgiris District with the pages of its publications!' (Zvelebil, 2000: p. 126). Moreover, into this rural mix have come Tamil immigrant communities which tend to occupy day-labouring roles within the District.

For our purposes, this Nilgiris' anthropology is significant inasmuch as the smallholder tea sector is the domain of the Badagas community. The precise origins of the Badagas people remain a point of academic debate, but linguistic affinities suggest strongly that these groups were plains-dwelling communities from Karnataka. Whilst a popular account is that the Badagas migrated up to the Nilgiris in the 1790s, during the time of Tipu Sultan's wars against the Maharaja of Mysore, Hockings (1997) dates the first migration earlier at around 1565 (during an earlier Moslem attack on the Vijayanagar Empire). Similarly, the 1908 *Nilgiris Gazetteer* (Francis, 2001) concludes that this migration must have occurred prior to 1600 based on the writings of an Italian Catholic priest who visited the Nilgiris in 1603. Daily village life in the 370 Badagas communities in the Nilgiris is dominated by the cycles of agricultural work, but like elsewhere in India, villagers are increasingly connected in complex ways to the global economy. Notably, remittance payments play a vital role in sustaining family incomes.[1]

The Badagas quickly adopted the cultivation of European vegetables following British settlement on the plateau, with Badagas tea plantations developing from around 1925 (Hockings, 1997: p. 156). However, it was not until the 1970s, when, encouraged by government supports under the *Crop Diversification Programme*, Badagas smallholders began to expand tea production significantly. The programme was promoted by a tripartite consortium of UPASI, the Government of India's national agricultural extension agency Krishi Vigyan Kendra (KVK) and the Tea Board of India, and was instigated in light of deepening problems in Nilgiris vegetable production, which included pest and disease threats, low prices and damage from the indigenous wildlife (including deer, bison and elephants) (Tea Board & U-KVK, 2001: p. 19). It was also argued at the time that tea plantings would mitigate erosion problems on steep hillsides, brought about by the intensified lopping of the district's primary forests. Tea cultivation quickly proved to be a popular alternative to vegetable production. Advantages included the relative absence of pests and disease, easier field maintenance, and the fact that regular plucking cycles generated a steady income stream. Importantly for Badagas villagers, local climatic conditions enable tea grown in the Nilgiris to be plucked throughout the entire year. When tea prices rose steeply in the mid-1990s, smallholder conversion to tea quickened pace. The total area under tea in the Nilgiris increased from 35,585 ha in 1996 to 60,427 ha in 2000 (Tea Board of India, 2003), so that it came to occupy 72 per cent of the Nilgiris cultivated land area (Swaminathan,

2002: p. 232). In the twenty-first century, tea still comprises the mainstay within the Badagas rural economy. A survey undertaken in 2001 indicated that 86 per cent of Nilgiris tea smallholders grew no other crops except tea (Tea Board & U-KVK, 2001: p. 40). Since that time there has also been strong growth of high-value horticulture and floriculture in the Nilgiris, but tea remains the dominant crop. At the turn of the twenty-first century, approximately 60,000 tea smallholders in the Nilgiris contributed some 40 per cent of total South Indian tea production.

## Nilgiris Smallholders and the Tea Crisis

The participation of Badagas smallholders in the tea industry caused a dramatic increase in production volumes in the 1990s, just at the time when Soviet buyers left South India and prices began to soften. With their lesser abilities to coordinate and credential their output, smallholders bore the brunt of price declines during the severe period of market free fall between 1998 and 2002. Since the end of the 1990s, roughly half of all tea production from South India has been exported (Tea Board of India, 2006). Consequent to this, since 2000 a widening gap has emerged between South and North India in terms of average export prices (Figure 7.1). South Indian teas have been in free fall whilst the higher-quality North Indian teas have been less exposed to these problems (Sanjith, 2004). Obviously, these outcomes relate to the situation, discussed in Chapter 3, whereby South India remains heavily reliant on export sales to low-price markets (in 2002 they received the low average export price of around Rs 62/kg) whereas in North India in the same year, more than 25 per cent of exports were destined to the European Union where they received an average export price of Rs 120.84/kg.

As producers of teas sold into low-priced export and domestic markets, Nilgiris tea smallholders are reliant on selling their product through auctions, and have received low prices because of quality detriments. The smallholder sector is associated with the cultivation of coarse and variable green leaf, and the factories which process this product operate to mixed standards. In light of global over-supply and the general trend for lead firms to demand that suppliers meet more stringently defined grades and standards (discussed earlier in this chapter), these growers have been downgraded to inhabiting an increasingly precarious status near the bottom rung of the global tea industry.

By way of illustration, in 1998 average green leaf prices were Rs 12/kg (approximately US$0.28/kg) (Ajjan and Raveendran, 2001: p. 29). Assuming an average annual yield of 8,850 kg/ha (Tea Board & U-KVK, 2001: p. 5) and annual production costs of Rs 40,000 per ha,[2] these prices would enable

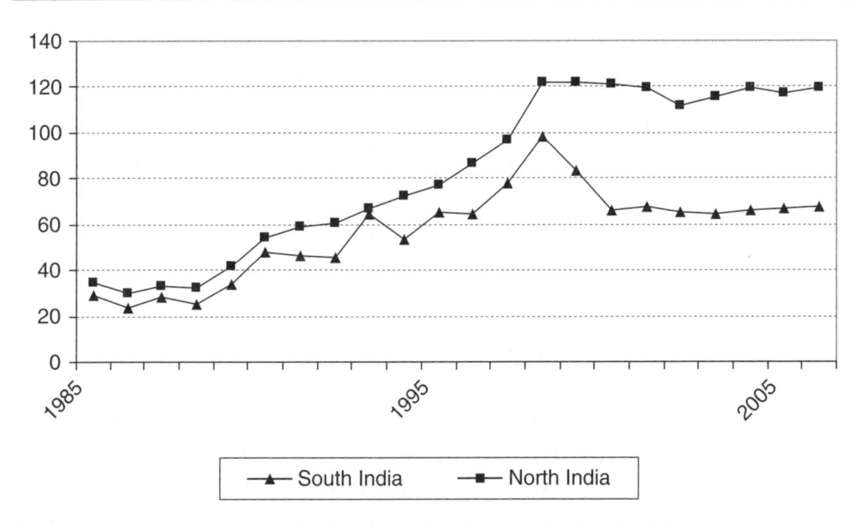


**Figure 7.1** Average tea export price (Rupees per kg), North and South India, 1985–2006.
*Source*: Tea Board of India (2006).

smallholders to generate an annual income of Rs 62,600 (or US$1,456) from one hectare of tea. By 2001, however, such incomes had evaporated. With average prices at just Rs 5.95/kg (approximately US$0.14/kg) for green leaf (Tea Board & U-KVK, 2005: p. 11), smallholders' annual net incomes would fall to only Rs 10,872, or approximately US$253. In short, a 50 per cent fall in green leaf prices generated a drop in smallholder income of around 83 per cent.[3]

Such losses to income were obviously unsustainable, and smallholders had little choice but to reduce production costs. The irony of this, however, was that it ricocheted back onto quality. As the biggest single cost item, plucking routines quickly deteriorated. Rather than pluck every 8–10 days during 'rush' cropping periods and 10–12 days during leaner periods (and thus capture new buds and fine leaf), many growers adopted a longer plucking cycle and replaced hand plucking with indiscriminative shear plucking. The net effect was to generate product in which coarse leaf predominated. The implication here was that low prices created a vicious cycle in which poorer quality became the norm, in turn placing growers under even more intense economic pressures. Aggravating this situation further, essential field practices such as regular pruning, application of fertilizers and pest and disease management were often abandoned. Moreover, in their desperate need to make up lost income, the felling of shade trees for sale as timber was not uncommon, again with obvious implications for the future productivity of tea gardens.

The spiralling problems affecting tea smallholders had severe social costs. These are brought into perspective in our field note summary of one Badagas tea smallholder we interviewed, living in a small village in Ooty Taluk:

> Nanjundan is aged in his 50s and lives in the village with his wife, two sons, one daughter-in-law and one grandson. He also has two daughters, both of whom are married and live with their in-laws. He owns 1.5 ha of land, of which 1 ha is planted to tea and the other 0.5 ha is planted to vegetables. He introduced tea to his land 15 years ago, encouraged by the Crop Diversification Scheme. Since the onset of crisis, however, he has been receiving very low prices, only Rs 2/kg to Rs 3/kg on average. His village is remote and with no local factory, he is at the mercy of local leaf agents who can bid down prices from desperate smallholders. With little incentive to comply with 'good practice', Nanjundan uses sickle plucking on his tea and plucks only once every two months. This generates a very low annual yield of around 2,000 kg of green leaf/hectare, or around 333 kg per plucking cycle. The coarse leaf plucking of his property is done in one day's work by a gang of five men and five women who are paid Rs 100 per day [approximately US$2.72] and Rs 50 per day [approximately US$1.36] respectively. On this basis it costs him Rs 750 to have his one hectare of tea plucked. If he gets only Rs 2/kg from the leaf agent he ends up losing money, but if he gets Rs 3 he makes a net gain of Rs 250 (approximately US$5.68); a paltry recompense for two months of tea growth. Essentially, at prevailing prices tea cultivation makes an infinitesimal addition to his material livelihood. He has reverted to a semi-subsistence lifestyle, feeding his family on home-grown foods plus access to the highly subsidized food security-based PDS [Public Distribution System].[4]

The collapse of tea prices sent shockwaves through the Badagas communities of the Nilgiris. Badagas villages typify the position of their inhabitants in the lower rungs of rural India. Housing is sturdy, if modest (Figure 7.2). An average family home typically consists of a sitting room, one or two bedrooms, a kitchen, a loft and possibly a toilet (although many villagers still rely on the communal village toilet). Cooking is performed with gas, kerosene or fuelwood. In the mid-1990s, returns from tea (nicknamed 'the currency bush' at the time) encouraged symbols of rising affluence to penetrate this social landscape. These included the purchase of consumer durables (including washing machines, new foam beds, and electric mixers), house extensions and an increasing number of 'two-wheelers' (motorcycles). These were the years when satellite dishes mushroomed from smallholders' houses, with the newly deregulated Indian television industry booming its 100+ channels into villages where, only a few years beforehand, populations crowded into the front rooms of the few owners of television sets to watch the latest cricket match or Bollywood sitcom.

**Figure 7.2** A typical Badagas village.
*Source*: Photo by the authors.

A decade later, the Badagas communities we interviewed were more circumspect about their future. A telling theme in many of our discussions with smallholders was grower exit and debt. In 2001, 60,389 tea smallholders were enumerated in the Nilgiris (Tea Board and U-KVK, 2001: p. 5). By 2006 the number of tea smallholders had fallen to 50,329.[5] Grower exit occurred through various means. For many members of the younger generation, continuing in their parents' footsteps as tea smallholders represented a poor option for economic advancement compared with educational and career opportunities elsewhere. On other occasions, grower exit occurred through more direct circumstances. Symptomatic of these changes has been the transfer of land through land-pawning arrangements. As described to us in one village, growers who urgently require cash would pawn their tea gardens for a fee of Rs 50,000 [approximately US$1,100] for a three-year period. If, after three years, the original owner was unable to repay the amount, the pawn-holder would work the fields for a further three years. If after this period, the owner was still unable to repay the original amount, the fields would be sold. In the village in which we conducted interviews on this subject, approximately 10 per cent of local smallholders had been forced into land-pawning arrangements since the onset of the tea crisis.[6]

In recent years, this large community of Badagas tea smallholders has provided a major political force in South India's plantation sector, with politicians of different hues seeking to curry their favour. Badagas tea smallholders have been active agitators, holding numerous protest marches and related activities seeking to advertise their plight. In 2000, a series of protests by smallholders gained national media attention. The protest was described by T.M. Kullan, self-styled vice-president of the 'Tribal Solidarity Association of the Nilgiris', in his polemic entitled *Toxic Tea Market in the Nilgiris, Nilgiris Profusely Bleeds!*:

> Badagas women clad in white, squatted on the roads while their men-folk pitched flags and shouted slogans. One could almost foretell the violence that entered the heels of the agitation. There were buses burnt, mass arrests, lathi charges and the unfortunate incident at Meekeri village, where cops went on rampage. (2006: p. 6)

Somewhat more restrained was International Tea Day 2006 (17 December), when, in Wyanad, workers' and smallholder representatives across India put out a call for the recognition of rights of livelihood and land, the need for a national federation of small tea growers, and an international commodity agreement on tea (*Labour File*, 2006). As a metaphor of the philosophical underpinnings of these arguments, growers initiated a 'pinch of soil' campaign whereby each sent 50 grams of soil to the State legislature in order to emphasize their plight.

## The Quality Upgradation Programme

As described in Chapter 3, the tea industry of South India is structured around an institutionally dense web of organizational interests. These arrangements provided a launching pad for a significant strategic response to the smallholder crisis in the Nilgiris, in the form of the 'Quality Upgradation Programme' (QUP). The prime mover in obtaining and implementing this programme was UPASI-Krishi Vigyan Kendra (U-KVK), which was introduced in Chapter 3 as a Coonoor-headquartered unit within India's nationwide KVK network of agricultural adaptive research agencies. U-KVK's staffing level fluctuates, but in 2006 employed ten scientists and 15 field staff; representing a vigorous repository of human capital in the district.[7] U-KVK is active in running seminars within the region, and works closely with Tamil Nadu Government Ministries (such as the Department of Horticulture) with cognate interests in rural development.

The QUP represented a coordinated strategy to address the position of smallholders in tea value chains (Neilson et al., 2006). This scheme had its origins in a pilot project undertaken in 1999 involving growers from Thummanatty village. Through a careful plucking regime and the processing of leaf in U-KVK's own factory at Coonoor, price realization of Rs 70/kg for made tea was obtained; translating to Rs 14/kg for the Thummanatty growers. This compared with an average price of Rs 8/kg being received by other small growers at this same time (Tea Board & U-KVK, 2005: p. 7; Ajjan and Raveendran, 2001: p. 29). This success signalled a strategic response to the (at the time emerging) tea crisis which would take the form of using *quality improvements* as the basis for improvements in grower livelihoods. Thus, the QUP embodied a quite different kind of response to the sort that has traditionally dominated the Indian rural sector. The usual response of Indian policy makers to rural hardship has been to implement price floors or direct subsidy payments, with the objective of securing farmers' livelihoods in tough times. These forms of policy intervention have played their part during the period of the tea crisis (discussed below) but, in general, as second fiddle to the more strategically oriented QUP. What was significant about the QUP was its premise that the solutions to the problems facing Nilgiris smallholders were to be found in the market. Rather than doling out subsidies or invoking the 'licence Raj' tradition of instituting price controls, the QUP operated on the assumption that the long-term viability of smallholders depended upon their capacity to grow a product that obtains a higher price. To this end, since commencing in July 2000 the Program sought to develop a set of incentive structures that align smallholders' interests to the promotion of improved quality.

The context in which the scheme was developed is described in the chain diagram (Figure 7.3). In the year 2000, smallholders contributed approximately three-quarters of the 95,194 tonnes of made tea produced in the Nilgiris. This was sold either to processing factories directly or via green leaf agents. The vast majority of smallholder processing was undertaken by bought-leaf factories (BLFs); however, the Indcoserve Cooperative accounted for approximately 10 per cent of Nilgiris production and a small amount was sold to estate factories through various outsourcing programmes. Most of this tea then passed through auction, with Coonoor traditionally constituting the major site where smallholder tea is traded. Because of its reliance on smallholder (as opposed to estate-grown) teas, prices at the Coonoor auction have historically been lower than those at Coimbatore and Kochi. A very small amount of BLF tea bypassed the auction system, being sold through private contract by the Nilgiris District Tea Producers' Marketing Cooperative (NILMA), an organization representing 36 BLFs.[8]

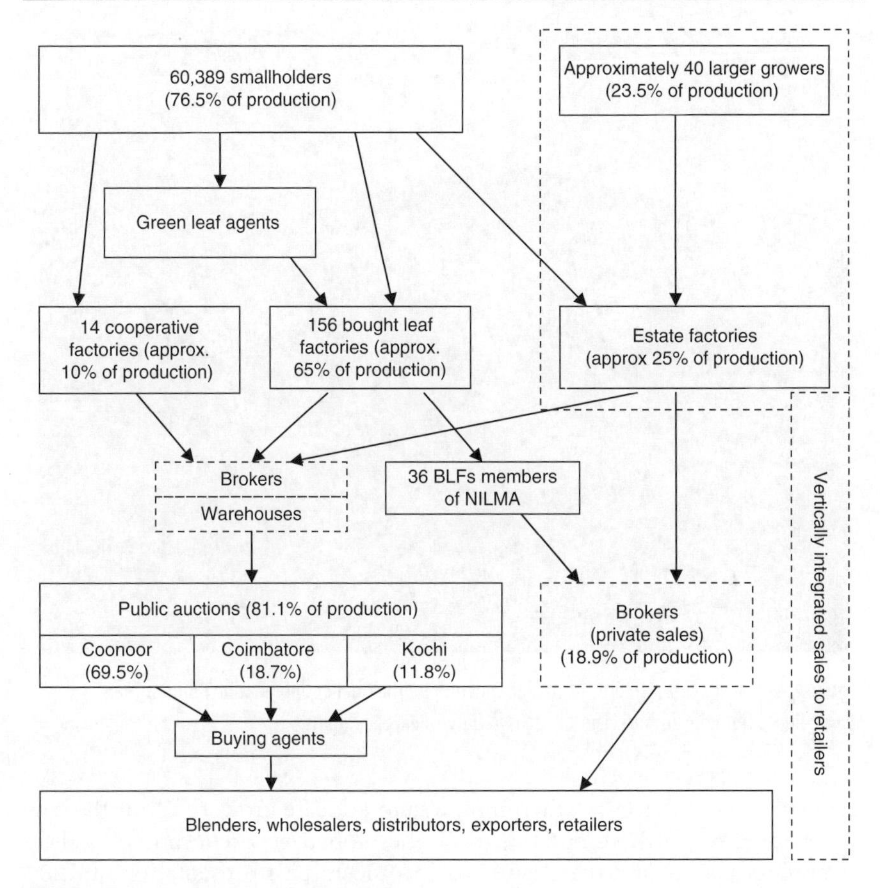


**Figure 7.3** Value chains for Nilgiris tea, 2000.
*Source*: Own work based on statistics in Mitra et al. (2003); Tea Board & U-KVK (2001) and field interviews.

The QUP sought to restructure these chain arrangements by instigating 'cultural change' within the smallholder sector, leading to improved field maintenance and the increased plucking of fine leaf grades. It then proposed to intervene in the interface between the grower and BLF sectors by challenging the role played by agents. Finally, it aspired to improve factory performance through the 'carrots and sticks' of training schemes and closer regulatory inspections. Through these means, the overall ambition was to generate improved price realization with assurances that smallholders would share in these gains.

The first two of these initiatives were highly inter-related, and thus are best considered in tandem. As already described, the collapse in prices for smallholders impacted directly and adversely on plucking routines. One of

**Figure 7.4** A village-awareness campaign meeting for the Quality Upgradation Programme.
*Source*: Photo from the authors. The village is in Ooty Taluk, the Nilgiris.

the first tasks of the QUP, therefore, was to educate growers about the role of leaf quality. To these ends, U-KVK developed an orchestrated 'Village Awareness Campaign' that saw its field officers give presentations to 203 grower communities during the 21-month period from July 2000 to March 2002 (Tea Board & U-KVK, 2005: p. 10). This was then followed up in a second phase involving regular village visits by 100 newly-employed field extension officers from the Tamil Nadu Department of Horticulture.[9] Village visits were coordinated with demonstration projects and also involved a number of local NGOs. This entire programme had the objective to increase the proportion of fine leaf grade plucking, and to reinstigate appropriate field maintenance procedures, notably pruning schedules (Figure 7.4).

These programmes aiming to educate and encourage growers about fine leaf plucking were coordinated with initiatives to provide quality-based economic incentives. The key intervention here involved the development of 'Quality Tea Procurement Centres' run by village-based women's Self-Help Groups (SHGs), in conjunction with the introduction of differential leaf pricing systems by BLFs. In 2002, the first year in which these schemes got underway, ten SHGs were established. By the end of 2004, a total of 70 were operational (Tea Board & U-KVK, 2005: p. 108). In terms of the local value chains for smallholder tea, the objective of the procurement centre/ SHG model was for village communities to take charge of the quality agenda by separating out fine leaf from coarse grades. They would do this by re-locating *shoot analysis* (a common index used to assess the percentage of finer growth in a batch of green tea leaf) from being undertaken in the factory, to being undertaken in the village. Effectively, the procurement centre/ SHG model enables leaf to be sent to factories already sorted, giving factories the scope to pay growers for their higher-quality product.

Discussion of this point needs to be contextualized by an appreciation of the role traditionally played by green leaf agents in the Nilgiris tea economy. There is divergence across the Nilgiris in terms of local supply arrangements from field to factory. Where there is a prevalence of factories and strong competition between them, many factories operate their own delivery systems. For example, this is evident in and around the small town of Aravenu, in Kotagiri Taluk, where there are 35 factories within a five-kilometre radius around the town.[10] In Ooty Taluk, by way of contrast, agents were responsible for transporting some 88 per cent of leaf to factories in 2001 (Tea Board & U-KVK, 2001: p. 51). Obviously, the latter type of arrangement adds an extra cost element to the supply chain, as agents require a fee for their services. Moreover, knowing that growers rely on them (green leaf should, ideally, be delivered to factories the day it is plucked) creates dependent power relations. As is commonly the case in such scenarios in developing countries, local agents can exercise various roles in communities, including the provision of credit. In exactly half of the 70 villages where the procurement centre/SHG model was implemented, green leaf agents were forced out of business. In the others, they lost trade and/or were forced to compete in terms more advantageous to growers (Tea Board & U-KVK, 2005: p. 110). Squeezing and/or eliminating these middlemen and the broader empowerment of SHGs, therefore, became a key outcome from the QUP.

There are evident economic advantages to growers from this model. By taking agents out of the local value chain and creating the conditions for the payment of price premiums on account of higher-quality product, significantly improved price realization was achieved. A survey of 20 procurement centres/SHGs undertaken by U-KVK found that, in 2004, this model assisted 1,065 tonnes of pre-sorted fine leaf grade to receive an average price premium of Rs 3.01/kg (Tea Board & U-KVK, 2005: p. 113). Our own interviews in grower communities corroborated this magnitude of positive outcome. In one village where we conducted interviews, growers received premiums between Rs 1 to Rs 4 per kg for fine grade leaf. This translates to a net transfer of Rs 3.2 million [approximately US$73,000] to grower communities, or about Rs 3,500 [approximately US$80] per participant. In some cases these monies have been distributed back to growers directly, while in others, they were used for dedicated community purposes including the purchase of school books, construction of storage sheds, health care, and festival funding (Tea Board & U-KVK, 2005: p. 131, and fieldwork).

## The role of women's Self-Help Groups and factory participation

One of the most important outcomes from the QUP intervention was that the scheme inserted women into this pivotal component of local value chains. Effectively, the scheme is predicated on the principle that local control and transparency in prices provide the best conditions for embedding the ethic of quality in communities. In the procurement centre/SHG model, there is a direct and visible link between local women sorting leaf prior to its delivery to factories, and the price premiums subsequently received. However, more than this, SHGs once established act as community-based institutions for revolving credit and for the financing of small-scale entrepreneurial activities. In this, the scheme adopted lessons from the wider SHG movement that has become a much-publicized socially progressive development success story across Tamil Nadu and other states of India (see, *inter alia*, Dash, 2003; Leach and Sitaram, 2002; Wilson, 2002; Lahiri-Dutt and Samanta, 2006).

In tea smallholder villages of the Nilgiris, once community members have set up SHGs they are registered with an NGO which, over a six-month period, audits the organization and provides advice and training, particularly with respect to financial and organizational management. Once this hurdle is passed, SHGs are then eligible to apply for bank loans. (The Government of Tamil Nadu and the Tea Board have intervened to encourage banks to extent credit facilities to SHGs.) In 2006, the interest rates charged on these funds were 9 per cent per annum; much lower than the alternative informal credit sector (often one-and-the-same as the leaf agents) which can charge around 5 per cent per month.[11] Members can generally borrow money within the group at a rate of 2 per cent per month. Minimal up-front collateral requirements are imposed on these loans, but, according to testimony from various field officers from U-KVK, the SHGs have a strong repayment record. It would seem that because these initiatives are grounded in the theme of collective community responsibility, there is a strong moral undercurrent that acts to enforce loan repayment.[12] Crucially, therefore, SHGs set up for leaf-sorting purposes have provided villages with enhanced institutional capacity for more general deployment.

The necessary other element to the success of the procurement centre/ SHG model is, of course, the willingness of factories to accept segmented leaf and pay a premium for its delivery. Differential leaf pricing was first introduced by the Kannavarai factory in Kotagiri Taluk, which in April 2002 announced it would pay its suppliers a premium for quality, which by October of that year was Rs 3.25 [US$0.07] per kg for fine green leaf (62 per cent above what it was paying for non-sorted leaf).[13] From growers' perspectives, of course, it makes little sense to pluck and separate fine leaf grade unless such price premiums exist. Whereas the use of shears can generate coarse leaf plucking rates of 30 kg/day (or more), plucking for fine leaf generally restricts daily output to 15–20 kg/day.

For factories, acceptance of fine leaf necessitates the establishment of dual processing lines and some other related investments (Ramamoorthy and Ramu, 2002). These needs led the Tea Board to institute the 2001 Factory Upgradation Programme (FUP). The FUP operates complementarily to the QUP, with both schemes administered by U-KVK. It offers factories with a subsidy of 50 per cent of the costs of upgrading equipment and gaining ISO9001 and HACCP certification (Tea Board, undated a). The programme had a considerable reach through the district, with a total of 88 BLFs in the Nilgiris given assistance through the FUP (Tea Board & U-KVK, 2005: p. 119). Operating in parallel to the FUP is a further subsidy of Rs 2/kg for conversion from CTC to orthodox processing,[14] and an Export Incentive Scheme which provides subsidies for 'better performing export companies' ranging from Rs 2/kg (for bulk orthodox tea) to Rs 10/kg (for value-added tea bag exports) (Kapur, 2002: p. 5).

The QUP and FUP acted to stratify the BLF sector in terms of price realization and profit. There is an emerging divergence between BLFs operating according in line with the linked factory–grower model, and those producing poorer-quality teas. An example of the former is IBS, which operates five tea-processing factories across the Nilgiris. IBS sources 80 per cent of its leaf from SHGs, has its own trucks to collect directly from them, and has established a premium label ('IBS plantation') to differentiate fine grade lead and seek higher price-realization.[15] In 2004, IBS purchased leaf from 20 procurement centres and, on average, received a price premium of Rs 15/kg for its fine grade tea (Tea Board & U-KVK, 2005: p. 137). The trend towards stratification marks the ongoing rationalization of the BLF sector. In the main, BLFs were established by the larger Badagas tea growers as an extension of their tea cultivation activities. Whilst the costs of establishing a factory are considerable, it is also a logical financial development for successful growers who have access to capital and already possess a good understanding of tea production systems and have existing social networks within the tea-growing communities.[16] Following an earlier period of expansion, the number of BLFs in the Nilgiris has declined steadily during the past decade, as those unable to achieve higher price realizations have closed. Accompanying these trends, some of the more successful enterprises have expanded through factory acquisition. Hence, the quality agenda is driving considerable restructuring of BLF ownership. BLFs producing poorer-quality teas have faced an increasingly difficult commercial environment. Since the advent of the FUP there has been extensive rationalization in the number of BLFs in the Nilgiris.

There is little doubt that the QUP provided a major injection of new ideas and practices into the Nilgiris tea smallholder sector. Yet despite these successes for its participants, its limitations also bear consideration. By the completion of the first stage of the project in 2005, the 70 villages which had implemented the procurement centre/SHG model represented a significant minority of the 370 tea-growing Badagas villages in the Nilgiris. Similarly, although 88 BLFs had obtained funding through the FUP, only 20 were linked formally into the procurement centre/SHG model. This patchy penetration was acknowledged by the Tea Board & U-KVK (2005: p. 104) in their evaluation of the project: 'Though there is only 20 to 25% improvement in the quality of green leaf throughout the district, the fact is that the entire small grower community has realized the importance of quality in tea to get remunerative green leaf price.'

By 2005, with problems of sub-economic pricing and poor quality still plaguing a large proportion of Nilgiris smallholders, the Tea Board determined that a stronger hand was required. Accordingly, in September it announced a ten-point set of directives that compelled the industry to adopt higher-quality standards (Figure 7.5):

1. The South Indian tea industry will operate from henceforward on the basis that at least 65 per cent of production must take the form of fine leaf grade.
2. Agents who buy leaf that is not at least 65 per cent fine leaf grade will have their licenses revoked.
3. Sickle harvesting is to be banned.
4. Factories processing leaf that does not meet the mandatory target of at least 65 per cent fine leaf grade will have their license revoked.
5. Brokers who sell leaf that does not comply with the minimum 65 per cent fine leaf grade standard will have their licenses revoked.
6. Actions will be taken against buyers who procure poor-quality leaf.
7. Actions should be taken against factories practicing adulteration of leaf grades in an attempt to evade these directives.
8. The optimal plucking interval (once every 12–15 days) should be maintained.
9. The Tea Board will support growers' arguments for minimum leaf prices only if improved quality standards are met.
10. The Tea Board seeks the cooperation of small growers.

**Figure 7.5** Tea Board poster mandating that smallholders are not to use sickles in tea plucking.
*Source*: Photo by the authors.

## Tea Board involvement in quality control

The Tea Board then commenced a 'second front' in its attack on poor-quality leaf by increasing the intensity of product inspections. The *Tea Marketing Control Order* vests extensive powers with the Tea Board to cancel or suspend the registrations of manufacturers, brokers or tea buyers if they are found to be violating aspects of the *Prevention of Food Adulteration (PFA) Act*. For many years the adulteration of tea has been an endemic problem in the industry; what has been lacking, however, has been the resources and resolve to instigate a tougher testing and enforcement regime. The most common practice has been for producers to add orange colouring agents with the objective of generating a fraudulently-obtained higher price at auction, along with typical adulterants such as sand, stone, and mineral salts. The *PFA Act* specifies that tea must contain not more than 17 per cent of crude fibre, and that it must satisfy certain tests relating to ash content.

From the middle of 2005 a series of inspections and product tests were implemented by the Tea Board which led to product confiscations and a media-oriented ceremonial public burning of 2,500 kg of smallholder tea that did not meet mandatory PFA requirements. As a result of this process, approximately 40 factories were closed. A number of these shut their doors after receiving 'show cause' letters from the Tea Board, whilst others took the decision to close their business in the face of escalating debts and new requirements emerging from the ten-point plan.[17] Because of these actions and the generally difficult commercial conditions facing the sector, as of September 2007 less than 70 BLFs were buying and processing green leaf (out of a total of 165 factories that had been built and were at some stage operational during the past decade).[18] Subsequent closures and consolidation in the bought-leaf factory sector have eliminated many of the more marginal and/or unscrupulous operators, with the effect that the industry now has less reason to fear that its teas fail to meet this most fundamental of quality benchmarks.

## Smallholder quality and estate outsourcing

With the improved quality performance of some smallholder production, some estates have been encouraged to commence outsourcing arrangements for green leaf. Such moves represent a significant break from past practices as, traditionally, the South Indian estate and smallholder sectors tended to operate as if in parallel universes. Estate factories would process only green leaf grown on the estate, and smallholder leaf would be processed exclusively in BLFs. However, with estates procuring smallholder tea for processing in their own factories, these distinctions are blurred. Comparable to schemes known elsewhere as 'out-grower projects' or 'nucleus estate' regimes (Eaton and Shepherd, 2001), these tendencies potentially have wide implications for the future economic organization of South Indian tea cultivation; issues we return to in the concluding section of this chapter.

There are no authoritative data on the magnitude of these organizational arrangements in the South Indian tea industry;[19] however, in our visits to estates and smallholder communities we encountered a number of instances where estate companies were strategically procuring smallholder leaf. Harrison Malayalam Ltd, for example, developed a smallholder procurement scheme for one of its CTC factories in Wyanad District. To ensure green leaf met a set of minimum quality requirements, it undertook an education programme relating to plucking techniques and pruning schedules.[20] Talamala Estates Ltd, also in Wyanad, took the decision to procure approximately 30 per cent of its throughput from smallholders, and thus make optimal use of its factory.[21] In the Nilgiris, the Darmona estate company has interests spread across a number of non-contiguous properties across the district, but its factory throughput relies 80 per cent on the purchase of smallholder leaf. Darmona's estate-grown tea, combined with some fine leaf bought from smallholders, is sold as premium branded product ('Darmona Estate') while the remainder is sold through lesser-priced brands or via the auctions.[22] In Gudalur Taluk, the Parry Agro and Silver Cloud estate factories have also linked with procurement centres to purchase fine leaf grown from smallholders (Tea Board & U-KVK, 2005: p. 128). The relative successes of the QUP appear to have convinced the estate sector of the possibilities of smallholder procurement, which has encouraged subsequent estate sector investment in supply chain development through outsourcing and quality improvement programmes at the farmer level. The decision by Silver Cloud to purchase smallholder fine leaf grade is significant, given that this estate provides an exemplar of high-price, high-quality tea production, with a state-of-the-art automated factory and some 85 per cent of its output being sold outside of the auction system.[23]

These and other observations lead us to suggest that the attempt to increase factory utilization and efficiency through smallholder leaf procurement is taking three emergent forms:

1. Ad hoc purchases, usually for the aim of ensuring optimal throughput of estate factories. In these cases, estates are keen to process and brand smallholder-cultivated leaf separately, because of fears its lower-quality could tarnish the estate's reputation;
2. Semi-permanent 'neighbourhood supply' arrangements, whereby estates purchase leaf from adjacent smallholders or medium-sized estates on a regular basis. Depending on leaf quality, these purchases could be bagged and branded as estate tea, or kept separate; and
3. Strategically formulated contract farming-type arrangements, whereby estates actively promote and monitor smallholder development on condition that they supply the estate factory. In these cases, smallholder production becomes a regular and necessary input that underpins the ongoing operation of a production system that transcends the individual estate.

In whatever form they take, the emergence of these kinds of estate outsourcing schemes complements the quality upgradation initiatives described earlier in this chapter. The willingness of estates to procure green leaf from smallholders widens the scope for the latter's engagement in higher-value segments of tea value chains. Moreover, the outsourcing of tea production in South India could be interpreted as a strategic response to enhanced demands for ethical accountability along the global value chain for tea. Due to the increasing costs of compliance to ethical standards within their permanent workforce, estates may be encouraged to outsource production beyond the current gaze of ethical audit regimes. The question that remains, of course, is whether such progress will provide sufficient additional economic benefits to substantially improve the livelihood prospects for this sector of the rural community. We now address this question.

## Is the Quality Agenda Sufficient to Improve the Lives of Smallholders? The Role of Government Supports

The Guest of Honour at the 2005 UPASI Annual Conference was the Union Minister for Commerce and Industry, the Hon. Kamal Nath. The minister was to arrive at the Conference, held in Coonoor, by helicopter. However, blustery conditions in this high-elevation location forced the helicopter pilot to avert the landing, and the minister had to cancel his appearance. Had the minister's helicopter been able to land, he would have been confronted by a massive street protest coordinated by tea smallholders. Along the main roads of upper Coonoor, where the UPASI Conference was being held, thousands of smallholders waved banners and chanted slogans which proclaimed their call for a statutory minimum tea price. Two years later, at the 2007 UPASI Annual Conference, the weather conditions were more benign, allowing the minister's helicopter to land. There was no street protest this time but, at the Conference, the minister's Congress Party colleague, the MP for Nilgiris R. Prabhu, made an impassioned call for minimum price supports. Each time he mentioned this issue his band of supporters would clap and cheer furiously, seeking as much as possible to leave a strong impression with the minister.

Calls for minimum prices or related forms of livelihood support payments have been a prevailing current in the Nilgiris since the onset of the crisis. A host of local organizations, including the Hill Area Development Program, the Tribal Solidarity Association of the Nilgiris and the Save the Nilgiris Committee, have weighed into the issue. This has been a fractious debate with claims and counter-claims frequently cross-cutting one another. Nevertheless, arguments for the livelihood needs of smallholders to be addressed have solidified into three kinds of programmatic response. The first of these corresponds to the obligation for smallholders to be treated fairly in their dealings with factory buyers. In the early 2000s, concerns were aired that the crash of tea prices was providing factories with opportunities to exploit their smallholder suppliers. Weaker conditions, it was suggested, provided an additional opportunity for factories to bid down the prices they paid for green leaf. In response, the Tea Board, in its *Tea Marketing Control Order* of 2003, instituted a mandatory price-sharing formula between growers and factories. Under these new rules, factories were obliged to pay growers the equivalent of at least 60 per cent of the price they realized upon the sale of their tea. So, for example, if a BLF realized a price of Rs 60/kg of made tea upon auction, it would be required to ensure that growers received at least 60 per cent of this amount, or Rs 36. Given that it takes 4.5 kg of green leaf to manufacture 1 kg of made tea, this would equate to a minimum green leaf payment to growers of Rs 8/kg. This scheme was modelled on comparable arrangements in Sri Lanka and Kenya, where a majority of the country's tea is cultivated by smallholders.[24]

The second type of programmatic response is direct subsidy payments. On various occasions during the crisis period subsidy payment schemes have been launched for the small grower sector. In 2000, the Tea Board executed an income assistance scheme which allocated Rs 262 million (US$5.7 million) to Nilgiris tea smallholders (Kapur, 2002: p. 4). This direct cash payment translated to an average of Rs 4,360 per smallholder (US$95). Then, upon the 2004 electoral success of the Congress Party-backed United Progressive Alliance (UPA), one of the first acts of the newly-appointed Kamal Nath was to approve an emergency Rs 2/kg subsidy payment for smallholders. This subsidy lasted until mid-2005, and upon its conclusion the government of Tamil Nadu stepped in with its own 12-month smallholder subsidy payment of Rs 4,400/hectare. Assuming a green leaf yield of 8,000 kg/hectare, this worked out to be the equivalent of Rs 1.8/kg, almost the same level as the previous scheme.[25] In February 2006, with state elections looming, the Tamil Nadu government topped this up with an additional one-off subsidy of Rs 3,080/hectare.[26] From the middle of 2006, once the election was over, the expiry of the Tamil Nadu government schemes meant that smallholders again were wholly reliant on the market. Yet this was perhaps less of a problem as it might otherwise have been, because rising tea prices from the first half of 2006, induced by the East African drought, provided a much-needed fillip to smallholder livelihoods.

Apart from the economic implications of these subsidies, these arrangements suffered from various administrative problems that impaired their efficiency. Subsidy payments from the Tea Board required growers to demonstrate their ownership of tea-planted land, which meant in turn that growers had to be in possession of land certificates. In some cases of legal uncertainty over land tenure (notably in Gudalur, discussed in Chapter 5), the absence of certificates denied growers access to payments. More generally, however, fees for certifying land can be expensive, and problems can arise in cases where land is formally held by family elders yet is worked by younger family members.[27]

The third kind of programmatic response to the livelihood struggles of tea smallholders relates to the debate on mandatory minimum prices for smallholder tea. As discussed above, the Tea Board and other Government of India agencies have been loath to return to the use of minimum prices in this sector, preferring instead to address the problems of smallholders through quality-raising initiatives. However, calls for mandatory minimum prices have been a continued refrain from various NGOs during the crisis period. Notably, in 2002 a legal petition was filed in the Madras High Court by a smallholder NGO, the Nilgiri Tea Growers' Protection Centre, demanding the institution of a minimum price scheme. Five years later, in July 2007 the Court issued writs that required the relevant parties (the government of Tamil Nadu, the Union Ministry of Commerce and the Tea Board) to find a solution to the problem of securing livelihoods for Nilgiris tea smallholders. Following the High Court decision, local growers' interests were quick to assert Rs 15/kg as a required minimum support price (*The Hindu*, 2007a), however the general consensus in the industry was that this was an ambit claim that, if implemented, would ruin the industry's international competitiveness.[28] Seeking to find middle ground, at the 2007 UPASI Annual Conference the MP for the Nilgiris R. Prabhu placed on the record his support for a minimum support price of Rs 10/kg for Nilgiris tea. However, in January 2008 the Executive Director of the Tea Board in South India, R.D. Nazeem, announced that the response to the Court's writs would be to amend the grower–factory price-sharing formula for smallholder tea (discussed above), from 60 to 65 per cent. This response, of course, neatly bypassed NGO demands for a mandatory minimum price in favour of a solution that continued to link growers' incomes to market prices. The president of the Nilgiris Small Tea Growers' Association was quoted in the media as indicating that although the proposed formula 'was not what the farmers had in mind, it was a step forward' (*The Hindu*, 2008).

## Conclusion

The interventions discussed in this last section of this chapter reveal an important characteristic of the Nilgiris tea sector. In thematic support with Chapter 5 (illustrating the partialness of global private regulation to alleviate problems of labour and livelihoods) and Chapter 6 (illustrating the limited scope of environmental codes of conduct to secure conservation outcomes in Kodagu), this chapter concludes by bridging the empirical narrative of Nilgiris smallholders, to the central arguments of the book.

Firstly, the value chain status of Nilgiris smallholders is contextualized by intersections of governance and institutions. The changing governance arrangements in tea GVCs at the end of the 1990s that served to marginalize Nilgiris smallholders enacted institutional responses in the form of quality-improvement programmes. These programmes were made possible by the particular institutional environment in which smallholders were embedded (and, in particular, the ways they were embedded within a supportive web of regional organizations such as UPASI-KVK and the SHG movement). In South India, political support for smallholder production systems is significantly improving the capacity of smallholders to engage with the complex and varied requirements of global value chains. The increased willingness of the estate sector to procure green leaf from smallholders through various kinds of outsourcing schemes further enriches the potential viability of smallholder production. However, as discussed in the last major section of this chapter, these quality-improvement programmes are themselves inserted within regional political, social and economic contexts. Agendas to improve growers' incomes through quality upgradation have had their share of success, but alone have not possessed the depth or scope to alleviate their ongoing dire economic circumstances.

In a robust democracy such as India, where NGO activism is ever-present, governments have felt obliged to augment such initiatives with the kind of responses discussed above. Quite obviously, the Government of India has responsibility towards the livelihood security of its citizenry, and courts are bound to rule on such matters when pressed. The story of Nilgiris smallholders' engagements with tea value chains is thus entwined within the governance arrangements into which they sell their tea, and the institutional environments that position them in particular ways to do this.

# Chapter Eight

# Making a Living in the Global Economy: Institutional Environments and Value Chain Upgrading

Late one afternoon in one of South India's tea-planting districts, we were interviewing an estate manager about labour management issues. He told us that every morning he liked to 'walk the rounds'; in part as an exercise regime, but also as an informal way of monitoring what was happening on the estate on a daily basis. Traditionally he'd take a different route every morning, so as to encounter workers unknowingly. With workers never really knowing when or where the manager would walk past, they were, in his words, 'kept on their toes'. But in the past year or so, he lamented, this strategy had gone awry. At first he noted that whenever he'd come across a group of pluckers, they seemed especially productive. But this didn't tally with his daily production figures, which were more or less unchanged. Only after a length of time did he cotton on to what was happening. The advent of mobile phones meant that pluckers could communicate with their workmates in different parts of the estate and forewarn them of the manager's imminent approach.

This humorous story makes a neat point of departure for this chapter. So far, we have emphasized the place-bound politics that ensue as economic actors embedded in *institutional environments* relate to the *governance* structures that determine social, economic and environmental outcomes. In the anecdote above, the pluckers' capacities to create a particular institutional environment via their usage of mobile phones shaped the governance arrangements which affected their lives. Through the previous three chapters, we have pursued comparably structured arguments with respect to labour and livelihoods in the tea industry, the environmental contexts of Kodagu coffee production, and the fate of Nilgiris tea smallholders. In this chapter, we bring a whole-of-industry perspective to these debates, and ask how the institutional environments of South Indian plantation sector *in toto* shape their prospects for *value chain upgrading*; the capacity of producers to improve the returns they obtain from value chain participation. Returning to themes we introduced in the Introduction to this book, we review whether and how South Indian tea and coffee producers are improving their livelihoods within contemporary globalization.

As discussed in Chapter 2, the concept of upgrading focuses attention on the economics and politics of global engagement. It asks questions about why and how global engagement can enable upstream producers to improve their prospects in some cases, but remain subordinated in others. In Gereffi's earlier work, the concept was applied to the question of how manufacturers in East Asian 'tiger economies' such as Taiwan were able to use their participation in global value chains to access the skills and technologies vital to national economic advancement (Gereffi, 1995). Other research has pointed to how particular governance structures inhibit upgrading. For example, research on African horticulture has highlighted how lead firms utilize tactics of control and coordination to subordinate upstream producers (Dolan and Humphrey, 2000; Humphrey and Schmitz, 2000; Humphrey et al., 2004). Evidently, therefore, the concept of upgrading reflects upon the potential double-edge sword of globalization.

Of direct relevance to the arguments of this book, moreover, the concept of upgrading provides a bridge that links the institutional and governance dimensions of the GVC approach. Referencing the notion of strategic coupling (Coe et al., 2004), we argue that the debate on upgrading is best articulated *not* from the (conceptually rigid) standpoint of how an external governance structure impacts upon place-bound actors, but from a *relational perspective* that understands governance arrangements and institutional formations as being co-produced. Hence, prospects for upgrading hinge on how the multi-scalar institutional formations into which economic actors are embedded interact with new governance arrangements frequently set in train by agents remote from their immediate environment.

The plantation industries of South India provide an apposite setting to deploy this concept. In Chapter 1, we discussed overarching tendencies in these industries. Many major consumer markets for these products have been consolidated through the forward integration of processing and trading firms. In both tea and coffee, a few multinational companies have accumulated a swathe of lucrative brands and now dominate this trade. These developments encase (existing and potential) upstream producers within restructured institutional environments. The enhanced size and global reach of lead firms in the tea and coffee industries gives these companies huge potential influence to reshape social, economic and environmental outcomes in upstream production sites. Yet in the same breath, we argue that this occurs not within a passive political landscape. The institutional environments of upstream producers shape both their capacity to participate in chains, and the economic benefits they obtain from such participation.

**Table 8.1** Upgrading initiatives in South Indian tea and coffee industries

| | | *Tea industry* | *Coffee industry* |
|---|---|---|---|
| **Process upgrading** (transforming inputs into outputs more efficiently by reorganizing the production system or introducing superior technology) | Efficiency-enhancing initiatives in tea and coffee cultivation and processing | • Plant breeding and other horticultural research; replanting schemes; mechanization and computerized plucking systems. | • Plant breeding; pests and disease management strategies; irrigation initiatives for small growers; use of biogas; streamlined processing. |
| | Initiatives with regards to market exchange | • Auction reforms; factory scheduling reforms and outsourcing of leaf procurement. | • Development of Indian-based coffee futures trading; e-trading; integration between curers and exporters. |
| **Product upgrading** (moving into more sophisticated product lines) | Ensuring tea and coffee meets/exceeds buyers' minimum standards | • Quality Upgradation Programme, MRL compliance, support for factory conversion/investment and ISO/HACCP certification, ETP compliance, meeting specialty tea buyer requirements (Harrod's). | • OTA compliance; changing chemical usage (e.g. Endosulfan and lindane); Utz Kapeh/4C compliance; meeting specialty tea buyer requirements (Illy, Starbucks). |
| | Generating premium prices from enhanced marketability and the development of niche quality attributes | • 'Golden Leaf' competition; GIs, trademarks and regional branding; single-estate tea, organic and fair trade certification. | • 'Cup of Excellence' competition; GIs, trademarks and regional branding; single-estate coffee, organic and fair trade certification. |
| **Functional upgrading** (acquiring new functions in the chain to increase the overall value content of activities) | In-country value-added processing | • Tea bag manufacturing and loose packaged tea. | • Instant coffee manufacturing (Unilever, Nestlé, Tata); roaster-retailer expansion (Coffee Day). |
| | Acquisition of brands | • Tata acquisition of Tetley's | • Tata acquisition of 'Eight O' Clock' coffee |
| **Inter-sectoral upgrading** (moving into different sectors) | Shift into estate tourism | • Kanan Devan Hills Plantation Company 'Tea Sanctuary' | • Tata Coffee bungalow rentals<br>• 'Homestay movement' in Kodagu. |

*Source*: Own work.

This chapter embellishes these points. As an organizational device, the discussion here follows the fourfold framework of different types of upgrading associated with the work of John Humphrey and Hubert Schmitz (Humphrey and Schmitz, 2002; Schmitz, 2004, 2006):

(i) *process upgrading* (transforming inputs into outputs more efficiently by reorganizing the production system or introducing superior technology);
(ii) *product upgrading* (moving into more sophisticated product lines);
(iii) *functional upgrading* (acquiring new functions in the chain, or abandoning existing functions, to increase the overall skill content of activities); and
(iv) *inter-sectoral upgrading* (using the knowledge acquired in particular chain functions to move into different sectors).[1]

As laid out in Table 8.1, the recent 'upgrading story' of South India's plantation sector encompasses a considerable number of initiatives under these four headings. Many aspects of this analysis have been discussed already within previous chapters, but some have not. Our purpose here is to consider them categorically, thereby drawing out their overarching characteristics and significance.

## Process Upgrading

Lower prices on international markets have spurred participants in the South Indian tea and coffee sectors to initiate a range of actions with the general intent of reducing costs and/or making production more efficient. To these ends, the institutional environment of South Indian tea and coffee production has provided a vitally important *enabling context* for such process upgrading to occur.

The recent history of process upgrading initiatives in South Indian tea and coffee sectors evinces two distinctive arenas where such agendas have been played out. First is the area of cultivation and processing, where R&D activities, extension services and strategic investments have enabled improved competitiveness. Examples of such upgrading described in this book include plant breeding programmes, improved processing methods and technologies, more effective pest management strategies, the introduction of irrigation, as well as the various attempts to minimize the wage bill, through mechanization, computerized harvest models, wage incentives for additional pluckings, and flexible labour regulation. The second is in the area of marketing and exchange, where recent years have witnessed intense debate over the imperative to streamline in-country systems of market exchange, and thus to improve transparency and reduce transaction costs within these sectors. In tea, the focus of these efforts has been related to improvements to the auction system. Although recent years have witnessed an increased propensity for South Indian tea to bypass auction channels (in 2002 fully 74.5 per cent of South Indian tea was sold via auction, but this had fallen to 59 per cent by 2006 [UPASI, 2007: pp. 1, 5]), auctions remain especially important for small growers and as a price discovery institution within the sector. Recent initiatives to reform the auction system are complex, and so are addressed separately in Appendix B. In coffee, it has involved the development of futures-based price transmission systems, the use of e-trading technologies such as plantersnet.com, and the integration of curer-exporters in the value chain.

The ways that producers are politically embedded within institutionally thick networks of national, regional and local organizations provides the platform from which these initiatives derive. As outside researchers, we were continually impressed by the abundance and availability of statistical data, the apparent quality and commitment of the extension services provided to producers, and the intensity and regularity of industry forums. Flagship events such as the UPASI Annual Conference are organized with a high degree of professionalism and sustained levels of industry commitment. It helps that many of industry actors, notably in the planters' class, are well-travelled and well-read; they possess an acute appreciation of how their industry is positioned in global markets and the issues it faces.

As discussed in Chapter 4, the tea and coffee industries have both been recipients of extensive R&D from government and industry organizations focusing on yield improvements, pest and disease mitigation, and, more latterly, sustainable cultivation. Indeed, a key feature of the evolution of these industries in India has been the sustained commitment of Colonial and post-Colonial administrations to R&D.[2] These arrangements, moreover, have been enacted in socio-cultural contexts whereby producers (in particular, the 'planter class') exhibit considerable professional pride and are thus enthusiastic recipients and implementers of R&D outputs. As noted with respect to planters in Kodagu in Chapter 6, moreover, the application of R&D activities can take place in complex ecological settings where local knowledge is paramount to their effectiveness, especially when it comes to issues relating to sustainable production and environmental conservation. Smallholders are targeted with purpose-specific schemes which seek to actively engage with these sectors (in coffee, the 'Support to Small Grower Scheme' described in Chapter 4; and in tea, the 'Quality Upgradation Program', described in Chapter 7).

The effectiveness of initiatives to improve the efficiency and transparency of market exchange is also intimately connected to the presence of institutionally thick organizational networks. This is seen most clearly in the case of coffee, where the development of e-trading and futures trading has involved joint actions by the Coffee Board, coffee industry associations and private companies. In tea, the restructuring of grower–processor supply arrangements (evidenced by the development of outsourcing schemes, as described in Chapter 7) is comparably premised on cooperative efforts amongst various parties, as estates will accept leaf only if minimum quality benchmarks are met, and this is expedited by smallholder support programmes from the key organizations of the Tea Board and U-KVK. The one area of market exchange in the tea industry where reforms have been forestalled is the auction system, which remains a point of intense conflict in the industry. However, during 2007, planters (though UPASI) began to take steps to develop tea futures trading that would augment/supplant the (allegedly flawed) price-discovery functions of the current auction system.[3]

Two conclusions come out of this overview of the diverse initiatives described above. Firstly, more intense competition on global markets give South Indian tea and coffee producers little option but to up the ante in terms of efficiency and productivity. These industries 'need to keep running to stand still'. The tea and coffee industries of South India have had been forced to invest in these efforts in order to ensure they continue to secure their (albeit often meagre) livelihoods for participants. Hence, these process-related initiatives have not so much been associated with 'upgrading', but have been necessary to avoid 'downgrading'.[4] Although the regional institutional environment provides an *enabling context* for driving efficiency gains, process upgrading alone is unlikely to radically transform these sectors for the better.[5]

Secondly, in terms of value chain dynamics, process upgrading occurs mainly through *horizontal* relations. Process upgrading in these industries fundamentally derives its strength from the acumen, energy and coordinating capacity of the industry itself. In the terminology of regional economics, it generally represents an *endogenous* source of transformation, linked to the ways that industries are politically and culturally embedded within their regional contexts. In an environment of excess supply, downstream lead firms have little incentive to intervene in ways that assist producers in regard to these initiatives, as resultant efficiency gains are captured by the producers themselves. Hence, in reference to earlier discussions on the relative importance of horizontal and vertical stimuli for upgrading, our assessment of process upgrading suggests it depends squarely on the former.

## Product Upgrading

Product upgrading refers to attempts to improve product quality in such a way that marketability is enhanced and a premium price is paid to producers.

Manifestations of product upgrading are best understood in terms of a 'ladder' of compliance and value-addition, with attainment of certain standards providing the basis for an ability to move 'to the next stage' and thus generate higher price-realization.

In South Indian tea and coffee industries, *de minimis* product standards require compliance with domestic food standard legislation (*Prevention of Food Adulteration Act*, especially important for tea) and, for exports, relevant benchmarks in consuming countries. As we discussed in the previous chapter, non-compliance with the basic domestic requirements of the PFA Act was historically a serious problem within the tea smallholder/BLF sector; however, interventions by the Tea Board in 2005–6 appear to have eliminated its worst excesses. The issue of export compliance is more complex and vexed, being entwined within the politics of international standard setting. In recent years, developed countries have demonstrated a propensity to adopt more stringent requirements with respect to import inspection and clearance. There is an overall shift in global agri-food trade towards more rigorous requirements for exporter registration and an onus on producers to give more thorough product assurances, which presents themselves through both state-based and private sector modes of regulation. A vital component of downstream lead firm governance within agri-food sectors is the deployment of 'technologies of audit' (Campbell and Le Heron, 2007: p. 133) which have the aim of intensifying monitoring of upstream suppliers (Harvey, 2007; Burch and Lawrence, 2007; Spriggs and Isaac, 2001; Hughes, 2005a, 2005b). State-driven change has been instigated to the greatest extent in the USA, where the *Bio-Terrorism Act* (2003) mandates that every food exporter selling into the USA must be registered with the US Food and Drug Administration and give prior notice of all shipments. An important manifestation of these requirements within the tea and coffee industries during recent years has been in respect to the setting of maximum residue levels (MRLs) for pesticides, an issue to which we will now turn.

The establishment of MRL limits is a complex multilateral exercise anchored around Codex Alimentarius, an international body jointly administered by the World Health Organization and the Food and Agricultural Organization. Codex Alimentarius is recognized by the World Trade Organization as an appropriate standard-setting body with regards to MRLs, but individual countries are not bound to follow Codex recommendations. The past decade has witnessed considerable debate relating to the roles of Codex and national standard-setting bodies in the global food system (Barling and Lang, 2005). On top of these requirements, of course, are private standards, which are often even more restrictive than Codex specifications. From developing countries' perspectives, the whispered suspicion is that excessively strict regulations imposed by western regulators and lead firms represent a *de facto* Technical Barrier to Trade (TBT)

(WTO, 2007). The claim is that excessively strict MRL ceilings in tea and coffee[6] have been written in mind of these countries' strong consumer lobbies and the linked economic imperative of multinationals to protect their valuable brand assets. This means, effectively, that producers bear considerable compliance costs for meeting standards to which downstream actors gain benefits. Regardless of the actual veracity of these claims, there is little doubt that the international politics of MRL compliance and enforcement illustrates the encasement of producer interests within a governance regime in which consumer country interests call the shots.

Whatever the scientific merits of particular food safety rules and regulations, the pertinent fact is that producers have been required to undertake considerable product upgrading efforts, just to ensure that their market access is retained. In this context, yet again, the strength of regional and national organizations at local levels has been vital. Both the Tea Board and the Coffee Board are highly active in these arenas, alongside industry associations. The discontent created out of these arrangements remains a volatile sticking point in tea, manifested in successive meetings of the IGGoT. Frustrated with the insubstantial progress made since the issue was first tabled at the Group's 13th Session in 2001, in August 2007 producer countries agreed to establish a new grouping – the International Tea Producers' Forum (ITPF) – to pursue these issues. Of note in the context of this book, Indian leadership was central in the ITRF getting off the ground. Indian producer organizations (especially UPASI representing the export-oriented South) provided the strategic vision to make the ITPF a reality.

Compliance with these statutory requirements then paves the way for producers to meet a more extensive and stringent set of market access criteria generally associated with private sector buyers. In general, these buyers' requirements are established either through company-specific Codes of Conduct (e.g., the Unilever SAI initiative, Starbucks' CAFÉ Practices program, etc) or collective private-sector arrangements (e.g., Global-GAP [tea and coffee], 4C [coffee] or the ETP [tea]). Schemes vary in terms of their scope, orientation, audit practices and engagement with civil society. Quasi-official process standards (notably HACCP and ISO systems) are often implicated within these schemes. Reflecting the strategic priorities of their authors, these forms of private regulation tend to mandate elevated (or at least, certified) compliance with specified quality, food safety, environmental and social standards.

Conforming to these private sector requirements enables access to affluent consumer markets that, in general, pay higher prices. At the same time, however, the costs of complying with these systems can be considerable. A persistent complaint of tea and coffee producers in South India (as with those in other parts of the world) is that although the cost of these private sector regulatory burdens seems to be ever-increasing, the price-realization from compliance does not. Indeed, the vignette that opened this book points to this very complaint. With respect to the 4C, in particular, the concern of producers is that the private sector authors of this scheme are keen to enshrine ethical practices into their upstream value chains, but are less keen to back this up 'by paying ethical prices'[7] (see also Neilson and Pritchard, 2007b).

A variation to these arrangements is in highly specialized gourmet product markets, where individual lead firms exercise considerable attention to 'bringing producers on board' with respect to particular quality requirements. An example of these private sector relations is provided by the experience of Illycaffe.[8] Based in Trieste, Italy, Illycaffe was founded in 1933 by Francesco Illy, who invented the first steam-powered coffee machine (predecessor of the modern espresso machine). Since then, the company has grown into one of the most successful coffee brands in the world, with a reputation for investing in research and innovation, and for supplying the upper-end high-quality segment of the market. With an annual turnover of €246 million in 2006, Illy Coffee is sold to 140 countries around the world and is the leading espresso brand in Italy (Businesswire, 2007). Illycaffe does not generally purchase coffee from international spot markets, but instead develops relationships with individual Arabica suppliers in producing countries. Along with Brazil, Guatemala and Columbia, India has emerged as an importance source region for the company in recent years (Urs, 2008). By 2002, the company was purchasing 2,400 tonnes of green coffee from India and whilst this is less than 1 per cent of India's total coffee production, it is almost 10 per cent of Plantation A grade coffee (Goswami, 2003). By 2007, Illy's purchases of Indian coffee had stabilized at around 2,500 tonnes of Arabica coffee per annum, with Indian coffee comprising up to 20 per cent of the company's espresso blends.

Illycaffe has very specific quality requirements and attempts to ensure these are met through partnerships with growers. Andreas Illy has proclaimed: 'We select growers, we transfer them our know-how and techniques and we pay a premium and sustainable price which ensures growers always make a profit' (Urs, 2008: not paginated). In late 2003, the company began organizing the 'India Coffee Quality Prize for Espresso', with winners taking home lucrative financial prizes and participating in visits to the company's facilities in Trieste. Then, in February 2007, Illycaffe announced the opening of a 'University of Coffee' campus in Bangalore, which will offer a variety of courses in agricultural practices as well as for retail-end baristas and other 'connoisseurs' (Kulkarni, 2007). These initiatives attempt not only to promote quality and coffee appreciation in a generic sense, but are also aimed at instilling Illy-defined quality conventions within source regions. Evidently, the company is taking a lead role in the establishment of quality conventions in accordance to its own priorities. This approach is consolidated at the farm level by ensuring price premiums for coffee meeting these standards. Coffee planters in Chikmagalur were receiving premiums of Rs 250 per bag (50 kg) for Illy-quality coffee (equivalent to about US$0.06/lb).[9] Price premiums then escalate at the export level where, according to one major exporter, during 2003–4 Illycaffe bought at an average FOB price of Rs 80/kg (equivalent to approximately US$0.80/lb when the ICO indicator price for Colombian Milds was around US$0.70/lb).[10] This compared to an average of Rs 45/kg for all other exports from India at the time. The active engagement of Illycaffe with upstream producers in India is, however, implemented through intermediaries (notably the large curing works/exporting firms) and not through direct purchasing agreements. Their ability to exert this influence through other value chain actors characterizes the role that roasting firms play in the governance of the global value chain for coffee. Whilst the upstream influence of Illycaffe in India is currently the most visible and far-reaching example of international roaster intervention in the sector, this may change in the future with the increased buying presence of Starbucks: in 2007 Tata Coffee announced that it had been selected by Starbucks as the first Indian supplier of green coffee beans to the company.

For producers, a vital feature of these schemes is that they represent 'upgrading within the chain': downstream buyers set the standards and/or facilitate the means for elevated product standards, and potentially higher levels of price-realization. However, the participation of producers in higher-paying market segments can also ensue through endogenous efforts to enshrine and promote regional quality attributes via geographical branding. In Chapter 4, we discussed the role of cooperative generic promotion efforts by public and private sector entities in building regional reputation (the *Golden Leaf* tea awards and *Flavor of India* awards in coffee).

Beyond this, however, place-based reputation can be codified as intellectual property via various legal forms. The simplest of this is the use of a corporate trademark to designate estate origins. In Chapter 4 we discussed how elite South Indian producers in tea (e.g., Chamraj Estates) and coffee (e.g., Palthorpe Estates) are increasingly branding their products in terms of single estate origins. Next in complexity come multiple-user trademarks, such as the 'Nilgiris mark' established by the Nilgiris Planters' Association (see Chapter 5). Finally, the most comprehensive system is that of Geographical Indications (GIs) registered under the aegis of the WTO's Trade-Related Intellectual Property Rights (TRIPS) Agreement. This is a form of product upgrading in that a respected geographical identifier provides a mark of exclusivity which can ensure that local producers control any benefits that arise from the attribution of geography in product marketing. In the Indian tea industry, the first attempt to establish geographically-based common-use intellectual property occurred with the 1998 trademark registration of a tea logo for Darjeeling (Damodaran, 1998) and the subsequent successful application of a GI for Darjeeling and Kangra (another production area in India's North-east). In recent years, the number of GIs registered in India has increased markedly. During 2007, the Government of India brought down favourable determinations for a slew of new GIs of relevance to the plantation sector including Assam orthodox tea, monsoon Malabar coffee, Tellicherry pepper, and Allepey green cardamom. Also in 2007, the Nilgiris Producers' Association finalized a GI application for Nilgiris orthodox tea, and at the time of writing, an official determination on this awaits. For producers, GIs can generate economic rents by promoting associations between the geographical origins and assumed notions of quality; and then retain exclusive control of these claims via the use of legal mechanisms. However, it should be noted that, despite the widespread interest in GIs across India, the actual economic benefits to producers of such protection are still being debated. Importantly, however, the establishment of GIs potentially threatens the capacity of downstream actors to control product marketing and, as a result, can be viewed as confrontational. Evidently, their capacity to achieve these ends rests on the institutional environment of capable producer associations, effective legal systems and supportive national policy mechanisms (notably, in the Tea Board and Ministry of Commerce). In this respect, India's enthusiasm and capacity to establish GIs is in contrast to other producing countries with very different institutional contexts, such as Indonesia (refer to Neilson, 2007). Should these goals be attained, producers in South India are provided with a platform for upgrading their position within global value chains.

This review of product upgrading underlines its diversity and significance for the tropical products sector at the current time. To chime again a point made in Chapter 1, gone are the days when tropical products' value chains were anchored by state marketing boards that arranged sales according to crude quality grades and operated price stabilization schemes. The contemporary tea and coffee sectors are marked by considerable price diversity based around product attributes. Climbing the ladders of compliance and reputation represents a vitally important aspect of livelihood enhancement for participants in these sectors. This being the case, it is apparent that different forms of product upgrading rely alternately on the mobilization of horizontal networks (e.g., ensuring compliance with food safety legislation or through applying for a GI) or are motivated by downstream lead firms seeking elevated compliance criteria (e.g., the ETP, 4C or Illycaffe's initiatives), or more commonly depend on the synergistic coupling of both.

## Functional Upgrading

'Functional upgrading' refers to attempts by value chain participants to *structurally* alter their positions within chains. For participants in South

Indian tea and coffee value chains, this logically involves downstream integration into value-added and/or branded products. The capacity to participate in these chain segments is beyond the capabilities of all but the largest corporate participants in the South Indian tea and coffee sectors. What we are talking about here is the capacity for South Indian tea and coffee producers to compete in the most hard-fought and lucrative parts of these sectors' global value chains. As Humphrey and Schmitz (2004: pp. 354–5) remind us, functional upgrading is a relative rarity amongst upstream developing country producers. In the South Indian tea and coffee sectors, there are only four corporate entities to which the discussion of functional upgrading meaningfully applies; the Indian operations of the world's two largest agri-food corporations, Nestlé and Unilever, and the two large Indian-owned corporate players, the Tata group and Amalgamated Bean Coffee (ABC).

These contexts raise an important question of definitional focus. GVC studies of functional upgrading are typically framed around a clear distinction between foreign lead firms, and subservient upstream (developing country) suppliers (e.g., Sturgeon and Lester [2004]). Their pivotal question is whether the latter can learn from their downstream customers and take functions off them. In the case at hand, however, we face a regional industry which includes two large foreign-owned multinationals (Nestlé and Unilever), and one local firm (Tata) with its own foreign subsidiaries. Also relevant here is the fact that although India remains a developing country, it is also rapidly emerging as a global economic powerhouse. With this come large affluent domestic markets, and thus a different set of political-economic contexts in which production, exchange and consumption takes place. Hence, the dualist distinction between downstream/foreign and upstream/local does not readily apply in this case. Therefore, the discussion that follows assesses functional upgrading in terms of both: (i) the extent to which the *regional economy* hosts activities that add value to these products compared with their basic tradable form (i.e., made tea and green coffee beans, respectively); and (ii) how these activities are connected to the repositioning of former plantation companies within the global tea and coffee value chains.

As illustrated in Table 8.2, the four major companies that are the focus of this discussion all have an extensive presence in various segments of the South Indian tea and coffee sectors. Strong growth in the Indian domestic market for branded/packaged tea has provided the basis for an overall expansion in processing facilities by Hindustan Unilever and Tata, the two companies which dominate this sector. Both of these companies have strategically divested from direct tea plantation ownership in order to focus on downstream activities. In the case of Hindustan Unilever, in 2005–6 the company sold its entire portfolio of six South Indian tea estates, which covered 7,000 ha and employed 18,000 persons (Unilever, 2006b: p. 22). The

**Table 8.2** Functional activities of major corporate groups in the South Indian tea and coffee industries

| *Company* | *Activities* |
|---|---|
| Nestlé India Ltd | *Production*. (none) |
| | *Processing*. Instant coffee factory at Nanjangud in Mysore; soluble tea processing factory at Choladi (Gudalur). |
| | *Branding*. Supplier of branded instant coffee to Indian market; *Coffee (instant)*: 'Nescafé', 'Sunrise'. |
| | *Links to overseas entities within the same corporate group*. Producer of soluble tea for export to other firms within the Nestlé group of companies. |
| Hindustan Unilever Ltd | *Production*. Owner of six South Indian tea estates until their divestment in 2005. |
| | *Processing*. Instant coffee and tea bagging and packaging factories at a number of locations Supplier of branded instant coffee, roasted coffee, packaged and bagged tea and instant tea to Indian market. |
| | *Branding*. Supplier of branded instant coffee to Indian market; *Coffee (instant)*: 'Bru' (domestic market), 'Bon' (exports to Russia); *Coffee (roasted)*: 'Bru Roast & Ground', 'Bru Malabar'; *Packaged/bagged tea*: 'Brooke Bond' ('Taj Mahal', 'Red Label', 'Taaza', '3 Roses'); 'Lipton' ('Yellow Label'); *Iced tea*: 'Lipton'. |
| | *Links to overseas entities within the same corporate group*. Producer of processed teas and soluble tea for export to other firms within the Unilever group of companies; Producer of instant coffee for export to Nestlé affiliates in Russia. |
| Tata | *Production*. Owner of 19 coffee estates; Owner of seven tea estates (in Munnar) until their divestment in 2005. |
| | *Processing*. Curing works in Karnataka and instant coffee processing factories in Hyderabad and Madurai; tea packaging and bagging and instant tea manufacture at Munnar, Kerala. |

| | |
|---|---|
| | *Branding*. Supplier of branded roasted coffee and packaged and bagged teas to Indian market; *Coffee (roasted)*: 'Mysore Gold', 'Mr Bean', 'Tata Coffee', 'Tata Kaapi', 'Coorg Pure', Valparai Gold'; *Packaged/bagged tea*: 'Tata Tea', 'Kanan Devan', 'Tetley', 'Agni', 'Gemini'; Supplier of roasted coffee to 'Barista' retailer-roaster franchise (part-owned until 2004).<br>*Links to overseas entities within the same corporate group*. Supply of tea to Tata-owned 'Tetley' operations in UK, Australia and elsewhere; Downstream acquisition of 'Eight O'Clock' coffee brand in US; Exploring potential for branding joint venture for the company's instant coffee for export to Russia. |
| Amalgamated Bean Coffee (ABC) | *Production*. Ownership of coffee estates in Karnataka.<br>*Processing*. Curing works in Karnataka.<br>*Branding*. Supplier of branded roasted coffee to Indian market via 'Coffee Day' retailer-roaster franchises and 'Perfect' and 'Fresh 'n Ground' brands.<br>*Links to overseas entities within the same corporate group*. (none) |

*Source*: Company websites and fieldwork.

Hindustan Unilever properties were packaged as Tea Estates India Ltd.[11] The media release justified the divestment thus:

> The Plantation Divisions do not fit in with the objective of HLL to focus on FMCG [fast moving consumer goods] businesses. The company therefore believes that it would be prudent to transfer them into separate subsidiaries with a view to providing clear focus to operations, both in terms of land productivity and manpower productivity to manage costs and restore economic viability ... Both the Divisions have gardens, which enjoy considerable equity both in the domestic and international markets. But they do not realize any premium for these equities, from captive supplies. It is believed that these equities can be better exploited in collaboration with an industry player, which is able to market garden teas in both domestic and international markets at considerable premium, while taking advantage of HLL's presence in Assam and Tamil Nadu, high quality plantation practices and harmonious relations with the work force. Also, operating experience over the last few years has demonstrated that there are very little synergies between the Plantations Divisions and the Packet Tea business. (Hindustan Lever Ltd, 2005: not paginated).

At around the same time Tata also sold off its tea estate interests. Since 2000, when the company successfully acquired the 'Tetley' brand, Tata had progressively become more focused on brand ownership rather than the ownership of tea plantations. In South India, Tata Tea's operations were focused on Munnar, Kerala, where it owned 24,000 ha, of which some 8,900 ha was planted to tea. As developed, these plans turned into what has been labelled 'the world's largest employee buy-out'. Details are provided in Appendix C, but the gist of the scheme involved the sale of the estates to a newly constituted company (Kanan Devan Hills Plantation Co. [KDHP]) which had been capitalized through a combination of loans and share purchases from local Tata Tea employees. Approximately 13,000 workers purchased shares, meaning that workers came to represent the majority voting block within the company's share registry. The 'Munnar model' signifies key elements of change in the South Indian tea industry. In relinquishing ownership of its upstream production activities, Tata has underlined the economic contradictions in operating relatively high-cost production arrangements within an environment of low price-realization among South Indian teas. Interpreted in the larger frames proposed by this book, it reiterates the interplay of forms of governance with the institutional environment. The fragile status of South Indian production in the global tea economy coupled with the inabilities of producers to gain lucrative price-realization on local auctions have provided an institutional environment unconducive to Tata's previous (vertically integrated) form of chain governance. The result has been a restructuring of organizational arrangements, leading to a re-localization of control, responsibility and risk within a company majority-owned by its plucker workforce.

In coffee, the growth of value-added market segments has effected a somewhat different response from the majors. The recent rise of retailer-roaster franchises in India has been connected with downstream investment by corporate coffee plantation owners. Based loosely on the 'Starbucks model' which, during the 1990s, revolutionized café décor in Anglo-American cultures, these cafés offer espresso-based coffee drinks, other beverages and snacks within air-conditioned premises. The context for these developments is the rapid rise of India's middle class, and the two companies at its centre are Tata Coffee and Amalgamated Bean Coffee (ABC), which bought into the 'Barista'[12] and 'Coffee Day' chains respectively. In 2004, Tata Coffee purchased a 34 per cent stake in Barista and began supplying the chain with coffee blends. The clear market leader at the current time, however, is ABC, which opened its first 'Coffee Day' on Brigade Road in Bangalore in 1996 and, by September 2007, had 450 'Coffee Day' cafes nationwide (*Indian Coffee*, 2007: p. 5). It also operates a comparable number of 'Fresh N Ground' and 'Coffee Day Xpress' kiosks for take-away customers only. The company's ownership of 'Coffee Day' enables it to boast on its website that it is India's only fully vertically-integrated coffee company.

The rapid expansion of chains such as 'Barista' and 'Coffee Day' has been assisted by the delayed entry of 'Starbucks' into India, which for some years was caused by restrictive FDI laws with respect to Indian retailing. Although foreign investment rules were substantially liberalized in India during the 1990s, restrictions remained on deemed 'nationally sensitive' industries, of which retailing was included (along with atomic energy, the lottery business, gambling and betting). Evidently, the prospect of large foreign-owned retailers out-competing the small business base of India's towns and cities was an extreme political risk for incumbent Indian administrations. Only in 2006, following US pressure, did the UPA Coalition Government of Prime Minister Manmohan Singh announce a retail liberalization package (Neilson and Pritchard, 2007a). Foreign investment restrictions gave breathing space for the Indian-owned 'Barista' and 'Coffee Day' to secure key real estate and gain invaluable brand-recognition among consumers. In the case of Coffee Day, moreover, it also paved the way for planned international expansion, with the company in 2007 announcing an intention to enter the European market. Meanwhile, in terms of foreign competition, the large UK-based chain 'Costa Coffee' opened its first Indian store only in 2006, and at the time of writing, retains only a marginal presence in the national industry. The entry of Starbuck's has been even more laggard, with press reports over the last few years frequently announcing the imminent opening of a Starbucks store in India (see, for example, Ghosh and Chatterjee, 2007). This was planned initially to occur through a joint venture arrangement with Kishore Biyani (owner of Pantaloon and Planet Retail Holdings), however in late 2007 this deal had collapsed and Starbucks had withdrawn its application to operate single brand retail stores in India (Rajghatta, 2007). At the time of writing (early 2008), Starbucks was still yet to open a store in India.

In both tea and coffee, these companies have made investments with the view of enhancing value-added exports. In tea, Nestlé and Hindustan Unilever now use South India as a sourcing site for soluble tea exports, destined for use as a base ingredient for these companies' flavoured iced tea brands (e.g., 'Nestea' and 'Lipton'). To this end, new processing facilities have been developed during the past few years. Comparably, instant coffee manufacturing facilities owned by Nestlé, Hindustan Unilever and Tata, developed initially to service Indo-Soviet trade, have been subject to new investment in the context of growth in domestic demand and continued trade with the former Soviet bloc.[13] In terms of the domestic market, instant coffee manufacturing is dominated by multinational beverage companies, including Hindustan Unilever ('Bru'), Nestlé ('Nescafé Classic' and 'Nescafé Sunrise') and Tata ('Tata Café' and 'Mysore Gold'), followed by smaller manufacturers such as Continental Coffee Products and the Narasu Coffee Company. Nestlé first established an instant coffee factory in India at Nanjangud (Karnataka) in 1989, whilst Tata has two instant coffee factories, near Hyderabad and Madurai, the latter of which has recently added a modern freeze-dried facility (*Indian Coffee*, 2006: p. 4). Paralleling their downstream move into tea branding, in 2006 the Tata Group purchased the US-based 'Eight O'Clock Coffee', the third-largest US instant coffee brand after 'Folgers' and 'Maxwell House' (*Indian Coffee*, 2006: p. 4). Additionally, Tata's 'Mysore Gold' instant coffee brand is popular in both Russia and Ukraine. Through the efforts of Tata, therefore, Indian-owned coffee brands are becoming key players in international markets.

However, the South Indian story of functional upgrading is clearly different from the much-heralded example of the Sri Lankan tea industry, where the pioneering efforts of a leading local-owned tea company (Dilmah), in conjunction with initial supports and incentives constructed by the Sri Lanka Tea Board and the Export Development Board of Sri Lanka, encouraged an overall shift in the industry towards value-adding (Ganewatta et al., 2005). The IGGoT (2005) reports that, in 2003, 41 per cent of Sri Lankan tea exports were in the form of value-added products (i.e. retail packs, tea bags, teas blended and packaged for retail sale, etc.). By comparison, in 2005 only 16 per cent of Indian tea was exported in packaged form.[14]

As the domestic Indian market has become more affluent, new opportunities have been opened in branded tea and coffee markets, and in the franchised retailer-roaster café sector. Other instances of functional upgrading (instant coffee sales to Russia; soluble tea exports) seem to reflect opportunistic ventures by these companies; being in the right place at the right time. It is clear that the considerable size of the Indian domestic market has provided an important rationale for both: (i) the location of value-added processing activities within South India; and (ii) the capacity of 'Indian-owned' companies to restructure their production portfolios towards brand-oriented activities in the global market. As argued by Schmitz (2006), however, functional upgrading is unlikely to occur from 'within the chain' as downstream lead firms protect their positions at key nodes of value-addition within the chain. Rather, it is often the regional and national institutional settings, including politically-motivated protectionism (such as the restriction of FDI in retailing allowing growth of domestically owned café chains) and the supportive institutional environment that have encouraged value-added functional upgrading in India.

## Intersectoral Upgrading

Compared to the categories of upgrading discussed in the three sections above, intersectoral upgrading does not really figure prominently in these industries. The very nature of tea and coffee as permanent tree crops and with estates having legal obligations with respect to land usage and labour, it is not easy to diversify production. Unless, of course, 'intersectoral upgrading' is used to refer to the adoption of generic plantation management skills in allied plantation commodities such as rubber, or to the application of brand promotion skills by companies such as Tata to a broad selection of products from automobiles to roasted coffee and packaged tea. Nevertheless, the extent that this form of upgrading has meaning within South Indian plantation districts is most evident in the growth of plantation-based tourism.

India's recent economic boom has provided a new set of contexts in which the tea and coffee sectors operate. The massive boost to urban middle-class incomes has pushed up the real estate value of tea and coffee lands, especially in amenity-rich environments such as Kodagu and the Nilgiris. Both of these districts are within a day's drive from Bangalore, and have become increasingly popular as locations for weekend breaks for that city's well-heeled professional elites. This has enticed many tea and coffee producers to augment their productive activities with tourism interests. 'Estate bungalow' homestay tourist rentals are now a prevalent feature of the Kodagu landscape (and have been embraced, in particular, by small and medium-sized coffee planters), and are becoming more widespread in the Nilgiris. To date, however, the most sophisticated expression of these trends towards a pluriactive rural landscape is in the vicinity of Munnar, Kerala, where the Kanan Devan Hills Plantation Co. has developed 'The Tea Sanctuary': a network of estate bungalows for high-priced tourist rental.

Such initiatives evidence a putative shift from plantation districts being wholly production-centred agrarian landscapes, to situations of multifunctionality; where production takes place in environments valued also for their consumptive (tourist and amenity) values.

## Summary: The Future of Upgrading for South Indian Tea and Coffee

The discussion in the sections above contextualizes the argument by Humphrey and Schmitz (2004) about the capacity of process and product upgrading, but the difficulty of functional upgrading. There is strong evidence that the ('horizontal' networks) of cooperative and supportive public and private sector organizations in South India are helping to generate considerable *process-related* efficiencies which are serving to sustain the participation of producers in world markets. There is also a robust case to be made that some producers, through a combination of 'horizontal' and 'vertical' processes, are working their way up the 'ladder' of *product upgrading*, and hence generate improved price-realization. Functional upgrading is taking varied forms, with uneven implications for industry participants, but is overwhelmingly dependent on horizontal rather than vertical supports. The observations presented in this chapter, then, provide further evidence of the complex interactivity between multi-scalar institutional settings (including here the internal governance structures of the value chain itself) and the life-chances of participants in the plantation industries of South India. In turn, these settings are constantly being altered through a series of place-based struggles, such as those presented in earlier chapters.

Identifying the longer-term effects of these transformations, however, is fraught with difficulty. As Schmitz (2006: p. 567) observes, 'Products and customers that appear minor in volume terms could be of major significance for the learning and eventual functional upgrading of the enterprise'. These industries are now undertaking more downstream processing and branding than they did in the past; however, at the present time it is unclear whether these activities will fundamentally alter the structural position of South Indian production within global value chains. We come to the conclusion, therefore, that the recent global restructuring of tea and coffee value chains is indeed a double-edged sword only partially within the rubric of local participants to control. The historical condition in the tropical products sector is one of excess supply, a sharper plane of price differentiation, and more rigorous specification of requirements by those lead firms who act as gatekeepers to potentially higher-paying markets. Producers face the need to comply with a succession of benchmarks with respect to food safety, product quality and credence attributes, if they are to move from the (often sub-economic) bottom rung of global markets. In conditions of information density and complexity, their capacity to do this is critically defined by the institutional environment in which they are situated. Geography has become more, rather than less, important in a globalized world marked by greater ease of transport and communication.

Our observations of South India lead us to emphasize the saliency of these arguments. As a producer of relatively low-value teas and coffees, the presence and strength of the institutionally thick web of industry and public sector networks represents a vital asset that enables the region's producers to 'stay in the game' of global competition. In the context of considerable present and future challenges to these sectors resulting from India's rapid economic transformation (labour supply problems, an appreciating currency, etc.), the tea and coffee industries of South India cannot be expected to remain viable if they simply continue to produce commodity-grade outputs. Global markets will continue to 'move on' to lower-cost suppliers. The future of upgrading within the South Indian plantation sectors will hinge on a combination of process-related initiatives to generate higher efficiencies, product differentiation to ensure access to high-end markets, the presence of downstream value-added industries, and successful diversification away from dependence on commodity production alone.

Critically, change is sweeping through the South Indian plantation districts. Confronted during the past decade by a coincidence of deregulated markets (both domestically and internationally), the shift in economic power to downstream lead firms, and the worldwide crash in tea and coffee prices, South Indian producers have undertaken an array of initiatives (based around process, product and intersectoral upgrading) that have sought to keep these sectors afloat. The next era in the industry would seem to require a more sophisticated set of responses still, in order to build an industry that transcends its legacy of Colonial-era mass production, and to become a producer of superior-quality, diversified value-added outputs. As this occurs, tea and coffee landscapes will be transformed. The amenity values of these upland areas (discussed briefly under the heading of intersectoral upgrading) will increasingly come into play. The key to developing a successful strategy to navigate through these opportunities and constraints will be to identify those aspects of the existing institutional environment that can be effectively coupled with opportunities offered within prevailing value chain structures.

# Chapter Nine

# Conclusion: What We Brewed

The geographical approach taken in this book has sought to situate value chain restructuring *in place*. We began in Chapter 1 by asking the question 'How is global value chain restructuring impacting upon South Indian tea and coffee producers?' Our frame of reference has not been to investigate tea and coffee chains as abstract structures, but their grounded manifestations *within* South India. Such an approach inevitably brings forth a complex and multilayered narrative of different sets of changes affecting regional economic actors in different kinds of ways. Whilst we can 'see through the trees' to offer generalized insights as to 'what has happened' (e.g., our analysis of upgrading presented in Chapter 8), our view is that those 'trees' are equally part of the story we tell. The abiding concern of this book has been to craft a reinvigorated geographical approach to GVC analysis; to complement the extensive research on GVC governance with a stronger focus on the institutional dimensions of chains that more effectively account for geographical variability. Through such an approach, we are then able to tease out critical insights into: (i) the dynamics of industry restructuring occurring within the global value chains for tea and coffee; and (ii) what this means for the future of regional production spaces such as South India.

This concluding chapter now distills the essence of arguments presented throughout the book. These culminating comments are submitted under two headings: (i) Institutions and governance in GVC theory (our contribution to an institutionally reinvigorated GVC model that adequately incorporates geographical diversity); and (ii) Struggles of upstream producers (we present industry restructuring in the GVCs for tea and coffee as the outcome of a constantly evolving series of struggles between multi-scalar institutions to enforce governance).

## Institutions and Governance in GVC Theory

The global tea and coffee industries exemplify a defining characteristic of the contemporary global economy, namely the intensifying upstream influence of multinational companies, consumer advocacy groups and international NGOs, enacted along the axes of global value chains, which is resulting in a re-regulation of trade with profound impacts on economic outcomes in supplier economies such as South India. Researchers concerned with these changes have increasingly turned to conceptual models such as the global value chain approach to explain the dynamics of industrial restructuring. The GVC approach provides a robust, and practically useful, conceptual toolkit that can be used to explain the role of 'lead firms' in dictating the terms of engagement for various economic actors in a globalized trade system. However, we further assert that the way global value chains are governed is equally determined by the multi-scalar institutional settings that have evolved throughout the value chain as a consequence of discrete place-based struggles over time. It is this assertion that compels us to put forward an alternative GVC model that more effectively accounts for such reiterative processes.

In broad terms, the orientation of our contribution echoes the concept of *global ethnography*, associated with the research of Michael Burawoy (2000a, 2000b), and Doreen Massey's (1993a, b, 2005) observations on geographical scale. The aim is to operationalize a multi-scalar *relational* framework in which observed outcomes at one scale are understood as manifestations of co-produced changes at a range of others. This perspective rejects 'globalization from above' accounts which posit individual places as passive in the face of a global 'other'. In this book, for instance, our interest has been on the iterative interplay between different geographical scales. Thus, for example, in Chapter 7 we narrated the story of depressed tea prices and a 'buyer's market' which raised the importance of quality parameters, and subsequently evoked a response in the form of the Quality Upgradation Programme in South India that aimed to re-engage smallholder producers with new market conditions. It is this kind of multi-scalar iteration of processes which creates and recreates economic geographies.

Applying these arguments to GVC analysis leads us to propose a revised approach that incorporates key insights from GPN theorists (*inter alia*, Bathelt, 2006; Coe et al., 2004, 2008b; Henderson et al., 2002; Hess and Yeung, 2006). In most GVC studies, governance issues are the central concern. Our approach recalibrates the emphasis towards the institutional environment. This shift in focus addresses the illusion (sometimes prevalent) that governance structures are rolled out in a spatial vacuum. In fact, global value chains are defined by a series of struggles which result in the co-production of governance structures and evolving institutional environments. These struggles determine how and to whose advantage economic actors participate in chains.

Successive chapters of this book identify processes at play in South India through these terms. In Chapter 4, we saw the path-dependent processes that have created institutionally thick economic geographies in South India's plantation districts. The coalition of planters' associations, commodity boards and political representation keenly defines the way this regional production complex engages with global markets. This complex presents an articulate and relatively coherent voice at international industry forums and intergovernmental groups and, as a result, plays an influential role in shaping the global institutional settings of the global value chains for both tea and coffee. This is evidenced through participation in debates at the ICO and the IGGoT (especially in terms of the MRL standard-setting processes) as well as resistance to industry codes such as the 4C, which is being re-formulated to respond to Indian industry concerns.

In Chapter 5, we saw how the pursuit of ethicality in tea production by lead firms in affluent western markets led to the systematic reproduction of uneven labour geographies. The emergence of these schemes in the first instance arose out of the fact that sites of tea production and consumption are inhabited by societies with highly disparate ethical values and labour conditions. This combination of settings at opposite extremes of the value chain created a new institutional norm in the form of the ETP, Fairtrade and other forms of ethical certification. The *modus operandi* of these schemes is, however, being reconfigured as a result of institutional struggles initiated by chain actors. For instance, the dominance of estate production in the global tea industry has influenced the rules and conditions under which Fairtrade tea certification can be obtained (i.e. not restricted to smallholders as is the case in many other commodities). The actual consequences for labour regulation in South India are emerging as the outcome of institutional struggles taking place between multi-scaled regulatory processes, and will continue to depend heavily on the constellation of local political and industry organizational structures.

In Chapter 6, we saw how externally authored sustainability agendas in the coffee forests of Kodagu clashed with, and unsettled, diverse pre-existing ways in which cultivators were socially, economically and environmentally embedded within these sites of production. Over the last few years, lead firms in the global coffee industry have sought to institute a mode of value chain coordination whereby producers would comply with a series of ethical and environmental baseline standards (the 4C initiative). The diverse ways that coffee producers are embedded within ecological, social and political systems worldwide has led to highly variable levels of compliance to the scheme. Indeed, the environmental governance structures within Kodagu, comprised of complex land tenure arrangements, highly specific cultural attitudes towards conservation, and notions of planter stewardship are unlikely to be brushed aside by globalized notions of sustainability from elsewhere in the chain. Instead, the overlapping institutions of environmental governance from various scales are interacting to generate new outcomes on the ground with ramification for how sustainability schemes are instigated globally.

In Chapter 7, we discussed the ways in which Nilgiri tea smallholders' abilities to engage (or not) with global markets were forged from their access to support agencies, such as the Tea Board and UPASI-KVK. Downward price pressures from international tea buyers was challenging the ability of South Indian estates to survive in the global market, and appeared to be driving a process of outsourcing leaf production to smallholders. At the same time, increasingly stringent quality and hygiene standards imposed by buyers were a threat to smallholder participation in the more lucrative value chains flowing into affluent western markets. The form of industry restructuring occurring in South India, however, has been preconditioned by the institutional environment within which actors are culturally, politically and economically embedded, resulting in responses such as the Quality Upgradation Programme.

It is through these interlocking sets of institutional norms that industry restructuring is taking place, punctuated by series of resistance, struggle and institutional evolution. How exactly, then, do we envisage that such recursive processes can be incorporated within a GVC framework? Due to the immense practical utility of GVC theory, and its capacity to articulate contemporary realities of corporate-led governance structures, we propose maintaining its central nomenclature (agreeing with Sturgeon et al. [2008] on this point). However, what we add to the mix is greater attention to the evolution of chain governance structures as the culmination of institutional struggles by place-based actors. This enlargement of focus allows space for a more forthright geographical perspective within the framework. It draws explicit attention to how governance structures are shaped from the institutional environments in which economic actors are embedded. These notions dovetail with recent advances within an explicitly relational approach to economic geography, where institutions are increasingly seen as multi-scalar manifestations. These in turn are reproduced from the way economic actors are socially embedded at different stages of the value chain.

This approach differs from that generally applied in GVC studies. The 'institution settings' of a GVC has been used conventionally as a convenient shorthand for an assessment of global rules of trade, as embodied in such institutions (for the beverage crops) as the International Coffee Agreements, WTO agreements and corporate schemes such as the Ethical Tea Partnership, and sometimes extending into the arena of national policy, manifested in tariff rates and taxation. However, the analytical framework articulated throughout this book proposes that the underlying social and cultural institutions that give birth to broader structures of governance and organizational form should be given an equal footing. The divergent global trade patterns of tea and coffee, where the former is increasingly sold into Eastern Europe and the Middle East and the latter is predominately sold into Western Europe and North America, has profound implications for the way chain governance mechanisms have evolved for each commodity. This, we argue, is due to the place-informed institutions that determine consumption and trade cultures in these different markets, characterized by a heightened concern for ethicality and product traceability in western markets (along with greater retail concentration). These influences are multi-scalar in the sense that the institutional setting at any particular node in the value chain is prefigured by local norms operating alongside national and global institutions. Why is such an approach helpful? By removing the unidimensional focus on global, buyer-driven regulation of value chains (the 'linearity myth' [Beck, 2000]), this approach helps to open up new opportunities for engagement, resistance and the creation of future possibilities.

Through the geographical perspective just outlined, we give emphasis to the ways place-bound actors are embedded (in territorial arrangements, networks, and societal structures), such that contemporary value chain restructuring is reflective of a series of *value chain struggles* as *institutions* negotiate the ability of *governance* structures to determine social, economic and environmental outcomes. This meso-scale frame of inquiry asserts the importance of grounded economic actors in configuring value chain arrangements. It suggests that an understanding of the outcomes from global change require an appreciation of the specificities of place; how and why particular economic actors relate to others in particular ways. For this reason, the substance of this book has concentrated on the nitty-gritty of recent episodes of change in South Indian producer communities. Such attentiveness brings forth grounded insights into who is winning and losing from contemporary globalization processes, and why.

## The Struggles of Upstream Producers: Surviving the Tropical Products Crisis

The chapters of this book reveal a complex and intricate story. But that, precisely, is the point. We empathize with Robert Wade, the influential development economist, in his landmark book on East Asian industrialization, *Governing the Market*, who quipped that: 'One distinguished development economist who has written at length on how Taiwan grew rich through applying nearly free market principles, when asked why none of his three visits to Taiwan had exceeded one week, replied, "If I stayed longer I'd just get confused". Exactly so, but it would be a more sophisticated kind of confusion' (Wade, 2004: pp. 50–1). In our case, there is no structural answer to the generic question of whether value chain restructuring is assisting or hindering livelihood conditions in South Indian plantation districts. Rather, these developments are reproducing geographies of access and denial; advantage and disadvantage; amenity and disrepair. These outcomes are the product of place-bound struggles, as we have reported, among economic actors embedded in various and manifold ways.

During the past decade, the fate of producer communities in tropical commodity sectors has been a touchstone for wider debates on developing countries, agriculture and globalization. Encouraged by the overarching ideology that has come to be known as the Washington Consensus, these formerly heavily-regulated sectors have undergone extensive liberalization. From the perspective of market liberalization advocates, these policy actions were said to be necessary to improve producers' livelihoods in the context of mismanaged and often-corrupt statutory marketing boards, supply retention schemes and, generally, the meddling by governments in markets. In this vision, liberalization advocates argued that allowing a freer hand for the market would give greater opportunities for individual producers to tailor their outputs to the nuances of supply and demand, thereby encouraging an overall improvement in producer livelihoods. Yet these views were not accepted by all. Through the 1990s and into the 2000s, researchers and NGOs emphasized how freer markets in these product sectors were advantaging some at the expense of others. The liberalization of these markets occurred hand-in-glove with rapid concentration of downstream components of these products' value chains. It was argued that the global reach of multinational firms enabled rent capture through their ability to play off one upstream supplier against another. The oppositional nature of this debate – addressing the question of whether or not contemporary forms of (liberal market) industry governance are acting to the benefit of producer communities – remains the prevailing way in which these industries are interpreted within multilateral organizations, NGOs and much research scholarship. Yet the story of institutional change and resistance with the tea and coffee sectors of South India presents an alternative interpretation of recent developments.

Just prior to the commencement of research for this book, world tea and coffee prices abutted historic lows. For coffee, the ICO composite indicator price fell from US$1.33/lb in September 1997, to US$0.41/lb in September 2001 (Figure 3.8). In tea, the FAO composite price fell from US$2.52/kg in February 1998, to US$1.44/kg in June 2002 (Inter-Governmental Group on Tea 2005: p. 3). This sharp collapse in prices indisputably caused massive levels of hardship and human suffering. During 2004–7, the years when this research was undertaken, tea and coffee prices recovered somewhat from these crisis levels, but, nevertheless, they still remain low in comparison to historic trends.

The fluctuation of commodity prices is a fact of life, but recent trends represent more than mere market wobbles. As reported in the 'territoriality' sections of Chapter 3, the past couple of decades have witnessed a profound geographical expansion in tea and coffee cultivation. New countries have entered these industries, and some existing producers have expanded their output. In conceptual terms, these processes reflect structural changes in the ways developing country agriculture is inserted within global markets. As part of their incorporation into the global economy, countries such as Viet Nam have aggressively increased their agri-exports. Comparably, existing producers such as Sri Lanka (tea) and Kenya (tea and coffee) have restructured domestic production and trade systems for these foreign-exchange earning sectors, resulting in an overall expansion of output. Tea and coffee production has become, in a word, more *globalized*; with an increased number of participants and more intense competition between them. These forces are inexorably encouraging a sustained decline in commodity prices. Frosts in Brazil or political upheaval in Kenya may momentarily cause prices to spike, but the long-term trend is downwards.[1]

In this context, there is a profound *development problem* that is central to the future of the tropical products sector. Despite a diversification of developing country agri-food exports (Humphrey, 2006), tropical products continue to provide an important source of foreign exchange earnings for many developing countries. More importantly, they underpin the employment base for large sections of the rural economy, especially in upland areas that may not have other obvious agri-export potentials. Maintaining the viability of tropical product sectors within the context of depressed world prices thus represents a key challenge within these industries. But what exactly has the recent era of liberal policy settings delivered? As Gibbon (2005) observes, this period of transformation in the tropical products sector has produced a more ambiguous set of outcomes than anticipated by liberalization proponents. With liberalization coincident with a period of lower world prices, the downside of free markets has weighed heavily on these industries. Bastardizing Winston Churchill's dictum about democracy being 'the worst form of government, except for all the others', apologists contend that liberal markets may provide the worst possible world for tropical producers, except for all of the other alternatives (Lindsey, 2004). Nevertheless, as the conditions facing producers have become increasingly severe, it is not surprising that increasingly robust calls have been made to develop new global mechanisms to (re-) regulate production and trade (Robbins, 2003; Koning and Robbins, 2005). This push gathered pace in 2006, when African countries petitioned the WTO to reintroduce a system of International Commodity

Agreements to manage demand and supply in tropical commodities (World Trade Organization, 2006).

Notwithstanding debate on the merit of these calls, the ideological and practical hold of market liberal principles within contemporary global governance makes it difficult to envisage a second coming of tropical product trade regulation of the form created through the erstwhile commodity agreements. Once the genie of market liberalism is out of its bottle, it is difficult to lock away again. In any case, the geographical diffusion of tea and coffee consumption to countries of the former Soviet bloc, the Middle East and East Asia, discussed in Chapter 3, would make any future international commodity agreement extremely difficult to negotiate and even harder to enforce. The International Commodity Agreements of the mid-twentieth century were undertaken in the context of a far simpler geography of production and consumption than that which exists today. Instead, new patterns of chain governance in these commodity sectors are emerging based around various modes of supply chain accountability, described in earlier chapters. Perhaps somewhat surprisingly, these patterns of non-state regulation are evolving out of the ways value chain actors are embedded within their respective environments and societies; market liberalization has actually accentuated the importance of underlying institutional influences.

Furthermore, assuming that the problems facing tropical products industries will not be addressed by a return to price-inflating international commodity agreements, the issue for producers is increasingly becoming one of how to position themselves optimally within global value chains. For researchers investigating the 'development problem' of tropical products industries, recent developments appeared to hold vexed prospects. The accumulated market power of downstream branded coffee roasters and tea blenders is certainly defining the scope for upstream chain activities, as lead firms instill buyer-driven forms of governance that dictate the terms of engagement to producers. These relations included the price and payment conditions for suppliers, the defined acceptability of grades and standards, and the monitoring and enforcement of credence claims relating to the environmental and social conditions under which production occurred. In short, the combination of the global market conditions described in the section above with these shifts to global value chain governance seemed to deliver a double-impact on producer interests; lower prices and more rigorous requirements for servicing markets. Again, as exemplified by our discussion of upgrading potentials in Chapter 8, the capacity of producing regions to respond to changing global market conditions is also fundamentally predetermined by the local institutional setting of production and exchange. It can be expected that, as in the case of estate abandonment in Central Travancore compared against Nilgiris smallholders, these processes will exacerbate territorial difference.

Much of the content of this book has been concerned with documenting and reviewing these developments in the South Indian context. In Chapter 1, we introduced the concept of *global private regulation* as a descriptor for the tendency for lead firm interests to author and monitor a set of requirements relating to product and process standards. Because these requirements tend to sit outside of the formal state-based rules of trade, it has been suggested that such institutional norms are indicative of a new phase in the international agri-food economy, where the private sector becomes the chief regulatory driver (Busch and Bain, 2004; Barling and Lang, 2005; Friedmann, 2005). The ascendency of such regulatory shifts reinforce the central theme of our book: that the institutional settings of any particular node along a global value chain is itself defined by the convergence of various multi-scalar institutions created through a series of place-based struggles.

Our contribution to debates over these issues is encapsulated in Figure 1.1. The box on the left of the diagram is illustrative of external pressures being placed on upstream producers via agendas of vertical coordination and supplier compliance. This is orchestrated through technologies of traceability and the widening domain of global private regulation. The geographical expression of these agendas is *universality*; the implicit objective behind schemes such as Global-GAP, the Ethical Tea Partnership (ETP) and the Common Code for the Coffee Community (4C) is for downstream lead firms to be able to deal with upstream suppliers *anywhere in the world* according to a common template. Fundamental to this objective is a commonly understood *lingua franca* specifying accepted benchmarks relating to the conformity of grades and standards, with a remit that covers such things as product specification, food safety and the ethical and environmental basis of production. Yet at the same time, as represented by the box on the right of Figure 1.1, the proposed enactment of these agendas occurs in a world of territorial difference. Across the world, upstream producers are embedded within local, regional, national and international economies in vastly differing ways. This variability shapes their capacities and willingness to comply with the external agendas discussed above. The result, as we suggest in the middle box, is an arena of struggle as upstream actors seek to negotiate their relations with buyers. The four bullet point 'struggles' identified in this box correlate to the four themes we addressed in Chapters 5–8 of this book; labour, environment, smallholders and upgrading. The conduct and outcome of these struggles is far from certain, and is indeed a constantly evolving and reiterative process.

Placing these themes within regional production contexts, moreover, sheds light on their inchoate and highly uneven penetration. Thus, whereas flagship schemes such as the ETP (in tea) have a presence in the South Indian industry, much production occurs beyond their frontier; for example, smallholder production sold at auction for markets in the domestic economy and for export to the Middle East, Pakistan, etc. In coffee, it is similarly the case that much of South India's production is destined for markets that are beyond the frontier of these considerations. Premium-priced speciality Arabica and Robusta destined for Western European markets accounts for less than 5 per cent of industry output. Almost one-quarter of the industry's production takes the form of cheaper Robusta sold to Italy (where it becomes an anonymous 'filler' bean for blends) and another quarter is converted to cheap instant coffees sold mainly to countries of the former Soviet bloc. The significance of tea and coffee markets 'beyond the frontier' of the lead firm/private regulation system raises an important point for scholarship and policy development relating to the tropical products sectors. Its limitations exert a powerful reminder of the continuing diversity of chain formations within these sectors. The world's tea and coffee is not sold exclusively into brand-conscious western markets. Whilst the integrity of provenance might be paramount to well-heeled European, American, Japanese and Australian consumers, it may barely figure in other marketplaces.

The fundamental point of this book, then, is that it is in the intricate detail of how and why place-specific actors respond (or not) to global market signals that we understand the path-dependent trajectories through which economic outcomes emerge. As we have shown, the status and upgrading potential of individual economic actors in South India is shaped by the complexities of such things as the land tenure histories of Gudalur, Badagas political representation in the Nilgiris, the *Jamma* rights of Kodagu coffee planters, and the legacies of British colonial planter culture. It is through these mediums that the complex story of value chain struggles in South India derives.

# Appendix A

# The Role of Managing Agents

The role of managing agents at the centre of these interlocking commercial networks was described in a satirical piece titled *Managing Agents Shorter Catechism* in an 1898 edition of *Planting Opinion* (a South Indian tea industry journal) from which we have extracted the following:

> *What are managing agents?* We are the middlemen between the Board of Directors in London and the Managers of the Tea Estate.
>
> *What is a Board of Directors?* A small body of kindly-disposed elderly gentlemen, who knowing nothing about the working of tea estates, and believing all planters are unbusiness-like and unreliable, wisely leave the control of the garden to us.
>
> *What is a manager?* He is a planter appointed by the Board to work and look after the gardens, but whose whole time is much better employed in supplying us [the managing agents] with information and statistics.
>
> *What is the chief aim of the Managing Agents?* Our chief aim, after taking care of DOWB,[1] is to impress the Board with the enormous amount of skillful supervising we bring to bear on the Manager.
>
> *How is this best accomplished?* By constantly inventing new forms of elaborate statistics to be supplied by the manager, which, republished and sent home [i.e., Britain] in neat, typewritten columns, causes the Board to feel how fortunate the company is to possess such able and zealous Agents.
>
> *Is it desirable that Managing Agents should understand the working of an Estate?* No, it is most undesirable.
>
> *Why?* Because such knowledge would seriously hamper the freedom of our criticizing.
>
> *What is a shareholder?* He is a man of no importance.
>
> (*Planting Opinion*, 1898; reproduced in Kariappa et al., 2004: pp. 42–4)

# Appendix B

# The Operation and Intended Reform of South India's Tea Auctions

South India's tea auction system plays a pivotal role in connecting tea producers with buyers, and ensuring price-discovery and market clearance. In India, the conduct and rules of individual tea auctions are laid down by tea trader associations, representing the interests of the brokers and buyers who participate in each auction place. The system as a whole, however, has oversight from the Tea Board of India, which is empowered to intervene in the tea industry through legal orders. In the South, the system as a whole works to a coordinated weekly schedule with auctions conducted in Kochi (Tuesdays, Wednesdays), Coonoor (Thursdays, Fridays) and Coimbatore (Fridays). Of these three auction sites, Kochi accounts for roughly 50 per cent of South Indian auction throughput. Coonoor then accounts for approximately 30 per cent and Coimbatore just 20 per cent (UPASI, 2007: p. 7). To some extent, these different auctions have a reputation for different teas and therefore different buyers will tend to participate to a greater or lesser extent in each depending on their needs. For example, the Kochi auction tends to offer more dust grades (it has been historically associated with the so-called 'red dust' teas of Kerala)[1] while much of the leaf grade tea produced by smallholders and destined for loose-tea markets in North India and some export markets is sourced out of Coonoor. Consequently, there can be a process of 'buyer matching' that involves selectively placing a tea in the correct auction to obtain the best price. These reasons help to account for the significant differences in average prices between these auction centres.[2]

If tea auctions are considered analogously to an orchestra, then brokers are their conductors. Brokers play a multifaceted role. Firstly, they liaise with producers on matters of market strategy (when, how and on what auction markets to offer their teas), and frequently act as informal financiers. Gaining access to institutional (bank) finance has been perennial problem for smaller rural enterprises in India, and brokers have often filled this breach through extending various short-term facilities to tea producers (A.F. Ferguson, 2002a: p. 23). Then brokers act as licensed agents for lodging product at auction. In line with auction rules, brokers cup-taste and value individual lots, before writing up auction catalogues and sending off samples to registered buyers. The art of cupping is a fine one, with all brokers employing specialist tasters whose job it is to assign the quality characteristics that will be used in describing the tea in auction catalogues. These descriptions, in turn, are made according to codified parameters relating to product characteristics, commensurate with the grades in Table 3.3. Catalogues are printed and samples are sent off to prospective buyers 13 days before auction. In South India, samples represent approximately 0.3 per cent of auction volume (A.F. Ferguson & Co, 2002a: p. 17). Finally, on the auction day itself, brokers act as auctioneers for the lots they are offering. For these functions, brokers receive a standard 1 per cent commission on the value of sales, paid by the seller.

In South India there are eight broking companies, six of which operate in all three auction centres.[3] All South Indian auctions specify that to purchase tea, a buyer must be registered with the organizers (the relevant tea traders' association) and be physically present in the auction room. Each of the South Indian auctions has more than 150 registered buyers, the majority of whom act as commission agents, buying tea on behalf of other parties (either on domestic or international markets). Once a buyer makes a successful bid, he (auction buying is a male domain in South India) must make payment within 13 days. On this basis, the complete auction cycle in South India has an approximate one-month duration from the delivery of tea to 'godowns' (warehouses), cupping and cataloguing, the holding of the auction, and buyers' payments and receipt of tea.

Notwithstanding the fact that South Indian tea auctions operate in accordance with an extensive and highly codified set of regulations, in literally dozens of interviews with tea sector participants across South India, our research tapped into a pervasive vein of discontent about the balance of these arrangements. According to this general producer consensus, the present structuring of auctions works to the advantage of buyers, not sellers. In his Presidential Address to the 112th Annual Meeting of the United Planters' Association of Southern India (UPASI), Anil Kumar Bhandari commented:

> For decades UPASI has striven to achieve a level playing field for the tea producers with regard to the auction process for tea. This system was introduced in the first place to provide a platform and a fair procedure for tea to be traded. But over time it was felt that the system as practiced did not provide all the necessary elements for true, fair and transparent price-discovery. (Bhandari, 2005: p. 3)

Producers claim that buyers collude in a host of informal ways to reduce the amount of competitive bids at any auction. Of course, it is difficult for researchers such as ourselves to assess the veracity of such claims. Rather obviously, buyers are less than keen to discuss these issues. Brokers, with an interest in perpetuating the auction trade, are also not keen to be drawn on these issues. Nevertheless, it is certainly the case that in South Indian tea auctions many lots are successfully bought on the basis of uncompetitive bidding (that is, one bid only). Under the 'open outcry' (manual) auction system, there is a maximum of 25 seconds per lot, and auctioneers (brokers) are expected to get through 2.5 lots per minute. Upon our observation, much of the time when a lot is announced a buyer will yell out a bid and then, virtually immediately, the broker will hammer confirmation. At face value this hardly takes the appearance of a highly competitive market.

Aggravating these problems is the issue of 'lot divisibility'. This is the process whereby one buyer bids for part of a single lot and is then joined by other parties on the same price basis. Effectively, this operates as a shared agreement; for example, a broker might put up for auction a 'lot' constituted by 20 bags – one buyer will bid for 10 at a certain price and then another chimes in to buy the other 10 at the same price. In effect, the allowance of divisibility represents authority to collude. Moreover, although widespread across the world's major tea auctions, local rules give greater leniency to this practice in South India than anywhere else: 'divisibility norms in South India are perhaps the most liberal in the world with lots of 35 packages [bags] and above being allowed to be shared by up to four buyers' (A.F. Ferguson & Co, 2002a: p. 9). And further to the point: 'Divisibility of any kind contradicts the basic auction principle since it permits some buyers to buy tea without bidding' (A.F. Ferguson & Co, 2002a: p. 9).[4]

In 2001 the Tea Board took the decision to commission the private consulting firm A.F. Ferguson & Co with a brief to enquire and make recommendations into the auction and post-auction systems (A.F. Ferguson & Co, 2002a, 2002b).[5] The consultant was unequivocal in its criticism of lot divisibility, calling for its abolition (2002a: p. 12). The firm also recommended that the period before auction when catalogues and samples are circulated be reduced to eight days (2002a: p. 15). The *Tea Marketing Control Order* (TMCO) *2003* that flowed from the A.F. Ferguson report, however, did not take up these calls; and as of September 2008, they still remain in place. The key thrust of the 2003 TMCO was to instil greater accountability in the system as a whole; a revamped regime for registering buyers, and rules requiring buyers to provide detailed accounts of to whom and at what price they sell their tea (Chattophayay, 2003a). The only concession to A.F. Ferguson's calls for the abolition of lot divisibility was a decision to impose a 5 per cent premium on the price of any lots that are divided, thus generating a disincentive for this practice. Nevertheless, lot divisibility remains sanctioned.

The introduction of these measures provoked the ire of buyers and producers alike. Buyers objected to the increased record-keeping burden and regulatory oversight attached to the 2003 Order. Producers, meanwhile, saw these moves as not going far enough. Yet these complaints were drowned out in the uptake of another of A.F. Ferguson & Co's key recommendations; namely, a proposal to introduce electronic trading to the entire tea auction system. According to this proposal, buyers would still be required to attend the auction hall, but would place bids electronically on dedicated computer terminals. A number of lots could be open to bidding at any one time (quickening the auction process) and bids could be registered and displayed anonymously, reducing the potential for collusive behaviour. Prices could be displayed in real-time through an Internet portal, allowing market analysts and participants in any location in the world to monitor developments and, potentially, give instructions via mobile phone to their buying agents in the auction hall.

The Tea Board embraced this recommendation. It seemed not only to provide the tea industry with a quantum leap into a modern, efficient future, but also to provide an answer to producers who were calling for reform of the auction regime. Soon after A.F Ferguson & Co tabled its report, then, the Tea Board began instituting plans to 'electronise' (the Board's term) India's tea auctions, engaging a global computing firm and an international corporate advisor to chaperone the project to completion. Electronic trading was trialled for leaf grades at the Coimbatore auction in the first half of 2004, and the system was rolled out nationally in 2005.

Electronic auctions, however, have had a troubled birth. Producer complaints were not allayed by the continuing allowance of lot divisibility and the fact that bidders' identities remained on display on terminals (in contravention to A.F. Ferguson's recommended model). As Bhandari commented:

> To arrive at true price discovery and prevent cartelization or private deals; it is vital to ensure that during the bidding for a specific lot the identities of the bid but not that of the bidder should be revealed. We had pointed out to the Tea Board that without discontinuing the division of lots and ensuring anonymity of bidding, the competitiveness and transparency at auctions would not be possible. (2005: p. 3)

As the body which instigated the electronic trading system yet failed to implement other recommendations from A.F. Ferguson & Co relating to matters such as lot divisibility, the Tea Board became a fount for much of this criticism. When, in February 2006 we asked the executive from one of South India's major tea producers whether the introduction of electronic auctions had improved the transparency of the system, the pithy response we received was: 'The Tea Board, I'm sorry, is a waste of time'.[6] Such views gained momentum soon after the introduction of the system due to a series of 'glitches' which interrupted trading. Then, in early 2007, the entire electronic auction regime suffered a fatal system crash and auctions went back to manual trading. The Tea Board hastened to take legal action against the companies it had hired just a few years earlier to set up the system, and within a few months, announced that a new electronic trading platform would be developed by the NSE-IT (National Stock Exchange IT) and Tata Consultancy Services.

# Appendix C

# Restructuring of Tata Tea's Munnar Operations

In 2005–6, a total of 96 per cent of the approximately 13,000 workers in the estates purchased shares in Kanan Devan Hills Plantation Co. (KDHP). The price of a single share in KDHP was Rs 100 (approximately US$2.20), with the median share purchase being Rs 3,500 per worker (about US$78). For a plucking workforce in which the average daily wage is around Rs 125 (US$2.78), these share purchases represent a significant financial commitment. To facilitate share purchases, the company offered financial advances to workers on the basis of afterwards deducting these costs from salaries over a period of one year. Shares have a lock-in period of three years; they are not tradable within this time. Afterwards, they are tradable only to other existing shareholders. In addition to share purchases by workers, local management made substantial investments in the new company (thus 'voting with their wallets' and displaying commitment to the new venture) and Tata also invested in the company, via a trust.

Through these arrangements, Rs 13.5 crore (US$3 million) of new equity was raised. KDHP used these funds as the basis for two loans (a primary loan from the ICICI Bank, worth Rs 25 crore [US$5.55 million], and a subordinated loan from Tata, worth Rs 35 crore [US$7.78 million]), providing total financial assets of Rs 73.5 crore (US$16.33 million). For these monies, a sub-lease was obtained on all Tata's estates in the Munnar leasehold area, plus all fixtures (factories, offices, bungalows, etc.). The leases held by Tata were perpetual; however, the sub-leasing deal with KDHP covered only 30 years. Hence, on 1 April 2025, Tata could legally resume its lands.

The 'Munnar model' presents a vital test case of relevance for the South Indian tea plantation sector as a whole. In Munnar, the enrolment of workers as part-owners of the enterprise becomes the vehicle for resolving the profit crisis of the tea industry. Moreover, this is achieved in the context where wages, on-costs and social welfare obligations as specified under the *Plantation Labor Act* are retained. For the otherwise highly vulnerable plucking workforce in a time of crisis in the industry, the importance of these actions cannot be underestimated. It is significant that in April 2007, representatives from the World Bank visited Munnar to document the model. As Tata rolled it out again in its North Indian estates, the World Bank took a stake in the initiative through its investment-arm subsidiary, the International Finance Corporation (*Food and Beverage News*, 2007: p. 16).

Yet whilst Tata and some business media have, inevitably, trumpeted these events as an illustration of the company's socially responsible business ethic, a closer reading of the events reveals other aspects that are also worthy of notation. As we discuss elsewhere (Neilson and Pritchard, 2008), the much vaunted take-up rate of shares by the estate workforce could be interpreted less as demonstrating their confidence in the enterprise, and more as a financial cost which the workforce deemed necessary to bear to save their futures, in the short term at least. From the Tata Group's perspective, moreover, the transaction has provided an exit from this loss-making enterprise and enables clearer focus on the company's stewardship of its highly valuable retail tea brands. Significantly, the transaction establishing KDHP did not involve any change in the rights to ownership over the 'Kanan Devan' trademark. Despite no longer operating the tea estates in Munnar, Tata still runs its blending operations just outside of the town where it produces and packages its 'Kanan Devan' retail teas. The continued possession of this trademark iterates the point that Tata's primary business interests in this field now lie in the operation of consumer brands, as opposed to tea production. Under a commercial agreement, Tata's blending factory at Munnar purchases tea from KDHP at a market rate, based on cupping tests and auction prices. For KDHP, however, these issues pose a contradiction; it is the steward of the Kanan Devan tea landscape, yet cannot translate this into a legal (trademark) framework from which it can extract commercial advantage.

# Notes

## Chapter One: Introduction

1 Eurep-GAP is the acronym of Euro-Retailer Produce-Good Agricultural Practices.
2 Gereffi initially used the terminology of 'Global Commodity Chains' (GCCs) whereas in this book, for consistency, the term 'global value chain' is used. As explained by Gibbon and Ponte (2005: p. 77) 'the latter [term] is thought to better capture a wider variety of products, some of which lack commodity features. As a result, the global commodity chain approach is now known as global value chain analysis.'
3 John Williamson (1990) coined the term 'Washington Consensus' at the conclusion of a 1989 conference sponsored by the Institute for International Economics in Washington, arguing there was a 'reasonable degree of consensus' amongst Washington's politicians and technocrats for ten policy instruments required for effective economic reform in Latin America following the crises of the 1980s. These ten instruments were fiscal discipline, allocation of public expenditure to education, health and public infrastructure and away from subsidies, tax reform (broadening the tax base), market-determined and positive interest rates, a competitive exchange-rate, trade liberalization, openness to foreign direct investment, privatization and deregulation.
4 In 1938, Unilever bought the US and Canadian arms of Sir Thomas Lipton's tea business, including the iconic 'Lipton's' brand (acquiring global rights over these brands in 1972). In 1980 Unilever acquired the group of tea brands associated with Brooke Bond. This company had its origins in a Manchester tea shop in 1845. Brooke Bond's most substantial and significant tea brands were 'PG Tips' and 'Red Rose', into which, yet again, Unilever has invested heavily.
5 Now known as Altria.
6 In passing, it is worthwhile to note that this book examines two major export crops (tea and coffee) located in a number of production districts within a major regional agrarian economy (South India). The use of this method beggars

consideration. Most of the published research on global value chains either takes the form of in-depth studies of one product at one site of production; or follows a product along a chain. In contrast, the distinguishing feature of this study is that it invokes the regional scale across two product sectors. There is no other GVC study that we know of that emulates this frame of reference. Perhaps the study with the closest parallel to this book is Gibbon and Ponte's (2005) examination of GVCs in African clothing and agriculture industries. However, Gibbon and Ponte's book uses GVC analysis to inform the debate on African development, while this study is focused on the situated contexts of GVC restructuring.

7 This mission, led by Lord George Macartney, was the first envoy of Great Britain sent to China, and had the goal of establishing a permanent diplomatic and trading presence in China (to which end it was unsuccessful).

8 The Chinese and Assamese teas were found to be the same species. Note that Sir Percival's son, John Griffiths, tells the same story in his 2007 book on tea (Griffiths, 2007: pp. 34–6).

9 Tharian (1984: p. 41) writes: 'During 1946–1951 tea topped all the 17 industries, excluding matches, in the matter of net profits after tax in relation to net worth. The profit on direct investment was 20.1 per cent and on the equity capital it was 12.67 per cent, which means that the total profits over the five-year period repaid the investment made until then; these profits were largely remitted'. This issue is discussed in further detail in Chapter 3.

## Chapter Two: Re-inserting Place and Institutions within Global Value Chain Analysis

1 In passing, the SoP has provoked considerable criticism since its inception, and its influence remains a point of dispute. It has been claimed that the SoP approach replicates arguments made earlier within the field (Watts, 1994; c.f. Fine, 2004: p. 332), and that the approach thereby neglects the cultural and symbolic properties that underlie the ways food choices are made (Guthman, 2002).

2 Reflecting her dualistic interests, Dixon coins her framework 'cultural economy'. Nevertheless, and importantly, her work remains explicitly positioned within structural political economy.

3 This includes the long-standing Professor of Geography at the University of Manchester (UoM) Peter Dicken; his former PhD student at UoM and now Professor of Geography at the National University of Singapore (NUS) Henry Wai-Chung Yeung; and UoM researchers Jeffrey Henderson, Martin Hess and Neil Coe. These five scholars were the researchers in the UK Economic and Social Research Council-funded project 'Making the connections: Global Production Networks in Europe and East Asia' (2000–3). This project provided the vehicle for the development and elaboration of the GPN approach.

4 However, this merging of terminology with Porter-inspired 'value chains' has not been unproblematic, as the latter was principally intended as a tool to diagnose competitive advantage between firms by disaggregating a firms activities

into discrete components. The analysis of a particular firms' various 'primary' and 'support' activities, such as inbound logistics, human resource management and technology development provides insights into firm decision-making (and by extension chain governance), but it cannot necessarily be equated with the broader political-economy concerns of GVC analysis.

5 This sequence included: primary commodity exports; export-processing assembly; component-supply subcontracting; original equipment manufacturing, and original brand-name manufacturing (Gereffi, 1995: pp. 120–33). Notably, the research was part-funded by research grants from Taiwan and Korea (Gereffi, 1995: p. 100).

6 Meyer-Stamer (2004) uses the terminology of value chain 'switching' in favour of 'inter-sectoral upgrading', as it denotes strategic change in the way that producers position themselves within alternative profit-making scenarios.

7 Sturgeon et al. (2008) argue: 'the value chain metaphor [c.f. the network metaphor] is the more useful heuristic tool for focusing research on complex and dynamic global industries' (p. 302) and 'the simplicity of the chain metaphor increases its usefulness as a research methodology and policy input' (p. 318).

8 Notwithstanding this relative neglect of this important concept, however, there is undoubted recognition of the need for its strengthening. In framing their model of GVC governance, Gereffi, Humphrey and Sturgeon (2005: p. 82) acknowledge that the real-world application of their ideas must necessarily 'bring in' geography: 'Clearly, history, institutions, geographic and social contexts, the evolving rules of the game, and path dependence matter; and many factors will influence how firms and groups of firms are linked in the global economy.'

9 'Old' institutional economics is quite different from new institutional economics (NIE) associated most closely with Oliver Williamson (1975, 2000) and its antecedents in the work of Ronald Coase (1937). The NIE ultimately derives from neo-classical economics and many of its applications (though not all!) have a much narrower frame of reference centering on transaction cost metrics. For our purposes, we see considerable limitations in such applications of NIE. Concurring with Powell and DiMaggio (1991: p. 33, cited in Amin and Thrift, 1994: p. 11): 'we are sceptical of arguments that assume that surviving institutions represent efficient solutions because we recognize that rates of environmental change frequently outpace rates of environmental adaptation. Because sub-optimal organizational practices can persist for a long time, we rarely expect institutions simply to reflect current political and economic forces. The point is not to discern whether institutions are efficient, but to develop robust explanations of the ways in which institutions incorporate historical experiences into their rules and organizing logics'. Furthermore, we are critical of tendencies within NIE analyses to de-contextualize space, history and circumstance (i.e., *geography!*). As Toboso (2001: p. 768) has suggested: 'Many new institutionalist scholars who are interested in investigating the changes in social conventions, legal rules, forms of organization among private companies, etc. are making explanatory analyses in which we see only abstract individuals (boundedly rational, etc.) who attempt to obtain income-favorable (usually efficient) legal rules/social norms in stylized situations that the analyst presents

as being characterized almost exclusively by the rule of competition and the existence of transaction costs.'

10 Note that we disagree with Amin and Thrift's terminology here. They use the term 'institutions' in this quote to refer to what we think of as 'organizations'. To recall the terminology of Douglass North, we draw a distinction between institutions (meaning rules of the game) and organizations (the players). To our mind, entities such as Chambers of Commerce are *organizations* which contribute to an *institutional* framework.

11 The first two of these are lodged on the GVC website maintained by Gary Gereffi, Timothy Sturgeon, and IDS researchers (www.globalvaluechains.org).

12 'Industry analyses' refers to studies by investment banks, etc.

13 As they say: 'Do not shut yourself into the library until you have read everything that has ever been written on your sector.'

## Chapter Three: How to Make a (South Indian) Cup of Tea or Coffee

1 In Indonesia, for example, smallholders are defined as cultivators in possession of less than 2 ha, while in India the threshold is 10.12 ha and in Sri Lanka, smallholders are defined as any cultivator with less than 35 ha. Furthermore, some smallholder systems operate on the basis of independent farm families with their own separate land tenure, while other systems largely operate in the form of mutually supportive systems characterized by the pooling of key resources and cooperative marketing arrangements.

2 The legitimacy of the MRLs was questioned by the Indian exporter who argued that Germany's MRL was out of line with international benchmarks, hence, drawing attention to the current inconsistencies in international regulations regarding pesticide levels. In the USA, for example, the comparable rate was 8 mg/kg (Saraswathy, 2002).

3 Tea bushes should be pruned for general maintenance once every five years, during which time no plucking is possible for a period of between four to six months.

4 In South India plucking occurs all year, but the cooler and dryer months of January to March produce slower growth which accentuates the polyphenol production and thus is claimed to deliver higher quality (Hudson, 2000).

5 Interview, plantation company owner, Kottayam, 23 February 2006.

6 Interviews, corporate estate managers, Wyanad, 14 September 2005.

7 Interview, Bought-Leaf Factory manager, Nilgiris District, 13 November 2004.

8 Blends can represent different combinations of tea alone, or tea blends with flavor additives. 'Earl Grey', for example, is blended tea with bergamot.

9 Interview, plantation company executive, Kochi, 17 February 2006.

10 It should be noted that whilst we have aggregated black and green tea in this discussion, there is in fact a notable difference in their production geographies. Green tea dominates production in China (84 per cent of total production) and

Japan (100 per cent), and is also important, though to a lesser extent, in Viet Nam (40 per cent) and Indonesia (23 per cent). Furthermore, the global market for green tea is currently more buoyant than is the case for black tea.

11 Known commonly as Robusta.

12 In July 2007, this 'discount' had narrowed to its lowest level in recent years, with the International Coffee Organization (ICO) weighted average for Robusta as much as 76 per cent of the ICO weighted average for 'Other Mild Arabicas', whereas in 2003 it had been less than 40 per cent (International Coffee Organization, 2007).

13 The 'Malnad' is the collective geographical term for the Western Ghats in Karnataka. In this book, it is used as a geographical descriptor of coffee growing districts.

14 Overbearing is when a coffee plant produces excess fruit which it cannot sustain and so actually causes damage to the plant.

15 Interview, head of Small Growers' Cooperative, Hassan District, 1 February 2005.

16 Interviews, coffee planters, Chikmagalur, 28 January 2005.

17 The PIC procedure is specified under the Rotterdam Convention, which promotes shared responsibilities in relation to importation of hazardous chemicals.

18 Interestingly, however, the use of Endosulfan is still common in Australian agriculture.

19 Interview, estate manager, Chikmagalur District, 28 January 2005.

20 The minimum for Robusta tends to be 75 kg/day. For Arabica it varies between estates. Interview, estate manager, Pollibetta, 25 January 2005.

21 Interview, coffee planter, Chettali, 24 January 2005.

22 This term refers to the period since 1980, when the Mandal Commission affirmed the legality of affirmative action programs assisting social groups that are known as scheduled tribes and other backward castes. The phrase *post-Mandal* India thus implies the period in which lower caste and tribal segments of the Indian population have more forthrightly asserted their place in society.

23 Interview, coffee planter, Chikmagalur, 31 January 2005.

24 Naxalism is a leftist revolutionary movement, originating in West Bengal, associated with violent attacks on government and establishment institutions.

25 This is apparently increasingly popular in India, as the conventional belief in the benefits of fermentation for cup quality is challenged by both planters and buyers. Another recent innovation is the so-called 'Pulped-Sun Dried' coffee which is dried immediately after pulping without any attempt at fermentation. Apparently, such coffee has become popular in the international specialty coffee sector, such that a separate category for it was established in the *Flavor of India – Fine Cup Awards* in 2007.

26 An alternative approach, adopted on some private estates and advanced by the Coffee Board research stations is the conversion of such waste to biogas. The 'bioreactor' can be used to capture gas released during treatment, which is then stored and used for household needs as well as fed back to the pulping station where, on one estate we visited, was able to reduce diesel consumption by 60 per cent. Interview, coffee planter, Chikmagalur, 29 January 2005.

27 One of the curers we interviewed believes that planters receive as much as 95 per cent of FOB prices in India, with local agents collect a commission of as little as Rs 25 per 50 kg bag (parchment coffee) and Rs 15 per 50 kg bag (cherries), once the cost of transport and processing are taken into account. (Note that one 50 kg bag of parchment is equivalent to approximately 42 kg of green beans [GBE], and that 50 kg of cherry is 26 kg of GBE.) In September 2005, agents were buying Arabica parchment at Rs 3,300/bag (equivalent to US$0.81/lb) at a time when the ICO 'Other Milds' category was US$0.99/lb. Planters were therefore receiving approximately 82 per cent of the international price and about 85 per cent of the FOB price (remembering also that Indian Arabica is generally sold equivalent, or at a slight discount, to New York prices). This suggests that in the liberalized era, the pre-export supply chain is operating relatively efficiently in Indian coffee, with a high degree of price transmission, transparency and competition.

28 In the interests of disclosing possible bias, it should be pointed out that these authors work for the Quality Management Division of the Coffee Board.

29 Observations at Kushalnagar Curing Works, 9 November 2004.

30 Note, however, that these data include all coffee exports, and that a considerable volume of instant coffee is manufactured and exported by branded multinationals such as Nestle, CCL, Tata and Unilever.

31 According to The ICO document WP Board No. 934/03, *Data concepts and variables used in the statistics of the Organization,* May 2003, available at www.ico.org.

32 Various interviews, including those with actors at the Bangalore auctions on 4 February 2005, and with a coffee broker at the North Coorg Club, 23 January 2005.

33 *Choupal* means 'gathering place' in Hindi.

34 As evidence of the globalized character of the South Indian coffee trade, there was intense speculation at the UPASI Annual Conference in September 2005 on the effects on coffee prices after Hurricane Katrina (in August that year) had cut a swathe through the coffee warehouses of New Orleans.

## Chapter Four: The Institutional Environment of the South Indian Tea and Coffee Industries

1 In addition to being considered the founding father of the Indian coffee industry, Baba Budan continues to be revered by Indian Muslims as a saint. His grave in the Baba Budan Giri range of Chikmagalur is an important pilgrimage site for many Muslims, and an annual festival is held to commemorate his death.

2 There is minimal historical scholarship on the managing agency system in the Indian coffee sector, compared to tea. Presumably this relates to the fact that during the Colonial period, exports of tea were much greater than coffee.

3 As opposed to 'Sterling companies' registered in London.

4 Talbot (1995: p. 151) writes: 'these actions by producers, along with the Cuban Revolution, convinced the Kennedy administration that cooperation with coffee producers was in the US interest. The administration was concerned that the continued decline in coffee prices would lead to economic instability and the spread of communism throughout Latin America'.

5 Interviews with representatives of the Coffee Board of India, Bangalore, 3 February 2005.

6 Prior to the Second World War, however, there were some international tea agreements. The first brief occasion this was attempted was in 1920–1, when the tea associations of India and Ceylon restricted supply in response to steep falls in global market prices. The second, and more substantial, attempt was in the 1930s in the context of declining world tea prices during the Great Depression. According to Sir Percival Griffiths (1967), tea prices in Calcutta for 1931–2 were 30 per cent lower than they were the previous year, and the lowest they had been since 1904. In March 1933, the then-Colonial territories of India, Ceylon and the Dutch East Indies (now Indonesia) launched a five-year International Tea Agreement through which each agreed to limit their exports by quota. This was an agreement between private associations, but was enforced by the colonial governments. In India, this was executed through a ban on new planting areas, which remained officially in force until 1948 (Tharian, 1984: p. 40). These voluntary restrictions served to raise tea prices almost to their pre-Depression levels, leading to a second Agreement coming into force in 1938. However, the Second World War caused this Agreement to go into abeyance and after the War it was not renewed. An informal voluntary retention scheme was initiated in 1969, (lasting only a year) and further examinations into an international tea agreement in 1974 were not concluded (Baffes, 2004: p. 24). Subsequently, the FAO instituted the Inter-Governmental Group on Tea (IGGoT), which has evolved to become the pre-eminent multilateral forum for the industry. The IGGoT has a generous mandate, but this does not extend to economic clauses relating to supply management.

7 The way this worked was that the Soviet Union's procurement authority would authorize various agents to bid on its behalf, with a general rider to maximize volumetric purchases given a certain budget. Interview, tea auctioneer, Coonoor, 20 September 2005.

8 The gendering of this expression is intentional: the social construction of a 'planter' assumes maleness.

9 Although some aspects of this are under challenge. By common agreement through the regions, the younger generation of plantation managers is less engaged with the trappings of plantation life. The bellwether indicator here is the level of attendance at the annual Planter's Ball, held at the Ooty Club. In 2005 and 2007, one of us had the perchance to be invited to this event. Few attendees of the event were under the age of 40, and participant numbers were dwindling year-on-year. A number of the Planter's Clubs across the planting districts of South India are now struggling in terms of membership.

10 Interview, UPASI-TRF regional officer, Gudalur, 16 September 2005.

11 Details as presented to the UPASI Annual Conference by N.K. Das, Chairman of the Tea Board, 19 September 2005.

12 Interviews, UPASI/Tea Board regional officer, Gudalur, 19 September, 2005; UPASI-KVK officials, Kotagiri, 19 June 2007.
13 Report to the UPASI Annual Conference, 15 September 2007. This project is being undertaken by the Centre for the Development of Advanced Computing (C-DAC), an Indian research institute headquartered in Pune. As of early 2008, this technology is at the trial stage with the expectation that it could become dominant within the industry in the space of five years.
14 SFTGFOP is a 'Special Fine Tippy Golden Flowery Orange Pekoe' orthodox whole leaf grade.
15 Interview, head of Small Growers Federation, Chikmagalur, 31 January 2005.
16 Interviews with Coffee Board staff and coffee growers, Wyanad District, 14–15 September 2005.
17 Interviews with Coffee Board officials, Hassan, 9 September 2005 and Kodagu, 24 January 2005.
18 Interview, estate managers, Pollibetta, 25 January 2005.
19 This focus for R&D reflects and supports ABC's extensive investments in the downstream Indian retailer-roaster industry, as discussed in Chapter 8.
20 Kenneth Davids is a well-known US-based coffee author, renowned cupper and co-founder of the Internet site: www.coffeereview.com.

## Chapter Five: Struggles over Labour and Livelihoods

1 It is necessary to distinguish between 'Fairtrade' (one word and capitalized, referring specifically to activities authenticated by the Fair Trade Labeling Organization [FLO]) and 'fair trade' as a generic description of these agendas.
2 Version 01.03.2007.
3 According to the *Generic Fairtrade Standards for Hired Labour Version 01.03.2007*, and the *Fairtrade Standards for Tea for Hired Labour Version 19.11.2007*.
4 www.flo-cert.net, accessed 18 December 2007.
5 Chamraj Estates, in the Nilgiris, provides an exemplar here.
6 Note that this estimate related to estates in Darjeeling. However, its conclusions are relevant for South India, as well.
7 Chamraj, noted above, grows approximately 400 tonnes of organically-certified green leaf. Also, the Peermade Development Society obtained organic certification in 2004, however, again, this was made possible from EU donor funding (Pereira, 2002) (Interview, manager, Peermade Development Society, Kuttikkanam, Kerala, 20 February 2006).
8 The three founding members were Unilever, Nestlé and Danone. As of September 2008, other SAI Platform members were Agrarfrost, Campina, Consorzio Interregionale Ortofrutticoli, Coca-Cola, Danisco, Ecom, Efisco, Elders, Fonterra, Friesland Foods, Kellogg, Kemin, Kraft, Lamb Weston/Meijer, McCain Europe, McDonalds, Neumann Kaffe Groupe, Sara Lee, Tchibo and Volcafe (SAI, 2008).
9 It is not possible to search the Global-GAP database on its website (www.globalgap.org, accessed 18 December 2007) for lists of certified producers, so it is unclear how many tea producers have already obtained certification worldwide.

10 Accord Services Ltd, All About Tea, Matthew Algie & Co Ltd, Bell Tea Company, Bettys & Taylors of Harrogate Ltd., DJ Miles & Co Ltd, Drie Mollen, Finlay Beverages Ltd, Imporient UK Ltd, Jing Tea, Keith Spicer Ltd, The Metropolitan Tea Company, Mother Parkers Tea & Coffee Inc, The Nambarrie Tea Company Ltd, Newby Tea, Republic of Tea, Sara Lee International, Tazo Tea, TEAZ Tea Boutique, The Tetley Group, R Twining & Company Ltd, Unilever (Europe), Williamson Fine Teas Ltd, The Windmill Tea Co Ltd.

11 Crucially, however, the scheme is self-audited by a NPA Committee. This lack of third-party audit evidently raises questions about the scheme. The politics of audit is a critical issue in the tea industry. *Prima facie*, the NPA scheme seems to offer a somewhat unsophisticated model for compliance verification. However, given that the scheme assures compliance with standards already under the audit gaze from other parties (for example, the *Plantation Labour Act* has its own regulatory regime) perhaps this is not fatal for the scheme's credibility.

12 Assuming an average plucking per worker of 30 kg/day but working only 260 days per year because of Sundays and public holidays, which leads to a year-round plucking rate of 20 kg/day. Also assumes a conversion rate of five kilogramme of green leaf to one kilogramme of made tea.

13 Field interview and witnessed financial data, Munnar, 29 May 2006.

14 This is based on interviews at Letchmi Estate, supported by sighted financial data. During the calendar year 2005, the company spent Rs 922,222 for 'lines' (housing) upkeep; Rs 162,573 for latrines/sanitation; Rs 176,786 for water supply (chlorination, tanks etc); and Rs 657,890 for medical facilities; not including the wages of doctors. This was for an estate with 1,646 permanent workers (field interviews, Munnar, 27 May 2006).

15 Arrapetta Estate, owned by Harrison's Malayalam Ltd. Field visit, 13 September 2005.

16 Interview, estate manager, Wyanad, 13 September 2005.

17 Interview, estate manager, Wyanad, 16 September 2005.

18 According to the FLO website, applicants for Fairtrade certification are required to pay an upfront fee of €250 and are also required to bear the costs of inspection and verification, which is levied at €350/day (FLO, 2007). An inspection would usually last three days for a small organization. Information within the FLO website at www.flo-cert.net/flo-cert/_admin/userfiles/file/Fees/PC%20FEESYSTEMSF%2019EN.pdf.

19 Interviews, 26–7 May 2006.

20 Interview, Tea Board staff, Peermade, 21 February 2006.

21 One company which abandoned its Travancore estates at the depths of the tea crisis was the RBT (Ram Bahadur Thakur) group. This company bought into the tea industry in 1976 on the basis of income earned from other diversified interests, notably including mining. When the company walked away from its estates in 2003, bush quality was poor, debts were owed to a large number of creditors (including the Government of Kerala), workers were owed wages, and a substantial amount of money which should have been deposited in the statutory Provident Fund for workers, was allegedly 'missing'. Subsequent legal action against the company, as of late 2007, had failed, to recover these monies.

22 The following narrative draws heavily on fieldwork in Gudalur in September 2005; interviews in Coimbatore, 21 September 2005, and the discussion on Janmam law provided in the *Nilgiris Gazeteer of 1908* (Francis, 2001).
23 Tipu Sultan's revolt against the pro-British Mysore Maharajah saw extensive population movement across our study areas, as communities fled the violence.
24 This was despite the fact that a later inquiry (the Escheat Enquiry of 1885) found that *Janmam* tenure in Wyanad-Gudalur had in fact been largely a British creation (Francis, 2001).
25 This legislative enactment by the Government of Tamil Nadu had the primary purpose to abolish and replace the zamindar with the *ryotwar* system. Through this legislation, the zamindari were compensated for their losses. The *ryotwari* system, meanwhile, ensured a direct legal relationship between the person who cultivated land, and the Government. Although the State retained ownership of these lands – the granting of ryotwari rights ('*ryotwari patta*') was not the equivalent of exclusive proprietary ownership – it held all the trappings of such. Effectively, it provided for cultivators of the land to have virtually all the equivalent rights of absolute possession, on the basis of annual taxes being payable.
26 The difference was that under the *Zamindari*, land tenure was held jointly with cultivators, whilst under *Janmam*, it was held exclusively by the *Janmi*.

## Chapter Six: Struggles over Environmental Governance in the Coffee Forests of Kodagu

1 The environmental consciousness of many consumers was also, of course, stimulated through the organic food movement, of which coffee was already an important part by the 1990s.
2 It should be pointed out that a widely accepted definition of 'Ecosystem Services' provided by the Millennium Institute (2003) in its *Ecosystems and Human Well-being: A Framework for Assessment*, emphasizes the anthropocentric nature of ecosystem services, which might actually disregard 'habitat protection' as such a service. However, in the sense that society may gain cultural satisfaction from habitat protection, and more importantly, may be willing to pay for this satisfaction, we are content to include it as such a service for the purposes of our discussion.
3 The 'environmental services' referred to in PES schemes include services, usually provided for free that fall outside conventional markets, such as watershed protection, erosion prevention, biodiversity conservation, landscape beauty and carbon sequestration. Wunder (2006) notes that a number of essentially similar schemes are described using associated terminology, such as Rewards for Environmental Services, Compensation for Environmental Services and Markets for Environmental Services.
4 Interviews with coffee planters, throughout 2005–7.
5 The Haleri dynasty of *Lingayat Rajas*, however, apparently consisted neither of Kodavas nor were they from Kodagu district itself. Instead, the Rajahs were of

the *Lingayat* faith (a Shivaite religion) in contrast to the nature-worshipping, pork-eating Kodavas, and appear to have descended from the great Vijayanagara empire of South India, following its fall to the Deccan sultanates in 1565 (Richter, 1870).

6 It should be noted, however, that those Kodagu planters campaigning for tree rights on *bane* lands vehemently deny that the granting of tree rights would result in such conversion, as evident in a heated debate during the *Ecosystem Services of the Landscapes of Kodagu* workshop at the Forestry College of Ponampet on 15–16 January 2008, attended by one of the authors.

7 The project *Connecting, enhancing and sustaining environmental services and market values of coffee agroforestry in Central America, East Africa and India* is coordinated by CIRAD (Centre de coopération Internationale en Recherche Agronomique pour le Développement) and, in India, is implemented together with the University of Agricultural Sciences Bangalore and the Coffee Board of India. The project commenced in 2007.

8 Such as through their *Promotion of Biodiversity Conservation within Coffee Landscapes* Project in El Salvador, and the *El Triunfo Biosphere Reserve: Habitat Enhancement in Productive Landscapes* Project in Chiapas, Mexico.

## Chapter Seven: Smallholder Engagement in Global Value Chains: Initiatives in the Nilgiris

1 Field interviews in the Nilgiris, various dates.

2 Production costs are limited to application of fertilizers such as urea and potash, herbicides during the establishment phase only, pruning and plucking. Tea plucking costs are easily the most important cost associated with smallholder production, and are frequently modified depending on market conditions. Interviews, tea smallholders in Kundah Taluk, 14 November 2004.

3 These numbers derive from official statistics and our fieldwork. According to growers we interviewed in a village in the Kotagiri district of the Nilgiri Hills, the highest price they received during the past decade was 18 Rs/kg for a short three-month period in 1998, and the lowest they received was 3 Rs/kg during 2003. The Tea Board & U-KVK estimate of average green leaf prices of Rs 5–6 between 2001 and 2002 is consistent with this village's experiences. During the period 2003–04 it was common for prices to range between Rs 7–11 per kg.

4 Interview, 21 September 2005. Name has been changed to protect anonymity.

5 The 2006 estimate is based on the number of reported beneficiaries in the Tamil Nadu Government's tea subsidy scheme (*New India Press*, 2006).

6 Interviews with villagers, Nilgiris, 13–14 November 2004.

7 Interview, UPASI-KVK Head Scientist, Coonoor, 14 February 2006.

8 Interview, executive with NILMA, Coonoor, 15 November 2004.

9 The 100 Department of Horticulture officers were employed for 12 months only.

10 Interview and field observations with UPASI-KVK Field Officer, Kotagiri Taluk, Nilgiris, 19 June 2007.

11 Interviews with SHG members and U-KVK field officers in Kundah Taluk, 15–16 February 2006.
12 Interviews with SHG members and U-KVK field officers in Kundah Taluk, 15–16 February, 2006.
13 The company was paying Rs 8.5/kg for fine green leaf, but only Rs 5.25 for normal leaf (Tea Board & U-KVK, 2005: p. 106).
14 Interview, UPASI/Tea Board regional officer, Gudalur, 19 September 2005.
15 Interview, IBS management, 13 November 2004.
16 Interviews, executive of NILMA, Coonoor, 15 November 2004.
17 Interview, Executive Director, Tea Board (South), Coonoor, 15 February 2006.
18 Presentation by R. Prabhu, MP for Nilgiris, UPASI Annual Conference, Coonoor, 16 September 2007.
19 Theoretically, such information is available from the Tea Board. 'Form E' in the compliance documentation attached to the *Tea Marketing Control Order* 2003 collects these data from estates. However, to date the Tea Board has not processed or published information collected from Form E.
20 Field visits, Wyanad, 13 September 2005.
21 Interview, Talamala Estate, Wyanad, 15 September 2005.
22 Interview, executive with Darmona Estate, 19 June 2007.
23 Interview, Silver Cloud estate management, 16 September 2005.
24 Interview with Tea Board of India officials, Kolkata, 17 March 2004.
25 Interview, U-KVK/Tea Board regional field officer, Gudalur, 16 September 2005; and presentation by R. Prabhu, MP for Nilgiris, UPASI Annual Conference, Coonoor, 16 September 2007.
26 Interview with UPASI-KVK Field Officer, Coonoor, 15 February 2006. As evidence of the explicitly political nature of these payments, cheques to each grower were enclosed within envelopes bearing a colour photograph of the Tamil Nadu Chief Minister up for re-election, J. Jayalalithaa. However, in a somewhat embarrassing (and revealing) incident, the payment of these subsidies was stopped by the Electoral Commission, which ruled that the Chief Minister's photograph constituted a political inducement. In one incident where smallholders were denied subsidy payments, a mob of 600 stormed a Tamil Nadu government office (New India Press, 2006).
27 Interviews with villagers, Nilgiris, 15 February 2006.
28 Remembering that more than 50 per cent of South Indian teas are exported, implementing this course of action would have dire consequences for the industry. It would most certainly backfire in terms of its avowed objective of supporting smallholders, because factories would go out of business and leave smallholders with no markets for their products.

## Chapter Eight: Making a Living in the Global Economy: Institutional Environments and Value Chain Upgrading

1 Meyer-Stamer (2004) uses the terminology of value chain 'switching' in favor of 'inter-sectoral upgrading', as it denotes strategic change in the way that producers position themselves within alternative profit-making scenarios.

2 This provides a key point of contrast between some of India's international competitors (notably in Africa and Southeast Asia), where generally weaker levels of R&D intensity during the Colonial era were not significantly built upon after Independence.
3 A representative of the Mumbai-based National Commodities and Derivative Exchange (NCDEX) spoke at the UPASI Conference of September 2007, with the view of examining the development of tea futures treading, notwithstanding the practical difficulties of such an endeavour, as discussed in Chapter 3.
4 Of note here are the problems of currency appreciation for South Indian tea and coffee producers. Between 2001 and 2008, the Indian rupee appreciated 25 per cent against the US dollar, and also appreciated against almost all of India's competitors in world tea and coffee markets: 15 per cent against the Indonesian rupiah, 33 per cent against the Vietnamese dong, 58 per cent against the Sri Lankan rupee, and 254 per cent against the Malawian kwacha. It remained steady against the Kenyan shilling, but political violence during January 2008 caused the Kenyan currency to collapse by 8.5 per cent against the Indian rupee within the space of a month. (All exchange rates sourced from oanda.com, accessed 27 January 2008.)
5 The incapacity of such measures to fundamentally improve the industry's position was brought into perspective during debate on the Tea Board Special Fund (see Chapter Three) which had the purpose of increasing tea yields (in made tea equivalent terms) from 1.7 tonnes to 2.4 tonnes per hectare. Questioning the contradictions of concentrating on efficiency gains in the tea sector, an iconoclastic UPASI Presidential Address in 2005 took aim at the logic behind this initiative: 'I am under the impression that a steep and sustained fall in tea prices caused by burgeoning surpluses was the cause of the [tea] crisis. There must be something skewed in our collective thought processes if our response to that type of crisis is to create within a decade an even bigger surplus' (Bhandari, 2005: p. 6).
6 As discussed in Chapter 3, this includes mandatory MRLs for tetradifon and ethion (notably in tea) and, for green coffee beans, the Prior Informed Consent (PIC) listing of Endosulfan and Lindane, and their potential complete banning. Additionally, the European Union has placed limits on Ochratoxin A (OTA) for roasted and soluble coffee imports.
7 Interview, UPASI Secretary-General, Coonoor, 18 June 2007.
8 Similar trends are evident within the speciality tea sector, such as the upstream activities of buyers such as Harrod's (the up-market UK retailer) to develop producer relationships according to specific requirements (Rahman, 1999). Although cut of a different cloth, some specialist niches in fair trade and organic sectors can comparably involve extensive interventions by downstream lead firms to facilitate compliance.
9 Interview, coffee planter, Chikmagalur District, 29 January 2005.
10 Interview, coffee exporter, 3 February 2005, Bangalore.
11 With little fanfare, as an aside, this transaction ended the last vestiges of major non-Indian ownership of the South Indian tea plantation sector.
12 Tata's shares in Barista, however, have been subsequently acquired by the Sterling Infotech Group.

13 The extensive instant coffee manufacturing facilities in South India is a point of significance at the global level. Talbot (1997b) reports that producer countries have struggled to successfully host these downstream processing activities. In a comparative assessment of producer country efforts to develop instant coffee manufacturing facilities, Talbot (2002b: pp. 714–15) finds that the provision of costly state subsidies has usually been required if producer countries were to stand a chance of competing with the high-tech facilities operated by multinationals in consuming countries. India provides an exception to this rule, however, with its status as a preferred supplier to the Soviet Union giving the local instant coffee industry a strategic kick-start.

14 Based on UNCOMTRADE data (www.comtrade.un.org), using HS category 0902030 (black tea in immediate packings of < 3kg) as a percentage of HS category 0902 (total black tea). Not that the figure of 16 per cent is for 'all India', and it would reasonably be assumed that the incidence of packaged tea exports would be higher in the North (Darjeeling, Assam, etc.) than in the South.

## Chapter Nine: Conclusion: What We Brewed

1 An alternative explanation, however, is that recent supply shortages in various agricultural commodities have been triggered by altered weather patterns in producing regions, prompted by human-induced climate change and many be symptomatic of more profound long-term shifts in market dynamics. For obvious reasons, this set of issues is beyond the capacity of this book to examine.

## Appendix A: The Role of Managing Agents

1 'Taking care of DOWB' is a peculiar, nineteenth-century British saying. It roughly translates as 'looking after oneself'. According to a *New York Times* article from 1860, its derivation is as follows: 'Everybody has heard of DOWB, that immortal type of nepotism in the administration of English affairs. DOWB, less widely known as DOWBIGGEN, had the felicity to be the nephew of Lord Pamure, the Secretary-at-War of Great Britain at the crisis of the Crimean struggle, and his affectionate uncle telegraphed to the British commander before Sebastopol, that whatever he did or did not do to maintain the honor of England, he must "not forget to take care of DOWB". The English were beaten ... But DOWB was made a Colonel: and his name has taken its place in history' (*New York Times*, 1860).

## Appendix B: The Operation and Intended Reform of South India's Tea Auctions

1 The term 'red dust' to describe dust-grade teas from Kerala was developed as a nickname reflecting the dominance of Communist political administrations in that state.

2 Interview, tea broker, Coonoor, 16 November 2004.

3 The six brokers operating in all three South Indian auctions are: Carrit Moran, J. Thomas, Forbes, Paramount, Contemporary Tea Auctioneers, and Best Tea Brokers. A seventh company, Global Tea Brokers, only operates in the Kochi and Coonoor auctions and an eighth, Imperial Tea Brokers, operates only in Coonoor.

4 In practice, auction rules which lend legitimacy to lot divisibility perpetuate a diminution of active and competitive bidding, and replace it with incentives for bidders to seek out one another prior to the auction date with a view to dividing up lots on offer. Further encouraging this state of affairs is the extended (13-day) time period between the sending out of catalogues and samples, and the actual auction taking place. This timing dates from the days when samples were air-shipped to London and local buyers eagerly awaited purchase orders via phone, telegram or fax. However, London buyers do not really figure at all in South Indian auctions nowadays and local commission agents can more than adequately peruse samples and catalogues in their offices usually located within a few kilometres (at most) from auction centres. The perpetuation of these anachronistic rules, in the opinion of producers, therefore now only serves the purpose of providing buyers with added market advantage. The significant forward notice provided to buyers via samples means they have not only the opportunity to consider the lots in great detail, but also a handsome period of 'deal-making time' to orchestrate the auction process to their advantage. Interview, executive with a leading tea plantation company, Kochi, 17 February 2006.

5 A.F. Ferguson also made numerous other recommendations, with the overall intention of making the Indian tea auction system consistent nationwide. The Tea Board implemented these recommendations through the *Tea Marketing Control Order* of 2003. Processes of standardization included, for example, rules to ensure all tea is sold in 20 kg packages, and that all tea offered at an auction must physically be present in warehouses at that place.

6 Interview, executive with a leading tea plantation company, Kochi, 17 February 2006.

# Bibliography


A.F. Ferguson & Co (2002a) *Study on Primary Marketing of Tea in India – Summary of Recommendations*, Tea Board of India/A.F. Ferguson & Co, Kolkata.

A.F. Ferguson & Co (2002b) *Study on Post-Auction Pricing Structure and Supply Chain for Tea*, Tea Board of India/A.F. Ferguson & Co, Kolkata.

Abernathy, F., Dunlop, J., Hammond, J., and Weil, D. (1999) *Stitch in Time: Lean Retailing and the Transformation of Manufacturing. Lessons from the Apparel and Textile Industries*, Oxford University Press, New York.

Accenture (2002) *Medium Term Export Strategy for Indian Tea 2002–07*, Tea Board of India, Kolkata.

ActionAid (2005) *Tea Break: A Crisis Brewing in India*, ActionAid, London.

Adams, T. (1989) Gudalur: A community at the crossroads, pp. 318–33 in Hockings, P. (ed.) *Blue Mountains: The Ethnography and Biogeography of a South Indian Region*, Oxford University Press, Oxford.

Aglietta, M. (1979) *A Theory of Capitalist Regulation*, New Left Books, London.

Ajjan, N. and Raveendran, N. (2001) Crisis in Nilgiris tea plantations. *Plant Horti Tech*, 3, pp. 29–31.

Akiyama, T. (2001) Coffee market liberalization since 1990, pp. 83–120 in Akiyama, T., Baffes, J., Larson, D. and Varangis, P. (eds) *Commodity Market Reforms: Lessons of Two Decades*, World Bank, Washington, DC.

Ali, R., Choudhry, Y.A., and Lister, D.W. (1998) Sri Lanka's tea industry: Succeeding in the global market, *World Bank Discussion Paper*, 368.

Alison, S. (2006) The 'branding law of corporate social responsibility' in the Australian processing tomato industry, unpublished B.LibStuds (Hons) thesis, available from the School of Geosciences, University of Sydney.

Allen, J. and Thompson, G. (1997) Think global, then think again: Economic globalization in context, *Area*, 29(3), pp. 213–27.

Altman, R. (2002) Tea export and transportation: Global trends, *Tea and Coffee Trade Journal*, 176(2), electronic journal, not paginated.

Altman, R. (2007) The art of grading tea, *Tea and Coffee Trade Journal*, 179(8), electronic journal, not paginated.

Amin, A. (1997) Placing globalization, *Theory, Culture and Society*, 14(2), pp. 123–37.

Amin, A. (2001) Moving on: institutionalism in economic geography, *Environment and Planning A*, 33, pp. 1237–41.

Amin, A. (2004) An institutionalist perspective on regional economic development, pp. 48–58 in Barnes, T.J., Peck, J., Sheppard, E. and Tickell, A. (eds) *Reading Economic Geography*, Blackwell, Oxford.

Amin, A. and Thrift, N. (1994) Living in the Global, pp. 1–22 in Amin, A. and Thrift, N. (eds) *Globalization, Institutions and Regional Development in Europe*, Oxford University Press, Oxford.

Amin, A. and Thrift, N. (1995) Globalisation, institutional 'thickness' and the local economy, pp. 91–108 in Healey, P., Cameron, S., Davoudi, S., Graham, S. and Madani-Pour, A. (eds) *Managing Cities: The New Urban Context*, John Wiley & Sons, New York.

Angel, D.P. and Rock, M.T. (2005) Global standards and the environmental performance of industry, *Environment and Planning A*, 37(11), pp. 1903–18.

Anupindi, R. and Sivakumar, S. (2007) Supply chain reengineering in agri-business: A case study of ITC's *e-Choupal*, in Lee, H.L. and Lee, C.-Y. (eds) *Building Supply Chain Excellence in Emerging Economies*, Springer Sciences + Business Media, London.

Arce, A. and Marsden, T. (1993) The social construction of international food: a research agenda, *Economic Geography*, 69, pp. 293–311.

Arindampal, J. and Somayaji, K.S. (2000) Coffee processing industry, *The Planters' Chronicle*, July issue, pp. 313–19.

Arjunan, M., Holmes, C., Puyravaud, J.-P. and Davidar, P. (2006) Do developmental initiatives influence local attitudes toward conservation? A case study from the Kalakad–Mundanthurai Tiger Reserve, India, *Journal of Environmental Management*, 79(2), pp. 188–97.

Arthurs, H. (2001) Reinventing labor law for the global economy: The Benjamin Aaron lecture, *Berkeley Journal of Employment and Labor Law*, 22(2), pp. 271–94.

Asia Pulse (2008) Tea review, *Asia Pulse* (online news database), 9 January.

Associated British Foods (ABF) (2007) *Annual Report*, ABF, London.

Auret, D. and Barrientos, S. (2006) Participatory Social Auditing: Developing a worker-focused approach, pp. 129–48 in Barrientos, S. and Dolan, C. (eds) *Ethical Sourcing in the Global Food System*, Earthscan, London.

Baffes, J. (2004) Tanzania's tea sector: Constraints and challenges, *World Bank Africa Region Working Paper*, The World Bank, Washington, DC.

Baffes, J. (2005) Reforming Tanzania's tea sector: a story of success, *Development Southern Africa*, 22(4), pp. 589–604.

Bair, J. (2005) Global capitalism and commodity chains: Looking back, going forward, *Competition and Change*, 9(2), pp. 153–80.

Bair, J. and Gereffi, G. (2001) Local clusters in global chains: the causes and consequences of export dynamism in Torreon's blue jeans industry, *World Development*, 29(11), pp. 1885–903.

Bair, J. and Gereffi, G. (2002) NAFTA and the apparel commodity chain: corporate strategies, inter-firm networks, and local development, pp. 23–50 in Gereffi, G., Spencer, D. and Bair, J. (eds) *Free Trade and Uneven Development: The North American Apparel Industry after NAFTA*, Temple University Press, Philadelphia, PA.

Bair, J. and Gereffi, G. (2003) Upgrading, uneven development, and jobs in the North American apparel industry, *Global Networks*, 3(2), pp. 143–69.

Bales, K. (1999) *Disposable People: New Slavery in the Global Economy*, University of California Press, Berkeley.

Barling, D. and Lang, T. (2005) Trading on health: cross-continental production and consumption tensions and the governance of international food standards, pp. 39–51 in Fold, N. and Pritchard, B. (eds) *Cross-continental Food Chains*, Routledge, London.

Barnes, T. (1999) Industrial geography, institutional economics and Innis, pp. 1–22 in Barnes, T.J. and Gertler, M. (eds) *The New Industrial Geography: Regions, Regulation and Institutions*, Routledge, London and New York.

Barrett, H.R., Ilbery, B.W., Browne, A.W. and Binns, T. (1999) Globalization and the changing networks of food supply: the importation of fresh horticultural produce from Kenya into the UK, *Transactions of the Institute of British Geographers*, 24, pp. 159–74.

Barrientos, S. and Dolan, C. (eds) (2006) *Ethical Sourcing in the Global Food System*, Earthscan, London.

Bates, R.H. (1997) *Open-Economy Politics: The Political Economy of the World Coffee Trade*, Princeton University Press, Princeton NJ.

Bathelt, H. (2006) Geographies of production: growth regimes in spatial perspective — towards a relational view of economic action and policy, *Progress in Human Geography*, 30(7), pp. 223–36.

Batt, P. (ed) (2006) Proceedings of the 1st international symposium on improving the performance of supply chains in the transitional economies, *Acta Horticulturae* 699.

Beck, U. (2000) *What is Globalization?*, Polity Press, Cambridge.

Bernstein, H. and Campling, L. (2006) Commodity studies and commodity fetishism I: Trading Down, *Journal of Agrarian Change*, 6(2), pp. 239–64.

Bhagwat, S.A., Kushalappa, C.G., Williams, P.H. and Brown, N.D. (2005a) A landscape approach to biodiversity conservation of sacred groves in the Western Ghats of India, *Conservation Biology*, 19(6), pp. 1853–62.

Bhagwat, S.A., Kushalappa, C.G., Williams, P.H. and Brown, N.D. (2005b) The role of informal protected areas in maintaining biodiversity in the Western Ghats of India, *Ecology and Society*, 10(1), article 8 (electronic journal, not paginated).

Bhagwati, J. (2005) *In Defense of Globalization*, Oxford University Press, Oxford.

Bhandari, A. K. (2005) Presidential address at the 112th Annual Conference of the United Planters' Association of Southern India, unpublished, available from UPASI, Glenview, Coonoor.

Bhowmik, S.K. (2003) Industrial relations in tea plantations, *Labour File* (Delhi), 1(3), pp. 31–4.

Bhowmik, S.K., Xaxa, V. and Kalam, M.A. (1996) *Tea Plantation Labour in India*, Friedrich Ebert Stiftung, New Delhi.

Biénabe, E., Boselie, D., Collion, M-H., Fox, F., Rondot, P., van de Kop, P. and Vorley, B. (2007) The internationalization of food retailing: Opportunities and threats for small-scale producers, pp. 3–17 in Vorley, B., Fearne, A. and Ray, D. (eds) *Regoverning Markets: A Place for Small-Scale Producers in Modern Agri-Food*

*Chains?*, Gower Publishing in association with Institute for International Development and Environment, Burlington, Vermont.

Biggart, N.W. and Beamish, T.D. (2003) The economic sociology of conventions: Habit, custom, practice, and routine in market order, *Annual Review of Sociology*, 29, pp. 443–64.

Blaikie, P. (2006) Is small really beautiful? Community-based natural resource management in Malawi and Botswana, *World Development*, 34(11), pp. 1942–57.

Boltanski, O. and Chiapello E. (1999) *Le Nouvel Esprit du Capitalisme*, Gallimard, Paris.

Boltanski, O. and Thévenot, L. (1991) *De la Justification: les Économies de la Grandeur*, Gallimard, Paris.

Bonanno, A., Busch, L., Friedland, W.H., Gouveia, L. and Mingione, E. (eds) (1994) *From Columbus to ConAgra: The Globalization of Agriculture and Food*, University of Kansas Press, Lawrence, KS.

Bourlakis, M. and Weightman, P. (eds) (2004) *Food Supply Chain Management*, Blackwell, Oxford.

Bramah, E. (2005) *The Bramah Tea and Coffee Walk Around London*, Christian Le Comte Publishing, London.

Brown, J. and Fraser, M. (2006) Approaches and perspectives in social and environmental accounting: An overview of the conceptual landscape, *Business Strategy and the Environment*, 15(2), pp. 103–17.

Burawoy, M. (1998) The extended case study method, *Sociological Theory*, 16(1), pp. 4–33.

Burawoy, M. (2000a) Introduction: Reaching for the global, pp. 1–40 in Burawoy, M., Blum, J.A., George, S., Gille, Z., Gowan, T., Haney, L., Klawitter, M., Lopez, S.H., O'Riaian, S. and Thayer, M. (eds) *Global Ethnography: Forces, Connections, and Imaginations in a Postmodern World*, University of California Press, Berkeley.

Burawoy, M. (2000b) Grounding globalization, pp. 337–50 in Burawoy, M., Blum, J.A., George, S., Gille, Z., Gowan, T., Haney, L., Klawitter, M., Lopez, S.H., O'Riaian, S. and Thayer, M. (eds) *Global Ethnography: Forces, Connections, and Imaginations in a Postmodern World*, University of California Press, Berkeley.

Burawoy, M., Blum, J.A., George, S., Gille, Z., Gowan, T., Haney, L., Klawitter, M., Lopez, S.H., O'Riaian, S. and Thayer, M. (eds) (2000) *Global Ethnography: Forces, Connections, and Imaginations in a Postmodern World*, University of California Press, Berkeley.

Burawoy, M., Burton, A., Ferguson, A.A. and Fox, K.J. (1991) *Ethnography Unbound: Power and Resistance in the Modern Metropolis*, University of California Press, Berkeley.

Burch, D. and Lawrence G. (eds) (2007) *Supermarkets and Agri-Food Supply Chains: Transformations in the Production and Consumption of Foods*, Edward Elgar, Melbourne.

Burki, S. and Perry, G. (1998) *Beyond the Washington Consensus: Institutions Matter*, The World Bank, Washington, DC.

Busch, L. and Bain, C. (2004) New! Improved? The transformation of the global agrifood system, *Rural Sociology* 69(3), pp. 321–46.

Busch, L. and Juska, A. (1997) Beyond political economy: Actor networks and the globalization of agriculture, *Review of International Political Economy*, 4, pp. 688–708.

Businesswire (2007) The Coca-Cola Company and Illy announce premium ready-to-drink espresso coffee joint venture, *Businesswire* (online news database), 15 October.

Buttel, F.H. and Newby, H. (eds) (1980) *The Rural Sociology of Advanced Societies: Critical Perspectives*, Allenheld Osmun, Montclair, NJ.

Callon, M. (1998) Introduction: the embeddedness of economic markets in economics, pp. 1–57 in M. Callon (ed.) *The Laws of the Markets*, Blackwell, Oxford.

Campbell, H. and Le Heron, R. (2007) Supermarkets, producers and audit technologies: The constitutive micro-politics of food, legitimacy and governance, pp. 131–54 in Lawrence, G. and Burch, D. (eds) *Supermarkets and Agri-Food Supply Chains: Transformations in the Production and Consumption of Foods*, Edward Elgar, Cheltenham.

Center for Education and Communication (CEC), FAKT and Traidcraft (2005) *Roundtable Consultative Meetings on Corporate Social Responsibility in the Tea Industry*, CEC, New Delhi. Available online http://www.cec-india.org accessed 2 February 2007.

Center for Education and Communication (CEC), FAKT and Traidcraft (2006) *Building a Business Case for Corporate Social Responsibility in the Indian Tea Industry: JustTea Project – the Final Review*, CEC, New Delhi. Available online http://www.cec-india.org accessed 2 February 2007.

Central Bank of Sri Lanka (2003) *Annual Report*, Central Bank of Sri Lanka, Colombo.

Chandrakanth, M.G., Bhat, M.G. and Accavva, M.S. (2004) Socio-economic changes and sacred groves in South India: Protecting a community-based resource management institution, *Natural Resources Forum*, 28(2), pp. 102–11.

Chari, S. (1997) Agrarian questions in the making of the knitwear industry in Tirupur, India: A historical geography of the industrial present, pp. 79–105 in Goodman, D. and Watts, M. (eds) *Globalising Food: Agrarian Questions and Global Restructuring*, Routledge, London.

Charveriat, C. (2001) *Bitter Coffee: How the Poor are Paying for the Slump in World Coffee Prices*, Oxfam International, London.

Chattopadhayay, S. (2003a) Crisis in the tea industry, *The Hindu* (online news database), 14 February.

Chattopadhayay, S. (2003b) Wailing from behind green bushes, *Labour File* (Delhi), 1(3), pp. 5–19.

Chattopadhayay, S. (undated) *Productivity and Decent Work in the Tea Industry of South India: A Report Prepared for the ILO Sub Regional Office-India*, ILO, New Delhi.

Chattopadhayay, S. and John, P. (2007) *Bitter Beans: An Analysis of the Coffee Crisis in India*, Partners in Change, New Delhi.

Cheesman, J., Bennett, J. W., Tran Vo Hung Son, Truong Dang Thuy and Vo Duc Hoang Vu (2007) The total economic value of sustainable coffee irrigation practices in Dak Lak, Viet Nam, Working Paper No 5 Australian Centre for International Agricultural Research, Project ADP/2002/015).

Christopherson, S and Lillie, N. (2005) Neither global nor standard: corporate strategies in the new era of labor standards, *Environment and Planning A*, 37, pp. 1919–38.

Christopherson, S. and Storper, M. (1986) The city as studio; the world as back lot: The impact of vertical disintegration on the location of the motion picture industry, *Environment and Planning D: Society and Space*, 4, pp. 305–20.

Cloke, P., Cook, I., Crang, P., Goodwin, M., Painter, J. and Philo, C. (2004) *Practising Human Geography*, Sage, London.

Coase, R.H. (1937) The nature of the firm, *Economica*, 4, pp. 386–405.

Cochin Chamber of Commerce and Industry (2007) *Tea Exports: Consolidated Annual Report 2005–2006*, Cochin Chamber of Commerce and Industry, Kochi.

Coe, N.M., Dicken, P. and Hess, M. (2008a) Introduction: global production networks — debates and challenges, *Journal of Economic Geography*, 8, pp. 267–9.

Coe, N.M., Dicken, P. and Hess, M. (2008b) Global production networks: realizing the potential, *Journal of Economic Geography*, 8, pp. 271–95.

Coe, N.M., Hess, M., Yeung, H.W.C., Dicken, P. and Henderson, J. (2004) 'Globalizing' regional development: a global production networks perspective, *Transactions of the Institute of British Geographers*, 29, pp. 468–84.

Coe, N.M., Kelly, P.F. and Yeung, H.W.C. (2007) *Economic Geography: A Contemporary Introduction*, Blackwell, Oxford.

Coffee Board of India (2003) Modalities on implementation of scheme on support to small grower sector, unpublished document, December, available from the Coffee Board of India, Bangalore.

Coffee Board of India (2004) *Karma and Quality*, Coffee Board of India, Bangalore.

Coffee Board of India (2005) *Action Programme to Combat the Coffee White Stem Borer*, Coffee Board of India, Bangalore.

Coffee Board of India (2006) *Database on Coffee*, Economic & Market Intelligence Unit of the Coffee Board, Bangalore.

Coffee Board of India (2007) *Database on Coffee May 2007*, Economic & Market Intelligence Unit of the Coffee Board, Bangalore.

Coffee Board of India (2008) *Self Help Groups in the Coffee Sector*, Coffee Board of India, Bangalore.

Common Code for the Coffee Community Association (CCCCA) (2008) Common Code for the Coffee Community Association, www.sustainable-coffee.net (accessed 31 January, 2008).

Conservation International (2008). Biodiversity Hotspots, at www.biodiversity hotspots.org (accessed 17 July 2008).

Cook, I. and Crang, P. (1996) The world on a plate: culinary culture, displacement and geographical knowledges, *Journal of Material Culture*, 1, pp. 131–53.

Cook, I., Crang, P. and Thorpe, M. (2004) Tropics of consumption: 'Getting with the fetish' of 'exotic' fruit, pp. 173–91 in Hughes, A. and Reimer, S. (eds) *Geographies of Commodity Chains*, Routledge, London.

Cooke, P. and K. Morgan (1993) The Network Paradigm: new departures in corporate and regional development, *Environmental and Planning D: Society and Space*, 11, pp. 543–64.

Cooke, P. and K. Morgan (1998) *The Associational Economy: Firms, Regions, and Innovation*, Oxford University Press, Oxford.

Coste, J. and Egg, J. (1996) Regional dynamics and the effectiveness of economic policies: the case of cereal markets in West Africa, in Benoit-Cattin, M., Griffon, M.

and Guillaumont, P. (eds) *Economics of Agricultural Policies in Developing Countries*, Editions de la Revue Française d'Economie, Paris.

Crang, P. (1996) Displacement, consumption and identity, *Environment and Planning A*, 28, pp. 47–67.

Damodaran, A. (1998) Darjeeling tea, Malabar pepper and complexity of geographic appellations, *The Planters' Chronicle*, November, pp. 503–5.

Damodaran, A. (2002a) Conflict of trade-facilitating environmental regulations with biodiversity concerns: The case of coffee-farming units in India, *World Development*, 30(7), pp. 1123–35.

Damodaran, A. (2002b) Economics of organic tea farming: A case study from Darjeeling, pp. 11–20 in UPASI & Tea Board (eds), *Tea Quality Upgradation Seminar on Organic Tea Status, Issues and Strategies*, UPASI Tea Research Foundation, Valparai.

Dankers, C. (2003) *Environmental and Social Standards, Certification and Labelling for Cash Crops*, Food and Agriculture Organization of the United Nations, Rome.

Dash, A. (2003) Strategies for poverty alleviation in India: CYSD's holistic approach to empowerment through the Self-Help Group Model, *IDS Bulletin*, 34(4), pp. 133–42.

Dash, K.C. (1999) India's International Monetary Fund loans: Finessing win-set negotiations within domestic and international politics, *Asian Survey*, 39(6), pp. 884–907.

Davids, K, (2004) *Kenneth Davids Portfolio: Ten Best Indian Coffees of 2004*, Coffee Board of India, Bangalore.

Daviron, B. and Fousse, W. (1993) *La compétitivité des Cafés Africains*, Ministère de al Coopération, Paris.

Daviron, B. and Ponte, S. (2005) *The Coffee Paradox: Global Markets, Commodity Trade and the Elusive Promise of Development*, Zed Books, London.

Davis, L.E and North, D.C (1971) *Institutional Change and American Economic Growth*, Cambridge University Press, Cambridge.

D'haeze D., Raes D., and Deckers J. (2005) Groundwater extraction for irrigation of Coffea canephora in Ea Tul watershed, Vietnam - a risk evaluation, *Agricultural Water Management*, 73(1), pp. 1–19.

Dicken, P. (2003) *Global Shift*, Sage, London.

Dicken, P., Kelly, P.F., Olds, K. and Yeung, H.W.-C. (2001) Chains and networks, territories and scales: towards a relational framework for analysing the global economy, *Global Networks*, 1, pp. 89–112.

Dicken, P. and Malmberg, A. (2001) Firms in territories: a relational perspective, *Economic Geography*, 77, pp. 345–63.

Directorate of Census Operations, Tamil Nadu (undated) *Census of India 2001* (CD-Rom).

Dirjen Bina Produksi Perkebunan (2004) *Statistical Estate Crops of Indonesia: Tea (2001–2003)*, Dirjen Bina Produksi Perkebunan (Directorate General of Estate Crops), Department of Agriculture, Jakarta.

Dixon, J. (1999) A cultural economy model for studying food systems, *Agriculture and Human Values*, 16(2), pp. 151–60.

Dolan, C. and Humphrey, J. (2000) Governance and trade in fresh vegetables: the impact of UK supermarkets on the African horticulture industry, *Journal of Development Studies*, 37(2), pp. 147–76.

Dolan, C. and Humphrey, J. (2004) Changing governance patterns in the trade in fresh vegetables between Africa and the United Kingdom, *Environment and Planning A*, 36(3), pp. 491–509.

Eaton, C. and Shepherd, A. (2001) Contract farming: Partnerships for growth, *FAO Agricultural Services Bulletin*, 145, FAO, Rome.

Elwood, S. and Martin, D. (2000) Placing interviews: Location and scales of power in qualitative research, *Professional Geographer*, 52(4), pp. 649–57.

Epstein, G. (ed.) (2005) *Financialization and the World Economy*, Edward Elgar, Amherst.

Eskes, A.B. and Leroy, T. (2004) Coffee selection and breeding, pp. 57–86 in Wintgen, J.N. (ed.), *Coffee: Growing, Processing, Sustainable Production*, Wiley-VCH, Weinheim (Germany).

Etherington, D.M. and Forster, K. (1993) *Green Gold: The Political Economy of China's Post-1949 Tea Industry*, Oxford University Press, Oxford.

Ethical Tea Partnership (2007) Ethical Tea Partnership, at www.ethicalteapartnership.org (accessed 18 December 2007).

Ethical Trade Initiative (ETI) (2005) Smallholder guidelines, at www.ethicaltrade.org (accessed 18 December 2007).

Ethical Trade Initiative (ETI) (2007) Ethical Trade Initiative, at www.ethicaltrade.org (accessed 18 December 2007).

Eurep-GAP (2003) Introducing EUREPGAP Coffee Reference Code – September 2003, http://www.eurepgap.org/coffee/publications.html (accessed 31 January 2008).

Fairtrade Labelling Organization (FLO) (2007) Fairtrade Labelling International, at www.fairtrade.net (accessed various dates, 2007).

Favereau, O. (2002) Conventions and regulation, pp. 312–19 in Boyer, R. and Saillard, Y. (eds) *Regulation Theory: The State of the Art*, Routledge, London and New York.

Ferraro, P.J., Uchida, T. and Conrad, J.M. (2005) Price premiums for eco-friendly commodities: Are 'green' markets the best way to protect endangered ecosystems?, *Environmental & Resource Economics*, 32(3), pp. 419–38.

Fine, B. (1993) Resolving the diet paradox, *Social Science Information*, 32(4), pp. 669–87.

Fine, B. (1994) Towards a political economy of food, *Review of International Political Economy*, 1(3), pp. 519–45.

Fine, B. (2002) *The World of Consumption: The Material and Cultural Revisited*, Routledge, London.

Fine, B. (2004) Debating production-consumption linkages in food studies, *Sociologia Ruralis*, 44(3), pp. 322–42.

Fine, B. and Leopold, E. (1993) *The World of Consumption*, Routledge, London.

Fine, B., Foster, N., Simister, J. and Wright, J. (1992) Consumption norms: A definition and an empirical investigation of how they have changed, 1975–1990, *School of Oriental and Asian Studies Working Papers*, 22.

Fine, B., Heasman, M. and Wright, J. (1996) *Consumption in the Age of Affluence: The World of Food*, Routledge, London.

Fisher, B.S. (1972) *The International Coffee Agreement: A Study in Coffee Diplomacy*, Praeger, New York.

Fitter, R. and R. Kaplinsky (2001) Who gains from product rents as the coffee market becomes more differentiated? A value-chain analysis, *IDS Bulletin-Institute of Development Studies*, 32(3), pp. 69–82.

Fold, N. (2008) Transnational sourcing practices in Ghana's perennial crop sectors, *Journal of Agrarian Change*, 8(1), pp. 94–122.

Food and Agricultural Organization (FAO) (2007) *FAOSTATS* (online database) http://faostats.fao.org (accessed on various dates).

Food and Agriculture Organization (FAO) (2006) *Enhancement of Coffee Quality through the Prevention of Mould Formation*, FAO, ICO and CFC, Rome.

Food and Agriculture Organization (FAO) (2004) *The State of Food Insecurity in the World 2004: Monitoring Progress Towards the World Food Summit and Millennium Development Goals*, FAO, Rome.

*Food and Beverage News* (2007) Tata Tea forms new company for plantation business, *Food and Beverage News* (Mumbai) 3(15) (3 March), p. 16.

Forster, K. (1993) China's tea industry in transition, *Tea and Coffee Trade Journal*, 167(4), electronic journal, not paginated.

Francis, W. (2001) *Madras District Gazetteers: The Nilgiris*, Asian Educational Press, New Delhi (First published in 1908 by the Superintendent, Government Press, Madras).

Friedberg, S. (2001) On the trail of the global green bean: Methodological considerations in multi-site ethnography, *Global Networks*, 1(4), pp. 353–68.

Friedland, W.H. (2005) Commodity systems: Forward to commodity analysis, pp. 25–38 in Fold, N. and Pritchard, B. (eds) *Cross-continental Food Chains*, Routledge, London.

Friedland, W.H., Barton, A. and Thomas, R. (1981) *Manufacturing Green Gold*, Cambridge University Press, Cambridge.

Friedman, T. L. (1999) *The Lexus and the Olive Tree*, Harper Collins, London.

Friedman, T.L. (2005) *The World is Flat: A Brief History of the Twenty-First Century*, Farrar, Straus & Giroux, New York.

Friedmann, H. (2005) From colonialism to green capitalism: Social movements and emergence of food regimes, pp. 227–64 in Buttel, F. and McMichael, P. (eds) *New Directions in the Sociology of Global Development*, Elsevier, Oxford.

Friedmann, H. and McMichael, P. (1989) Agriculture and the state system: The rise and decline of national agricultures, 1870 to the present, *Sociologia Ruralis*, 29, pp. 93–117.

Ganewatta, G., Waschik, R., Jayasuriya, S. and Edwards, G. (2005) Moving up the processing ladder in primary product exports: Sri Lanka's "value-added" tea industry, *Agricultural Economics*, 33, pp. 341–50.

Garcia, C. and Pascal, J.-P. (2005) Sacred forests of Kodagu: Ecological value and social role, pp. 1999–2029 in Cederlöf, G. and Sivaramakrishnan, K. (eds) *Ecological Nationalisms: Nature, Livelihoods and Identities in South Asia*, Permanent Black, New Delhi.

Garcia, C., Marie-Vivien, D., Kushalappa, C.G., Chengappa, P.G., and Nanaya, K.M. (2007) Geographical Indications and biodiversity in the Western Ghats: Can labelling benefit producers and the environment in a mountain agroforestry landscape?, *Mountain Research and Development*, 27(3), pp. 206–10.

Gereffi, G. (1994) The organization of buyer-driven global commodity chains: how US retailers shape overseas production networks, pp. 95–122 in Gereffi, G. and Korzeniewicz, M. (eds) *Commodity Chains and Global Capitalism*, Praeger, Westport, CT.

Gereffi, G. (1995) Global production systems and third world development, pp. 100–42 in Stallings, B. (ed.) *Global Change, Regional Response: The New International Context of Development*, Cambridge University Press, New York.

Gereffi, G. (1996) Global commodity chains: new forms of coordination and control among nations and firms in international industries, *Competition & Change*, 1(4), pp. 427–39.

Gereffi, G. (1999) International trade and industrial upgrading in the apparel commodity chain, *Journal of International Economics*, 48(1), pp. 37–70.

Gereffi, G. and Korzeniewicz, M. (eds) (1994) *Commodity Chains and Global Capitalism*, Greenwood Press, Westport, CT.

Gereffi, G., and Korzeniewicz M. (1990) Commodity chains and footwear exports in the semi-periphery, pp. 45–68 in Martin, W.G. (ed) *Semiperipheral States in the World-Economy*, Greenwood Press, Westport, CT.

Gereffi, G., Garcia-Johnson, R. and Sasser, E. (2001) The NGO–industrial complex, *Foreign Policy*, 125, pp. 56–65.

Gereffi, G., Humphrey, J. and Sturgeon, T. (2005) The governance of global value chains, *Review of International Political Economy*, 12(1), pp. 78–104.

Gereffi, G., Humphrey, J., Kaplinsky, R. and Sturgeon, T. (2001) Introduction: globalisation, value chains, and development, *IDS Bulletin*, 32(3), pp. 1–8.

Gereffi, G., Korzeniewicz, M. and Korzeniewicz, R.P. (1994) Introduction, pp. 1–16, in Gereffi, G. and Korzeniewicz, M. (eds) *Commodity Chains and Global Capitalism*, Greenwood Press, Westport, CT.

Gereffi, G., M. Martínez and J. Bair (2002) Torreon: the new blue jeans capital of the world, pp. 203–23 in Gereffi, G., Spencer, D. and Bair, J. (eds) *Free Trade and Uneven Development: The North American Apparel Industry after NAFTA*, Temple University Press, Philadelphia.

Ghosh, A. and Chatterjee, S. (2007) Starbucks to open in India, *International Herald Tribune*, 14 January.

Gibbon, P. (2001a) Upgrading primary production: a global commodity chain approach, *World Development*, 29(2), pp. 345–63.

Gibbon, P. (2001b) Agro-commodity chains: An introduction, *IDS Bulletin*, 32(3), pp. 60–8.

Gibbon, P. (2003) Value-chain governance, public regulation and entry barriers in the global fresh fruit and vegetable chain to the EU, *Development Policy Review*, 21(5–6), pp. 615–25.

Gibbon, P. (2005) Commodity policy in an era of liberalized global markets, pp. 153–80 in Lines, T. (ed.) *Agricultural Commodities, Trade and Sustainable Development*, International Institute for Environment and Development (IIED) and International Centre for Trade and Sustainable Development (ICTSD), London and Geneva.

Gibbon, P. and Ponte, S. (2005) *Trading Down: Africa, Value Chains and the Global Economy*, Temple University Press, Philadelphia.

Gilbert, C. (1996) International commodity agreements: an obituary note, *World Development*, 24(1), pp. 1–19.

Gilbert, C. (2008) Value chain analysis and market power in commodity processing with application to the cocoa and coffee sectors, pp. 5–34 in Food and Agriculture Organization (ed.) *Commodity Market Review 2007–08*, FAO, Rome.

Giovannucci, D. and Ponte, S. (2005) Standards as a new form of social contract? Sustainability initiatives in the coffee industry, *Food Policy*, 30, pp. 284–301.

Goodman, D. and Watts, M. (1994) Reconfiguring the rural or fording the divide? Capitalist restructuring and the global agro-food system, *Journal Of Peasant Studies*, 22(1), pp. 1–49.

Goodman, D. and Watts, M. (eds) (1997) *Globalizing Food*, Routledge, New York.

Goswami, U.A. (2003) Illy to outsource Arabica coffee from India, *The Economic Times* (online news database) 29 November.

Graham, D. and Woods, N. (2006) Making corporate self-regulation effective in developing countries, *World Development*, 34(5), pp. 868–83.

Granovetter, M. (1985) Economic action and social structure: The problem of embeddedness, *American Journal of Sociology*, 91(3), pp. 481–510.

Granovetter, M. and Swedberg, R. (1992) *The Sociology of Economic Life*, Westview Press, Boulder.

Gresser, C. and Tickell, S. (2002). *Mugged: Poverty in Your Coffee Cup*, Oxfam International, Oxford.

Griffiths, J. (2007) *Tea: The Drink that Changed the World*, André Deutsch, London.

Griffiths, P. (1967) *The History of the Indian Tea Industry*, Weidenfeld and Nicolson, London.

Grootaert, C. (1997) Social Capital: The Missing Link?, in World Bank (ed.) *Expanding the Measure of Wealth: Indicators of Environmentally Sustainable Development*, The World Bank, Washington, DC.

Grote, M.H. and Täube, F.A. (2006) Offshoring the financial services industry: implications for the evolution of Indian IT clusters, *Environment and Planning A*, 38(7), pp. 1287–305.

Gupta, B. (1997) Collusion in the Indian tea industry in the Great Depression: An analysis of panel data, *Explorations in Economic History*, 34, pp. 155–73.

Guthman, J. (2002) Commodified meanings, meaningful commodities: Rethinking production-consumption links through the organic system of provision, *Sociologia Ruralis*, 42(4), pp. 295–311.

Haraway, D. (1991) Situated knowledges: The science question in feminism and the privilege of partial perspective, pp. 183–201 in Haraway, D. (ed.) *Simians, Cyborgs and Women: The Reinvention of Nature*, Routledge, London.

Harvey, D. (2001) *Spaces of Capital: Towards a Critical Geography*, Routledge, New York.

Harvey, M. (2007) The rise of supermarkets and asymmeties of economic power, pp. 51–73 in Lawrence, G. and Burch, D. (eds) *Supermarkets and Agri-Food Supply*

*Chains: Transformations in the Production and Consumption of Foods*, Edward Elgar, Cheltenham.

Hayami, Y. and Damodaran, A. (2004) Towards an alternative agrarian reform: Tea plantations in South India, *Economic and Political Weekly*, 4 September, pp. 3992–7.

Henderson, J., Dicken, P., Hess, M., Coe, N. and Yeung, H.W.-C. (2002) Global production networks and the analysis of economic development, *Review of International Political Economy*, 9, pp. 436–64.

Hess, M. (2004) Spatial relationships? Towards a reconceptualization of embeddedness, *Progress in Human Geography*, 28(2), pp. 165–86.

Hess, M. and Coe, N.M. (2006) Making connections: global production networks, standards, and embeddedness in the mobile-telecommunications industry, *Environment and Planning A*, 38(7), pp. 1205–27.

Hess, M. and Yeung, H.W.-C. (2006) Guest editorial: Whither global production networks in economic geography? Past, present, and future, *Environment and Planning A*, 38(7), pp. 1193–204.

*Hindu Business Line* (2006) Indian Robusta losing Italian market to Vietnam, *Hindu Business Line* (online news database), 2 October.

*Hindu, The* (2005a) Coffee Board intimated about package for growers, *The Hindu* (online news database), 20 June.

*Hindu, The* (2005b) Move to 'modernise' Plantation Labour Act, *The Hindu* (online news database), 17 September.

*Hindu, The* (2006) Meeting on tree rights a farce, says committee, *The Hindu* (online news database), 12 November.

*Hindu, The* (2007a) Panel to be formed to look into fixing the floor price for tea, *The Hindu* (online news database), 6 September.

*Hindu, The* (2007b) Move to form panel for tackling tea growers' problems hailed, *The Hindu* (online news database), 19 September.

*Hindu, The* (2008) New price sharing formula comes into force in the Nilgiris, *The Hindu* (online news database), 9 January.

Hindustan Lever Ltd (2005) HLL proposes to transfer Doom Dooma and Tea Estates Plantation Divisions to subsidies, Media Release, 8 April, at http://www.hll.com/mediacentre/release.asp?fl=2005/PR_HLL_040705plant.htm, accessed 10 January, 2008.

Hirsch, P.M. and Lounsbury, M.D. (1996) Rediscovering volition: the institutional economics of Douglass C. North, *Academy of Management Review*, 21(3), pp. 872–84.

Hirst, P. and Thompson, G. (1999) *Globalization in Question: The International Economy and the Possibilities of Governance*, 2nd edition, Polity Press, Cambridge.

Hobbs, J.E. and Kerr, W.A. (2006) Consumer information, labelling and international trade in agri-food products, *Food Policy*, 31(1), pp. 78–89.

Hockings, P. (ed.) (1997) *Blue Mountains Revisited: Cultural Studies on the Nilgiri Hills*, Oxford University Press, New Delhi.

Hodgson, G.M. (1993) *Economics and Evolution: Bringing Life Back into Economics*, Polity Press, Cambridge.

Hopkins, T. and Wallerstein, I. (1977) Patterns of development of the modern world-system, *Review*, 1(2), pp. 11–145.

Hopkins, T. and Wallerstein, I. (1986) Commodity chains in the world economy prior to 1800, *Review*, 10(1), pp. 157–70.

Hudson, J.B. (2000) Raw material and quality of tea (not paginated), in UPASI-KVK (ed.) *Tea Factory Manual*, UPASI-KVK, Coonoor.

Hughes, A. (2001) Global commodity networks, ethical trade and governmentality: organizing business responsibility in the Kenyan cut flower industry, *Transactions of the Institute of British Geographers*, 26, pp. 390–406.

Hughes, A. (2005a) Responsible retailers? Ethical trade and the strategic re-regulation of cross-continental food supply chains, pp. 139–52 in Fold, N. and Pritchard, B. (eds) *Cross-Continental Food Chains*, Routledge, London.

Hughes, A. (2005b) Corporate strategy and the management of ethical trade: the case of the UK food and clothing retailers, *Environment and Planning A*, 37, pp. 1145–63.

Hughes, A. and Reimer, S. (2004) Introduction, pp. 1–16 in Hughes, A. and Reimer, S. (eds) *Geographies of Commodity Chains*, Routledge, London.

Hughes, A., Wrigly, N. and Buttle, M. (2008) Global production networks, ethical campaigning, and the embeddedness of responsible governance, *Journal of Economic Geography*, 8, pp. 345–67.

Humphrey, J. (2005) *Shaping Value Chains for Development: Global Value Chains in Agribusiness*, Deutsche Gesellschaft für Technische Zusammenarbeit (GTZ), Eschborn (available at www.gtz.de).

Humphrey, J. (2006) Horticulture: responding to the challenges of poverty reduction and global competition, *Acta Horticulturae* 699, pp. 1–34.

Humphrey, J. and Schmitz, H. (2000) Governance and upgrading: linking industrial cluster and global value chain approach, *IDS Working Paper*, 120.

Humphrey, J. and Schmitz, H. (2001) Governance in global value chains, *IDS Bulletin*, 32(3), pp. 19–29.

Humphrey, J. and Schmitz, H. (2002) How does insertion in global value chains affect upgrading in industrial clusters?, *Regional Studies*, 36(9), pp. 1017–27.

Humphrey, J. and Schmitz, H. (2004) Chain governance and upgrading: taking stock, pp. 349–81 in Schmitz, H. (ed) *Local Enterprises in the Global Economy: Issues of Governance and Upgrading*, Edward Elgar, Cheltenham.

Humphrey, J., McCulloch, N. and Ota, M. (2004) The impact of European market changes on employment in the Kenyan horticulture sector, *Journal of International Development*, 16(1), pp. 63–80.

*Indian Business Insight* (2007) Great guzzler: Tata Tea, *Indian Business Insight*, 3 June, not paginated.

*Indian Coffee* (2006) In the news, *Indian Coffee*, 70(8), p. 4.

*Indian Coffee* (2007) In the news, *Indian Coffee*, 71(9), p. 5.

*Indian Food and Industry* (2006) Indian tea market – what is brewing?, *Indian Food and Industry*, 28 February, p. 44.

Inter-Governmental Group on Tea (IGGoT) (2005) *Upgrading in the International Tea Sector: A Value Chain Analysis*, Paper presented to the Sixteenth Session, Committee on Commodity Problems, Inter-Governmental Group on Tea, Bali, 20–2 July, Document CCP: TE 05/04.

Inter-Governmental Group on Tea (IGGoT) (2006) *Current Market Situation and Medium-term Outlook*, Paper presented to the Seventeenth Session, Committee on

Commodity Problems, Inter-Governmental Group on Tea, Nairobi, Document CCP:TE 06/2.

International Coffee Organization (2007) Codex Alimentarius: Work related to food safety in Coffee, ED 2015/07 (Executive Director's Communication), International Coffee Organization, London, available at http://www.ico.org/documents/ed2015e.pdf (accessed 8 February 2008).

International Coffee Organization (ICO) (2007) *Trade Statistics*, online database, accessed various dates.

International Union of Foodworkers (2003) What will unions do next? *Asian Food Worker* www.asianfoodworker.net/india/tea.htm (accessed 24 February 2004).

J. Thomas & Co. (2004) *Tea Market Annual Report and Statistics*, J. Thomas & Co., Kolkata.

Jackson, P. (1999) Commodity cultures: The traffic in things, *Transactions of the Institute of British Geographers*, 24(1), pp. 95–108.

Jackson, P. (2002) Commercial cultures: Transcending the cultural and the economic, *Progress in Human Geography*, 26, pp. 3–18.

Jackson, P., Ward, N. and Russell, P. (2006) Mobilising the commodity chain concept in the politics of food and farming, *Journal of Rural Studies*, 22, pp. 129–41.

Jacob, S. (2003) A bitter brew in the high ranges, *The Hindu* (online news database), 28 September.

Jansen, A. (2005) *Plant Protection in Coffee: Recommendations for the Common Code for the Coffee Community-Initiative*, Common Code for the Coffee Community, Geneva (www.sustainable-coffee.net).

Jayarama and D'Souza, M.V. (2006) Shade trees for sustained coffee production, *Indian Coffee*, 70(5), pp. 9–11.

John, J. (2003a) There is blood in the tea we drink, *Labour File* (Special Issue on Tea) (Delhi), 1(3).

John, J. (2003b) Interview: We should have better access to international market, *Labour File* (Delhi) 1(3), pp. 71–6.

Jones, G. and Wale, J. (1998) Merchants as business groups: British trading companies in Asia before 1945, *Business History Review*, 72, pp. 367–408.

Jones, J. (1998) Apparel sourcing strategies for competing in the US market, *Industry, Trade, and Technology Review*, December issue, pp. 31–40.

Joseph, A.E. (2003) Neglected exports and unfair demands on employers, *Labour File* (Delhi) 1(3), pp. 65–6.

Jothikumar, P.D. (2002) Practices and constraints in cultivation of organic tea in South India, pp. 1–3 in UPASI & Tea Board (eds) *Tea Quality Upgradation Seminar on Organic Tea Status, Issues and Strategies*, UPASI Tea Research Foundation, Valparai.

Kalantaridis, C. and Bika, Z. (2006) Local embeddedness and rural entrepreneurship: case-study evidence from Cumbria, England, *Environment and Planning A*, 38(8), pp. 1561–79.

Kaplinsky, R. and Morris, M. (2001) *A Handbook for Value Chain Research*, online resource available at www.globalvaluechains.org.

Kapur, V. (2002) Tea situation in South India, pp. 1–9 in Tea Board and UPASI-KVK (eds) *Tea Quality Upgradation in Small Sector: A Dialogue on South Indian Tea Industry*, Tea Board and UPASI-KVK, Coonoor.

Kariappa, C.P. (ed.) (2002) *The Connoisseur's Book of Indian Coffee*, Macmillan India, Coffee Board and UPASI, Bangalore.

Kariappa, C.P., Dhunjeebhoy, H.D., Ramaswamy, V., and Datta, A. (eds) (2004) *Planting Times*, Macmillan and UPASI, Bangalore.

Karkkainen, B. C. (2004) Post-sovereign environmental governance, *Global Environmental Politics*, 4(1), pp. 72–96.

King, K. and McGrath, S. (2004) *Knowledge for Development. Comparing British, Japanese, Swedish and World Bank Aid*, Zed Books, London.

Kjosavik, D.J. and Shanmuganratnam, N. (2007) Property rights dynamics and indigenous communities in highland Kerala, South India: An institutional-historical perspective, *Modern Asian Studies*, 41(6), pp. 1183–260.

Kling, B.B. (1966) The origin of the managing agency system in India, *Journal of Asian Studies*, 26(1), pp. 37–47.

Kolk, A. (2005) Corporate Social Responsibility in the coffee sector: The dynamics of MNC responses and code development, *European Management Journal*, 23(2), pp. 228–36.

Konefal, J., Mascarenhas, M. and Hatanaka, M. (2005) Governance in the global agro-food system: Backlighting the role of transnational supermarket chains, *Agriculture and Human Values*, 22(3), pp. 291–302.

Koning, N. and Robbins, P. (2005) Where there's a will there's a way: Supply management for supporting the prices of tropical export crops, pp. 181–200 in Lines, T. (ed.) *Agricultural Commodities, Trade and Sustainable Development*, International Institute for Environment and Development (IIED) and International Centre for Trade and Sustainable Development (ICTSD), London and Geneva.

Krivonos, E. (2004) The impact of coffee market reforms on producer prices and price transmission, *World Bank Policy Research Working Paper* 3358.

Kulkarni, J., Mehta, P., Boominathan, D. and Chaudhuri, S. (2007) *A Study of Man–Elephant Conflict in Nagarhole National Park and Surrounding Areas of Kodagu District in Karnataka, India*, Envirosearch, Pune.

Kulkarni, V. (2007) Illy opens coffee varsity in Bangalore, *Hindu Business Line* (online news database), 23 February.

Kullan, T.M. (2006) *Toxic Tea Market in the Nilgiris, Nilgiris Profusely Bleeds*, Hill Teq Prints, Coimbatore.

Kushalappa, C.G. and Bhagwat, S. (2001) Sacred Groves: Biodiversity, Threats and Conservation, pp. 21–9, In Shaanker, R.U., Ganeshaiah, K.N. and Bawa, K.S. (eds) *Forest Genetic Resources: Status, Threats and Conservation Strategies*, IBH Publishing and Oxford, New Delhi.

Kydd, J., Pearce, R. and Stockbridge, M. (1996) The economic analysis of commodity systems: environmental effects, transaction costs and the francophone filière tradition, unpublished paper presented at the ODA/NRSP Socio-economics Methodology Workshop, ODI London, 29–30 April.

*Labour File* (2006) Tea workers and small growers demand protection of right to life: Thousands march in Gudallur [sic], *Labour File*, 17 December. www.labourfile.org/newsMore.aspx?Nid=97 (accessed 16 July 2007).

Lahiri-Dutt, K. and Samanta, G. (2006) Constructing social capital: Self-Help Groups and rural women's development in India, *Geographical Research*, 44(3), pp. 285–95.

Latour, B. (1987) *Science in Action: How to Follow Scientists and Engineers Through Society*, Open University Press, Oxford.

Lauret, F. (1983) Sur les études de filières agro-alimentaires, *Économies et Sociétés*, 17(5), pp. 721–40.

Law, J. (1987) Technology and heterogenous engineering: the case of Portuguese expansion, pp. 111–34 in Bijker, W., Hughes, T. and Pinch, T. (eds) *The Social Construction of Technological Systems: New Directions in the Sociology and History of Technology*, MIT Press, Cambridge, MA.

Leach, F. and Sitaram, S. (2002) Microfinance and women's empowerment: a lesson from India, *Development in Practice*, 12(5), pp. 575–88.

Lee, R. and Wills, J. (eds) (1997) *Geographies of Economies*, Arnold, London.

Leslie, D. and Reimer, S. (1999) Spatializing Commodity Chains, *Progress in Human Geography*, 23, pp. 401–20.

Levy, D.L. (2008) Political contestation in global production networks, *Academy of Management Review*, 33.

Lewin, B., Giovannucci, D. and Varangis, P. (2004) Coffee markets: New paradigms in global supply and demand, *Agriculture and Rural Development Discussion Paper* 3, World Bank, Washington, DC.

Lindsey, B. (2004) *Grounds for Complaint: 'Fair Trade' and the Coffee Crisis*, Adam Smith Institute, London.

Lingle, T. (1993) *The Basics of Cupping Coffee*, Specialty Coffee Association of America, Long Beach, CA.

Lockie, S. (2002) 'The invisible mouth': mobilizing 'the consumer' in food production-consumption networks, *Sociologia Ruralis*, 42(4), pp. 278–94.

Lockie, S. and Goodman, M. (2006) Neoliberalism and the problem of space: Competing rationalities of governance in Fair Trade and mainstream agri-environmental networks, *Research in Rural Sociology and Development*, 12, pp. 95–117.

Lockie, S. and Kitto, S. (2000) Beyond the farm gate: Production-consumption networks and agri-food research, *Sociologia Ruralis*, 40(1), pp. 3–19.

Loh, A.T., Kam, B.H. and Jackson, J.T. (2003) Sri Lanka's plantation sector: A before-and-after privatization comparison, *Journal of International Development*, 15, pp. 727–45.

Mabbett, T. (2007) Inner strength of soluble coffee on the outer edges of Europe, *Tea and Coffee Trade Journal*, 179(2), electronic journal, not paginated.

Macdonald, K. (2007) Globalising justice within coffee supply chains? Fair Trade, Starbucks and the transformation of supply chain governance, *Third World Quarterly*, 28(4), pp. 793–812.

Macfarlane, A. and Macfarlane, I. (2003) *Green Gold: The Empire of Tea*, Ebury Press, London.

Macleod, G. (2001) Beyond soft institutionalism: accumulation, regulation, and their geographical fixes, *Environment and Planning A*, 33, pp. 1145–67.

Madeley, J. (ed.) (2001) *Robbing Coffee's Cradle*, Actionaid, London.

Mahanty, A. (2002) Conservation and development interventions as networks: The case of the India eco-development project, Karnataka, *World Development*, 30(8), pp. 1369–86.

Maizels, A. (1997) *Commodity Supply Management by Producing Countries: A Case-study of the Tropical Beverage Crops*, Clarendon Press, New York.

Markusen, A. (1996) Sticky places in slippery space: a typology of industrial districts, *Economic Geography*, 72, pp. 293–313.

Markusen, A. (2003) Fuzzy concepts, scanty evidence, policy distance: the case for rigour and policy relevance in critical regional studies, *Regional Studies*, 37 (6 and 7), pp. 701–17.

Marsden T.K., Munton, R., Ward, N. and Whatmore, S. (1996) Agricultural geography and the political economy approach: a review, *Economic Geography*, 72(4), pp. 361–75.

Marsden, T. and Wrigley, N. (1996) Retailing, the food system and the regulatory state, in Wrigley, N. and Lowe, P. (eds) *Retailing, Consumption and Capital: Towards the New Retail Geography*, Longman House, Harlow.

Martin, R. (2000) Institutionalist approaches in economic geography, pp. 77–94 in Sheppard, E. and Barnes, T. (eds) *A Companion to Economic Geography*, Blackwell, Oxford.

Martin, R. (2002) *The Financialization of Daily Life*, Temple University Press, Philadelphia.

Massey D. (1993a) Power-geometry and a progressive sense of place, pp. 59–69 in Bird, J., Curtis, B., Putnam, T., Robertson, G. and Tickner, L. (eds) *Mapping the Futures: Local Cultures and Global Change*, Routledge, London.

Massey, D. (1993b) Questions of locality, *Geography*, 78(2), pp. 142–9.

Massey, D. (2005) *For Space*, Sage, London.

McCormick, D. and Schmitz, H. (2001) *Manual for Value Chain Research on Homeworkers in the Garment Industry*, online resource at www.globalvaluechains.org.

McCormick, R. (2006) A qualitative analysis of the WTO's role on trade and environment issues, *Global Environmental Politics*, 6(1), pp. 102–24.

Meadowcroft, J. (2002) Politics and scale: some implications for environmental governance, *Landscape and Urban Planning*, 61(2–4), pp. 169–79.

Meerman, J. (1997) *Reforming Agriculture: The Bank Goes to Market*, The World Bank, Washington, DC.

Menon, S (2004) The aromatic coffees of India Part 1, *Tea and Coffee Trade Journal*, 176(9), electronic journal, not paginated.

Menon, S. (2002) Stepping stones to tasteful success, *The Planters' Chronicle*, Feb.–March issue, pp. 79–84.

Menon, S. (2003a) An ode to the women in tea plantations, *Labour File* (New Delhi), 1(3), pp. 51–6.

Menon, S. (2003b) Starving to death, *Labour File* (New Delhi), 1(3), pp. 67–70.

Menon, U. (2005) Foresee 4c Foxy (editorial), *The Planters' Chronicle*, 101(10), p. 1.

Messner, D. (2004) Regions in the 'world economic triangle', pp. 20–52 in Schmitz, H. (ed.) *Local Enterprises in the Global Economy: Issues of Governance and Upgrading*, Edward Elgar, Cheltenham.

Meyer-Stamer, J. (2004) Paradoxes and ironies of locational policy in the new global economy, pp. 326–48 in Schmitz, H. (ed.) *Local Enterprises in the Global Economy: Issues of Governance and Upgrading*, Edward Elgar, Cheltenham.

Miles, M. and Crush, J. (1993) Personal narratives as interactive texts: collecting and interpreting migrant life-histories, *Professional Geographer*, 45(1), pp. 84–94.

Millennium Institute (2003) *Ecosystems and Human Well-Being: A Framework for Assessment*, Island Press, Washington, DC.

Mintel (2007) *Tea and Herbal Tea: UK*, Mintel, London.

Mitra, S.K., Rajan, A., Dasgupta, C., Pal, R., Mukherjee, D.K. and Malakar, N. (2003) *Techno-Economic Survey of Tea Industry in Nilgiris*, Tea Board of India, Kolkata.

Mohammad, R. (2001) 'Insiders' and/or 'outsiders': positionality, theory and praxis, pp. 101–17 in Limb, M. and Dwyer, C. (eds) *Qualitative Methodologies for Geographers: Issues and Debates*, Arnold, London.

Moppert, B. (2000) The elaboration of the landscape. In Ramakrishnan, P.S., Chandrashekara, U.M., Elouard, C., Guilmoto, C.Z., Maikhuri, R.K., Rao, K.S., Saxena, K.G. and Shankar, S. (eds) *Mountain Biodiversity, Land Use Dynamics and Traditional Ecological Knowledge*, Oxford and India Book House, New Delhi.

Mortimore, M. (2002) When does apparel become a peril? On the nature of industrialization in the Caribbean Basin, pp. 287–307 in Gereffi, G., Spencer, D. and Bair, J. (eds) *Free Trade and Uneven Development: The North American Apparel Industry after NAFTA*, Temple University Press, Philadelphia, PA.

Moxham, R. (2003) *Tea: Addiction, Exploitation, and Empire*, Carroll & Graf, New York.

Muradian, R. and Pelupessy, W. (2005) Governing the coffee chain: The role of voluntary regulatory systems, *World Development*, 33(12), pp. 2029–44.

Murdoch J. (1995) Actor-networks and the evolution of economic forms: combining description and explanation in theories of regulation, flexible specialisation, and networks, *Environment and Planning A*, 27(5), pp. 731–58.

Murdoch, J., Marsden, T.K. and Banks, J. (2000) Quality, nature, and embeddedness: Some theoretical considerations in the context of the food sector, *Economic Geography*, 76(2), pp. 107–25.

Murdoch, J. (1997) Towards a geography of heterogenous associations, *Progress in Human Geography*, 21, pp. 321–37.

Murdoch, J. (2000) Networks – A new paradigm of rural development, *Journal of Rural Studies*, 16(4), pp. 407–19.

Mutersbaugh, T. (2005a) Fighting standards with standards: harmonization, rents and social accountability in certified agrofood networks, *Environment and Planning A*, 37(11), pp. 2033–51.

Mutersbaugh, T. (2005b) Just-in-space: Certified rural products, labor of quality, and regulatory spaces, *Journal of Rural Studies*, 21(4), pp. 389–402.

Myers, N., Mittermeier, R.A., Mittermeier, C.G., da Fonseca, G.A.B. and Kent, J. (2000) Biodiversity hotspots for conservation priorities, *Nature*, 403, pp. 853–8.

Nadvi, K. (2008) Global standards, global governance and the organization of global value chains, *Journal of Economic Geography*, 8, pp. 323–43.

Naqvi, W. (2002) Experience in organic tea production in South India, pp. 4–7 in UPASI & Tea Board (eds) *Tea Quality Upgradation Seminar on Organic Tea Status, Issues and Strategies*, UPASI Tea Research Foundation, Valparai.

National Coffee Association (2002) *Coffee Facts and Figures*, National Coffee Association, New York.

Neilson, J. (2007) Institutions, the governance of quality and on-farm value retention for Indonesian specialty coffee, *Singapore Journal of Tropical Geography*, 28(2), pp. 188–204.

Neilson, J. and Pritchard, B. (2006) Traceability, supply chains and smallholders: Case studies from India and Indonesia, *Report to the 17th Session of the Committee on Commodity Problems/Intergovernmental Group on Tea*, Food & Agricultural Organization of the United Nations (FAO), Nairobi (Kenya) Document CCP:TE 06/4.

Neilson, J. and Pritchard, B. (2007a) The final frontier? The global roll-out of the retail revolution in India, pp. 219–42 in Burch, D. and Lawrence, G. (eds) *Supermarkets and Agri-Food Supply Chains: Transformations in the Production and Consumption of Foods*, Edward Elgar, Melbourne.

Neilson, J. and Pritchard, B. (2007b) Green Coffee? The contradictions of global sustainability initiatives from the Indian perspective, *Development Policy Review*, 25(3), pp. 311–31.

Neilson, J. and Pritchard, B. (2008) Big is not always better: Global value chain restructuring and the crisis in South Indian tea plantations, in Le Heron, R. and Stringer, C. (eds) *Agri-food Commodity Chains and Globalising Networks*, Ashgate, Aldershot.

Neilson, J. and Pritchard, B. (forthcoming) Fairness and ethicality in their place: The regional dynamics of fair trade and ethical sourcing agendas in the plantation districts of South India, *Environment and Planning A*.

Neilson, J., Pritchard, B. and Spriggs, J. (2006) Implementing quality and traceability initiatives among smallholder tea producers in Southern India, *ACTA Horticulturae*, 699, pp. 327–34.

Neven, D. and Reardon, T. (2004) The rise of Kenyan supermarkets and evolution of their horticulture product procurement systems: Implications for agricultural diversification and smallholder market access programs, *Development Policy Review*, 22(6), pp. 669–99.

*New India Press* (2006) Tea subsidy runs into rough weather, *New India Press* (online news database), 8 March.

*New York Times* (1860) Patronage and the pickaxe: Have we a DOWB among us?, *New York Times*, 9 May.

Nguythitiencu and Nguyentanphong (2005) *Country Report: Vietnam*, International Tea Day 2005 reports, www.teaday.org/contents/Vietnam.pdf.

Nicholls, A. and Opal, C. (2005) *Fair Trade: Market-Driven Ethical Consumption*, Sage, London.

Ninan, K.N. (2006) *The Economics of Biodiversity Conservation: Valuation in Tropical Forest*, Earthscan, London.

Ninan, K.N. and Sathyapalan, J. (2005) The economics of biodiversity conservation: a study of a coffee growing region in the Western Ghats of India, *Ecological Economics*, 55(1), pp. 61–72.

Nissanke, M. and Aryeetey, E. (eds) (2003) *Comparative Development Experiences of Sub-Saharan Africa and East Asia: An Institutional Approach*, Ashgate, Aldershot.

North, D.C. (1990) *Institutions, Institutional Change, and Economic Performance*, Cambridge University Press, Cambridge.

North, D.C. (1995) The New Institutional Economics and Third World development, pp. 17–26 in Harriss, J., Hunter, J. and Lewis, C. (eds) *The New Institutional Economics and Third World Development*, Routledge, London.

North, D.C. (1998) Economic performance through time, pp. 242–61 in Brinton, M. and Nee, V. (eds) *The New Institutionalism in Sociology*, Russell Sage Foundation, New York.

North, D.C. and Thomas, R.P. (1973) *The Rise of the Western World: A New Economic History*, Cambridge University Press, Cambridge.

O'Brien, R. (1992) *Global Financial Integration: The End of Geography*, Royal Institute for International Affairs, London.

O'Brien, T. and Kinnaird, M.F. (2003) Caffeine and conservation, *Science*, 300 (5619), p. 587.

O'Rourke, D (2005) Market movements: Nongovernmental organization strategies to influence global production and consumption, *Industrial Ecology*, 9(1–2), pp. 115–28.

O'Rourke, D. (2006) Multi-stakeholder regulation: Privatizing or socializing global labour standards? *World Development*, 34(5), pp. 899–918.

Ohmae, K. (1995) *The End of the Nation State: The Rise of Regional Economies*, Harper Collins, London.

Organization for Economic Cooperation and Development (OECD) (2006) *Private Standard Schemes and Developing Country Access to Global Value Chains: Challenges and Opportunities Emerging from Four Case Studies*, OECD, Paris.

Organization for Economic Cooperation and Development (OECD) (2007) *Economic Outlook*, OECD, Paris.

Oxfam (2001) *Bitter Coffee: How the Poor are Paying for the Slump in Coffee Prices*, Oxfam International, Oxford.

Oxfam (2002a) *Rigged Rules and Double Standards*, Oxfam International, Oxford.

Oxfam (2002b) *Mugged: Poverty in Your Coffee Cup*, Oxfam International, Oxford.

Pempel, T.J. (1999) *The Politics of the Asian Economic Crisis*, Cornell University Press, Ithaca.

Pendergrast, M. (2001) *Uncommon Grounds: The History of Coffee and how it Transformed the World*, Texere, New York.

Pereira, S. (2002) Scope for organic farming in small growers' sector, pp. 8–10 in UPASI & Tea Board (eds) *Tea Quality Upgradation Seminar on Organic Tea Status, Issues and Strategies*, UPASI Tea Research Foundation, Valparai.

Pesticides Action Network (2001) Lindane: A chemical of the past persists in the future (Update of article originally published in 1995), *Pesticides News*, 28(12), pp. 28–9.

Philo, C. and Parr, P. (2000) Institutional geographies: Introductory remarks, *Geoforum*, 31, pp. 513–21.

Philpott, S.M., Bichier, P., Rice, R. and Greenberg, R. (2007) Field-testing ecological and economic benefits of coffee certification programs, *Conservation Biology*, 21(4), pp. 975–85.

*Planting Opinion* (1898) Managing agents' shorter catechism, *Planting Opinion*, 6 August. Reproduced in Kariappa et al. (2004: pp. 40–4).

Polanyi. K. (1957) *The Great Transformation*, Farrar & Rinehart Inc, New York; first published 1944.

Ponte, S. (2002a) The 'latte revolution'? Regulation, markets and consumption in the global coffee chain, *World Development*, 30(7), pp. 1099–122.

Ponte, S. (2002b) Brewing a bitter cup: Quality and the re-organization of coffee marketing in East Africa, *Journal of Agrarian Change*, 2(2), pp. 248–72.

Ponte, S. and Gibbon, P. (2005) Quality standards, conventions and the governance of global value chains, *Economy and Society*, 34(1), pp. 1–31.

Porter, M.E. (1990) *The Competitive Advantage of Nations*, The Free Press, New York.

Praveen, C. (2006) Tiger on the prowl, *Coffeeland News* (Madikeri), 28 July, p. 1.

Pritchard, B. (1999) Switzerland's billabong? Brand management in the global food system and Nestlé Australia, pp. 23–40 in Goss, J., Burch, D., and Lawrence, G. (eds) *Restructuring Global and Regional Agricultures: Transformations in Australasian Agri-Food Economies and Spaces*, Aldershot, Avebury.

Putnam, R.D. (1995) Bowling alone: America's declining social capital, *The Journal of Democracy*, 6(1), pp. 65–78.

Raghuramulu, Y. (2007) Impact of diversification initiatives in Indian coffee, *Indian Coffee*, 71(1), pp. 22–9.

Rahman, H. (1999) Specialty tea buying, *The Planters' Chronicle*, October, pp. 469–73.

Raikes, P., Jensen, M.F. and Ponte, S. (2000) Global commodity chain analysis and the French *filière* approach: comparison and critique, *Economy and Society*, 29(3), pp. 390–417.

Rainforest Alliance (2006) Yuban and Other Kraft Coffee Brands Bring Rainforest Alliance Certified Beans to Mainstream, 11 May, www.rainforest-alliance.org/news (accessed 11 December 2007).

Rainforest Alliance (2007a) Rainforest Alliance, www.rainforest-alliance.org (accessed various dates).

Rainforest Alliance (2007b) Sustainable Coffee Production Finds a Growing Number of Fans in Germany, September 28, www.rainforest-alliance.org/news (accessed 11 December 2007).

Rajghatta, C. (2007) Starbucks and India's brewing beverage bubble, *The Times of India*, 16 August.

Raji (2004) *Bitter Coffee: Crisis of the Indian and Global Coffee Industry*, Vimukthi Prakashana, Bangalore.

Ramakrishnan, P.S., Chandrashekara, U.M., Elouard, C., Guilmoto, C.Z., Maikhuri, R.K., Rao, K.S., Saxena, K.G. and Shankar, S. (eds) (2000) *Mountain Biodiversity, Land Use Dynamics and Traditional Ecological Knowledge*, Oxford and India Book House, New Delhi.

Ramamoorthy, G. and Ramu, S. (2002) Interim Report on the impact of tea quality upgradation project on manufacturing technology in small sector, *The Planters' Chronicle*, April issue, pp. 141–7.

Ramaswamy, V. (2003) A strong role for tea auctioneers, *Hindu Business Line*, 22 September, p. 33.

Rappole, J.H., King, D.I. and Rivera J.H.V. (2003) Coffee and conservation, *Conservation Biology*, 17(1), pp. 334–6.

Ravindranath, N.G. (1999) Liberalisation of Indian coffee: Does it mean prosperity?', *Indian Coffee*, 63(1).

Raynolds, L. (2002) Consumer/producer links in fair trade coffee networks, *Sociologia Ruralis*, 42(4), pp. 404–24.

Raynolds, L., Murray, D. and Wilkinson, J. (eds) (2007a) *Fair Trade: The Challenges of Transforming Globalization*, Routledge, London and New York.

Raynolds, L.T., Murray, D. and Heller, A. (2007b) Regulating sustainability in the coffee sector: A comparative analysis of third-party environmental and social certification initiatives, *Agriculture and Human Values*, 24, pp. 147–63.

Reardon, T. and Berdegué, J. (2002) The rapid rise of supermarkets in Latin America: Challenges and opportunities for development, *Development Policy Review*, 20(4), pp. 317–34.

Reardon, T. and Farina, E.M.M.Q. (2002) The rise of private food quality and safety standards: Illustrations from Brazil, *International Food and Agricultural Management Review*, 4, pp. 413–21.

Reardon, T. and Timmer, C.P. (2007) Transformation of markets for agricultural output in developing countries since 1950: How has thinking changed? in Evenson, R.E., Pingali, P. and Schultz T.P. (eds) *Handbook of Agricultural Economics: Agricultural Development: Farmers, Farm Production and Farm Markets*, Elsevier, Amsterdam.

Reardon, T., Timmer, C., Barret, C. and Berdegué, J. (2003) The rise of supermarkets in Africa, Asia and Latin America, *American Journal of Agricultural Economics*, 85(5), pp. 1140–6.

Reardon, T., Berdegué, J. and Farrington, J. (2002) Supermarkets and farming in Latin America: Pointing directions for elsewhere? *Overseas Development Institute, Natural Resource Perspectives (monograph)*, 81.

Reardon, T., Timmer, P. and Berdegué, J. (2004) The rapid rise of supermarkets in developing countries: Induced organizational, institutional and technological change in agri-food systems, *Electronic Journal of Agricultural and Development Economics*, 1(2), pp. 168–83.

Renard, M.C. (2005) Quality regulation, certification and power in Fair Trade, *Journal of Rural Studies*, 21(4), pp. 419–31.

Ribbentrop, B. (1900) *Forestry in British India* (reprint 1989), Indus Publishing Company, New Delhi.

Rice, R.A. and Ward, J.R. (1996) *Coffee, Conservation and Commerce in the Western Hemisphere: How Individuals and Institutions can Promote Ecologically-sound Farming and Forest Management in Northern Latin America*, Natural Resources Defence Council and Smithsonian Migratory Bird Center, Washington, DC.

Richter, G. (1870) *Gazetteer of Coorg: Natural Features of the Country and the Social and Political Condition of its Inhabitants*, Low price Publications, New Delhi. (2002 Reprint of the original 1870 publication.)

Robbins, P. (2003) *Stolen Fruit: The Tropical Commodities Disaster*, Palgrave Macmillan, London.

Sachs, J. (2005) *The End of Poverty: How We Can Make it Happen in our Lifetime*, Penguin, London.

Salais, R. and Thévenot L. (eds) (1986) *Le Travail: Marchés, Règles, Conventions*, INSEE-Economica, Paris.

Sands, P. (2005) *Lawless World: America and the Making and Breaking of Global Rules*, Allen Lane, London.

Sanjith, R. (2004) Tea prices: North–South divergence, *The Planters' Chronicle*, June issue, pp. 24–9.

Saraswathy, H. (2002) 'Your cup of tea', *Indigenous Agriculture News* (New Delhi), 1(1), pp. 1–3.

Satish, B.N., Kushalappa, C.G. and Garcia, C. (2007). Land tenure systems: A key to conserve tree diversity in coffee-based agroforestry systems of Kodagu, Central Western Ghats, *Second International Symposium on Multi-strata Agroforestry Systems with Perennial Crops,* September 2007, CATIE, Turrialba, Costa Rica.

Schmitz, H. (2004) Globalized localities: introduction, pp. 1–19 in Schmitz, H. (ed.) *Local Enterprises in the Global Economy: Issues of Governance and Upgrading*, Edward Elgar, Cheltenham.

Schmitz, H. (2006) Learning and earning in global garment and footwear chains, *The European Journal of Development Research*, 18(4), pp. 546–71.

Scott, A. (1988) Flexible production systems and regional development: the rise of new industrial spaces in North America and western Europe, *International Journal of Urban and Regional Research*, 12(2), pp. 171–86.

Sibanda, L. (2006) Auto-control of Ochratoxin A levels in green coffee, *Indian Coffee*, 70(3), pp. 11–12.

Sivaram, B. (1996) Productivity improvement and labour relations in the tea industry in South Asia, *International Labor Organization Sectoral Assistance Program Working Papers*, SAP 2.54/WP.101.

Smith, S. and Dolan, C. (2006) Ethical trade: What does it mean for women workers in African horticulture?, pp. 79–96 in Barrientos, S. and Dolan, C. (eds) *Ethical Sourcing in the Global Food System*, Earthscan, London.

Smithsonian Migratory Bird Center (SMBC) (2007) Migratory birds, nationalzoo.si.edu/ConservationAndScience/MigratoryBirds/Coffee (accessed 28 November 2007).

Smithsonian Migratory Bird Center (SMBC) (not dated) *Shade Management Criteria for Smithsonian 'Bird-friendly' Coffee*, Smithsonian Migratory Bird Centre, National Zoological Park, Washington, DC.

SOMO, ProFound and India Committee of the Netherlands (2006) *Sustainabilitea: The Dutch Tea Market and Corporate Social Responsibility*, SOMO-Centre for Research on Multinational Corporations, Amsterdam.

Specialty Coffee Association of America (SCAA) (1999) *Coffee Market Summary*, Specialty Coffee Association of America, Long Beach, California.

Spriggs, J. and Isaac, G. (2001) *Food Safety and International Competitiveness: The Case of Beef*, CABI Publishing, Oxford.

Srinivas, M.N. (1965) *Religion and Society among the Coorgs of South India*, Asia Publishing House, Bombay.

Stiglitz, J. (2002) *Globalisation and its Discontents*, Norton & Co, New York.

Storper, M (1999) The resurgence of regional economies, pp. 23–53 in Barnes, T.J. and Gertler, M. (eds) *The New Industrial Geography: Regions, Regulation and Institutions*, Routledge, London and New York.

Storper, M. (1997) *The Regional World: Territorial Development in a Global Economy*, Guilford Press, New York.

Storper, M. and Walker, R. (1989) *The Capitalist Imperative: Territory, Technology and Industrial Growth*, Blackwell, Oxford.

Sturgeon, T.J. (2001) How do we define value chains and production networks?, *IDS Bulletin*, 32(3), pp. 9–18.

Sturgeon, T.J. (2002) Modular production networks: A new American model of industrial organization, *Industrial and Corporate Change*, 11(3), pp. 451–96.

Sturgeon, T.J. (2003) What really goes on in Silicon Valley? Spatial clustering dispersal in modular production networks, *Journal of Economic Geography*, 3, pp. 199–225.

Sturgeon, T.J. (2006) Conceptualizing integrative trade: The Global Value Chains framework, paper prepared for the CTPL conference 'Integrative Trade between Canada and the United States – Policy Implications', 6 December 2006, online resource www.carleton.ca/ctpl/conferences/documents/Sturgeon-Conceptualizing IntegrativeTrade-Paper.pdf (accessed 11 June 2008).

Sturgeon, T.J. (2007) How globalization drives institutional diversity: the Japanese electronics industry's response to value chain modularity, *Journal of East Asian Studies*, 7(1), pp. 1–34.

Sturgeon, T.J. and Lester, R.K. (2004) The new global supply-base: New challenges for local suppliers in East Asia, pp. 35–88 in Shahid, Y.M., Anjum, A. and Nabeshima, K. (eds) *Global Production Networking and Technological Change in East Asia*, The World Bank, Washington, DC.

Sturgeon, T.J., van Biesebroeck, J. and Gereffi, G. (2008) Value chains, networks and clusters: reframing the global automotive industry, *Journal of Economic Geography*, 8, pp. 297–321.

Sustainable Agriculture Initiative (SAI) (2008) SAI Platform, at www.saiplatofrm.org (accessed 26 September, 2008).

Swaminathan, P. (2002) Quality upgradation in the bought leaf sector, *The Planters' Chronicle*, June issue, pp. 232–4.

Swyngedouw, E. (1997) Neither global nor local: 'glocalisation' and the politics of scale, pp. 137–66 in Cox, K. (ed.) *Spaces of Globalization: Reasserting the Power of the Local*, Guilford Press, New York.

Swyngedouw, E. (2000) Authoritarian governance, power, and the politics of rescaling, *Environment and Planning D: Society and Space*, 18, pp. 63–76.

Talbot, J. (1995) The regulation of the world coffee market: Tropical commodities and the limits of globalization, pp. 139–67 in McMichael, P. (ed.) *Food and Agrarian Orders in the World Economy*, Praeger, New York.

Talbot, J. (1996) Regulating the coffee commodity chain: Internationalization and the coffee cartel, *Berkeley Journal of Sociology*, 40, pp. 112–49.

Talbot, J. (1997a) The struggle for control of a commodity chain: instant coffee from Latin America, *Latin American Research Review*, 32(2), pp. 117–35.

Talbot, J. (1997b) Where does your coffee dollar go? The division of income and surplus along the coffee commodity chain, *Studies in Comparative International Development*, 32(1), pp. 56–91.

Talbot, J. (2002a) Information, finance and the new international inequality: The case of coffee, *Journal of World-Systems Research*, 8(2), pp. 214–50.

Talbot, J. (2002b) Tropical commodity chains, forward integration strategies and international inequality: coffee, cocoa and tea, *Review of International Political Economy*, 9(4), pp. 701–34.

Talbot, J. (2004) *Grounds for Agreement: The Political Economy of the Coffee Commodity Chain*, Rowman & Littlefield, Lanham MD.

Tallontire, A. (2006) The development of alternative and fair trade: Moving onto the mainstream, pp. 35–48 in Barrientos, S. and Dolan, C. (eds) *Ethical Sourcing in the Global Food System*, Earthscan, London.

Tea Board of India & United Planters' Association of Southern India-Krishi Vigyan Kendra [U-KVK] (2001) *Tea Quality Upgradation in Small Sector: Field Technical Report*, U-KVK, Coonoor.

Tea Board of India & United Planters' Association of Southern India-Krishi Vigyan Kendra [U-KVK] (2005) *Reflections: Five Years of Quality Upgradation Programme in Small Tea Sector in Nilgiris*, U-KVK, Coonoor.

Tea Board of India (2003) *Tea Digest*, Tea Board of India, Kolkata.

Tea Board of India (2006) India tea portal, online database (accessed 22 June 2006).

Tea Board of India (undated a) *Quality Upgradation and Product Diversification Scheme: For the Tenth Plan Period*, Tea Board of India, Kolkata.

Tea Board of India (undated b) *Tea Plantation Development Scheme: For the Tenth Plan Period*, Tea Board of India, Kolkata.

Tea Board of Kenya (2006) Tea statistics, online database www.teaboard.or.ke/statistics.asp, (accessed 22 June 2006).

*Tehelka Newswatch* (2004) Disaster brews in the land of coffee, *Tehelka Newswatch*, 11 September.

Tewari, M. (1999) Successful adjustment in Indian industry: the case of Ludhiana's woollen knitwear cluster, *World Development*, 27(9), pp. 1651–72.

Tharian, G.K. (1984) Historical roots of the crisis in the South Indian tea industry, *Social Scientist*, 12(4), pp. 34–50.

Tharian, G.K. and Joseph, J. (2005) Value addition or value acquisition? Travails of the plantation sector in the era of globalisation, *Economic and Political Weekly*, 25 June, pp. 2681–7.

Thomsen, L. (2007) Accessing global value chains? The role of business–state relations in the private clothing industry in Vietnam, *Journal of Economic Geography*, 7(6), pp. 753–76.

Timmer, C.P. (1986) *Getting Prices Right: The Scope and Limits of Agricultural Price Policy*, Cornell University Press, Ithaca.

Titus, A. and Pereira, G.N. (2006) Eco-friendly Indian coffee: A profile, *Indian Coffee*, 70(11), pp. 12–16.

Toboso, F. (2001) Institutional individualism and institutional change: the search for a middle way mode of explanation, *Cambridge Journal of Economics*, 25(6), pp. 765–83.

Ton, G., Bijman, J. and Oorthuizen, J. (eds) (2007) *Producer Organisations and Market Chains: Facilitating Trajectories of Change in Developing Countries*, Wageningen Academic Publishers, Arnhem (The Netherlands).

Traidcraft (2007) *A Fair Cup?*, Traidcraft in association with DfID, London.

Tsing, A.L. (2005) *Friction: An Ethnography of Global Connection*, Princeton University Press, Princeton.

Turok, I. (1993) Inward investment and local linkages: how deeply embedded is 'Silicon Glen', *Regional Studies*, 27(5), pp. 401–17.

Twinings (2005) Twinings, at www.twinings.co.uk (accessed 1 March 2005).

Twinings (2007) Twinings, at www.twinings.co.uk (accessed 19 December 2007).

Typhoo Tea (2007) Tea Sourcing, at www.typhoo.com/teasourcing.php (accessed 19 December 2007).

Ukers, W.H. (1935). *All About Coffee*, The Tea and Coffee Trade Journal Company, New York.

Umweltbundesamt (UBA) (2007) *Endosulfan: Draft Dossier Prepared in Support of a Proposal of Endosulfan to be Considered as a Candidate for Inclusion in the Annexes to the Stockholm Convention*, German Federal Environment Agency – Umweltbundesamt (UBA), Dessau, February 2007, available at http://www.pops.int.

Unilever (2006a) *Annual Review*, Unilever, London.

Unilever (2006b) *Sustainable Development Report*, Unilever, London.

Unilever (2007) Unilever commits to sourcing all its tea from sustainable ethical sources, Media Release 25 May 2007, www.unilever.com/ourcompany/newsandmedia/pressreleases/2007/sustainable-tea-sourcing.asp.

United Nations Conference on Trade and Development (UNCTAD) (2003) *Report of the Group of Eminent Persons on Commodities*, UNCTAD, Geneva.

United Nations Conference on Trade and Development (UNCTAD) (2005) Commodity policies for development: A new framework for the fight against poverty, *Trade and Development Board, Commission on Trade in Goods and Services, and Commodities*, Document TD/B/COM.1/75.

United Nations Conference on Trade and Development (UNCTAD) (2006) Enabling small commodity producers and processors in developing countries to reach global markets, *Trade and Development Board, Commission on Trade in Goods and Services, and Commodities, Expert Meeting on Enabling Small Commodity Producers and Processors in Developing Countries to Reach Global Markets*, Document TD/B/COM.1/EM.32/2.

United Nations Conference on Trade and Development (UNCTAD) (2007) *Challenges and Opportunities Arising from Private Standards on Food Safety and Environment for Exporters of Fresh Fruit and Vegetables in Asia: Experiences of Malaysia, Thailand and Viet Nam*, United Nations, New York and Geneva.

United Nations Education and Social Commission (UNESCO) (2007) *World Heritage Centre Tentative List*, Reference No. 2203, whc.unesco.org/en/tentativelist (accessed 2 February 2007).

United Planters' Association of Southern India (UPASI) & Tea Board of India (2007) *The Golden Leaf India Awards: Southern Tea Competition* (awards brochure), UPASI, Coonoor.

United Planters' Association of Southern India (UPASI) (2007) *Plantation Statistics*, UPASI, Coonoor.

United Planters' Association of Southern India (UPASI)-Krishi Vigyan Kendra (KVK) (2000) *Tea Factory Manual*, UPASI-KVK, Coonoor.

United Planters of Southern India (UPASI) (2005) *Year Book and Annual Report*, UPASI, Coonoor.

United States Department of Labor/Bureau of International Labor Affairs (1997) *By the Sweat & Toil of Children (Volume IV) Consumer Labels and Child Labor*, www.dol.gov/ilab/media/reports/iclp/sweat4/tea.htm#top1, (accessed 16 July, 2007).

Urs, A. (2008) Illycaffe takes India to level of South America, *Business Standard* (online news database) 24 January.

Uthappa, K.G. (2004) *Land Tenure, Land Holding, and Tree Rights of Kodagu*, Kodagu Model Forest Trust, Ponnampet.

Utz Kapeh (2007) *Good Inside Newsletter*, November.

van der Wal, S. (2008) *Sustainability Issues in the Tea Sector: A Comparative Analysis of Six Leading Producing Countries*, SOMO (Centre for Research on Multinational Corporations), Amsterdam. (Report available at www.somo.nl.)

van Dijk, J.B., van Doesburg, D.H.M., Heijbroek, A.M.A., Wazir, M.R.I.A., and de Wolff, G.S.M. (1998) *The World Coffee Market*, Rabobank, Utrecht.

Varangis, P., Siegel, P., Giovannucci, D., and Lewin, B. (2003) Dealing with the coffee crisis in Central America: Impacts and strategies, *World Bank Policy Research Working Paper*, 2993.

Veblen, T. (1919) *Place of Science in Modern Civilisation and Other Essays*, B.W. Huebsch, New York.

Vellema, S. and Boselie, D. (eds) (2003) *Cooperation and Competence in Global Food Chains*, Shaker Publishing, Maastricht.

Venkatachalam, L. (2005) Perspectives on sustainability and globalization and the challenges for the coffee sector, *Proceedings of the 2nd World Coffee Conference*, 23–5 September, Salvador, Brazil www.ico.org/event_pdfs/wcc2/presentations/venkatachalam.pdf.

Venkatesh, K. and Basavaraj, K. (1999) Quality of Indian coffee, *The Planters' Chronicle*, 2.

Venu, M.K. (2007) India–ASEAN FTA likely to be finalised by May 2008, *The Economic Times*, 22 November.

Vorley, B., Fearne, A. and Ray, D. (2007) *Regoverning Markets: A Place for Small-Scale Producers in Modern Agri-Food Chains?*, Gower Publishing in association with Institute for International Development and Environment, Burlington, Vermont.

Wade, R. (2004) *Governing the Market: Economic Theory and the Role of Government in East Asian Industrialization*, Princeton University Press, Princeton, NJ (first published 1990).

Wallerstein, I. (1974) *The Modern World System*, Academic Press, New York.

Watts, M. (1994) What difference does difference make? *Review of International Political Economy*, 1(3), pp. 563–70.

Weatherspoon, D. and T. Reardon (2003) The rise of supermarkets in Africa: Implications for agrifood systems and the rural poor, *Development Policy Review*, 22(5), pp. 333–55.

Webster, A. (2006) The strategies and limits of gentlemanly capitalism: the London East India agency houses, provincial commercial interests, and the evolution of British economic policy in South and South East Asia 1800–50, *Economic History Review*, 59(4), pp. 743–64.

Weller, S. (2006) The embeddedness of global production networks: the impact of crisis in Fiji's garment export sector, *Environment and Planning A*, 38(7), pp. 1249–67.

Werther, W.B. and Chandler, D. (2005) Strategic corporate social responsibility as global brand insurance, *Business Horizons*, 48(4), pp. 317–24.

Whatmore, S. and Thorne, L. (1997) Nourishing networks: alternative geographies of food, pp. 287–304 in Goodman, D. and Watts, M. (eds) *Globalising Food: Agrarian Questions and Global Restructuring*, Routledge, London.

Wilkinson, J. (1997) A new paradigm for economic analysis? *Economy and Society*, 26, pp. 305–39.

Williamson, J. (1990) *What Washington Means by Policy Reform. Latin American Adjustment: How Much Has Happened*, Institute for International Economics, Washington, DC.

Williamson, O.E. (1975) *Markets and Hierarchies: Analysis and Antitrust Implications*, The Free Press, New York.

Williamson, O.E. (2000) The New Institutional Economics: Taking stock, looking ahead, *Journal of Economic Literature*, 38(3), pp. 595–613.

Wilson, K. (2002) The new microfinance: An essay on the Self-Help Group Movement in India, *Journal of Microfinance*, 4(2), pp. 217–46.

Wilson, K.C. (1999) *Coffee, Cocoa and Tea*, CABI Publishing, Wallingford.

Wood, A and Valler, D. (2001) Guest Editorial: Turn again? Rethinking institutions and the governance of local and regional economies, *Environment and Planning A*, 33(7), pp. 1139–44.

World Bank (2002) *World Development Report 2001/2002: Institutions for Markets*, The World Bank, Washington, DC.

World Bank (2007a) *World Development Report 2008: Agriculture for Development*, The World Bank, Washington, DC.

World Bank (2007b) *Social Capital* (electronic resource) web.worldbank.org/WBSITE/EXTERNAL/TOPICS/EXTSOCIALDEVELOPMENT/EXTTSOCIALCAPITAL, accessed 29 July 2007.

World Trade Organization (WTO) (2006) *Modalities for Negotiations on Agricultural Commodity Issues: Proposal Submitted by the African Group to the Special Session of the Committee on Agriculture*, Geneva, WTO, Document TN/AG/GEN/18.

World Trade Organization (WTO) (2007) Private voluntary standards and developing country market access: Preliminary results, Committee on Sanitary and Phytosanitary Measures, Document G/SPS/GEN/763.

Wrigley, N., Coe, N.M. and Currah, A. (2005) Globalizing retail: conceptualizing the distribution-based transnational corporation (TNC), *Progress in Human Geography*, 29(4), pp. 437–57.

Wunder, S. (2006) Are direct payments for environmental services spelling doom for sustainable forest management in the tropics?, *Ecology and Society*, 11(2), p. 23.

Wunder, S. (2007) The efficiency of Payments for Environmental Services in tropical conservation, *Conservation Biology*, 21(1), pp. 48–58.

Yeung, H.W.C. (1998a) The social-spatial constitution of business organizations: a geographical perspective, *Organization*, 5(1), pp. 101–28.

Yeung, H.W.C. (1998b) *Transnational Corporations and Business Networks: Hong Kong Firms in the ASEAN Region*, Routledge, London.

Yeung, H.W.C. (2005) Rethinking relational economic geography, *Transactions of the Institute of British Geographers*, 30, pp. 37–51.

Yeung, H.W.C. and Lin, G.C.S. (2003) Theorizing economic geographies of Asia, *Economic Geography*, 79, pp. 107–28.

Zukin, S. and DiMaggio, P. (eds) (1990) *Structures of Capital: The Social Organisation of the Economy*, Cambridge University Press, New York.

Zvelebil, K.V. (2000) Blue Mountains revisited: cultural studies on the Nilgiri Hills: book review, *Journal of the American Oriental Society*, 120(1), pp. 126–9.

# Index